D1526767

PHYSICAL VAPOR DEPOSITION OF THIN FILMS

PHYSICAL VAPOR DEPOSITION OF THIN FILMS

JOHN E. MAHAN
Colorado State University

A Wiley-Interscience Publication
JOHN WILEY & SONS, INC.
New York · Chichester · Weinheim · Brisbane · Singapore · Toronto

TS
695
M34
2000

A NOTE TO THE READER
This book has been electronically reproduced from digital information stored at John Wiley & Sons, Inc. We are pleased that the use of this new technology will enable us to keep works of enduring scholarly value in print as long as there is a reasonable demand for them. The content of this book is identical to previous printings.

This book is printed on acid-free paper. ∞

Copyright © 2000 by John Wiley & Sons, Inc. All rights reserved.

Published simultaneously in Canada.

No part of this publication may be reproduced, stored in a retrieval system or transmitted in any form or by any means, electronic, mechanical, photocopying, recording, scanning or otherwise, except as permitted under Sections 107 or 108 of the 1976 United States Copyright Act, without either the prior written permission of the Publisher, or authorization through payment of the appropriate per-copy fee to the Copyright Clearance Center, 222 Rosewood Drive, Danvers, MA 01923, (508) 750-8400, fax (508) 750-4744. Requests to the Publisher for permission should be addressed to the Permissions Department, John Wiley & Sons, Inc., 605 Third Avenue, New York, NY 10158-0012, (212) 850-6011, fax (212) 850-6008, E-Mail: PERMREQ@WILEY.COM.

For ordering and customer serivce, call 1-800-CALL-WILEY.

Library of Congress Cataloging-in-Publication Data:

Mahan, John E.
    Physical vapor deposition of thin films / John E. Mahan.
       p.   cm.
    "A Wiley-Interscience publication."
    ISBN 0-471-33001-9 (cloth : alk. paper)
    1. Vapor-plating. 2. Thin films. I. Title.
  TS695.M32   2000
  671.7'35—dc21   99–21926

Printed in the United States of America.

10 9 8 7 6 5 4 3 2 1

To Linda, Peter, and Wesley

CONTENTS

PREFACE xv

I Introduction to Physical Vapor Deposition 1

    I.1 Physical Vapor Deposition Technologies and Their Basic Physical Science, 1

        Overview, 1
        Kinetic Theory, 5
        Adsorption and Condensation, 8
        High Vacuum, 12
        Sputtering Discharges, 14

    I.2 Summary of Principal Equations, 16
    I.3 Mathematical Symbols, Constants, and Their Units, 17
        Reference, 18

II The Kinetic Theory of Gases 19

    II.1 Statistics, 20

        The Boltzmann Distribution, 20
        Characteristic Particle Speeds, 22

    II.2 Collisions, 23

        Impingement Rate and Incident Flux Angular Distribution, 23
        The Ideal Gas Law, 26
        Mean Free Path, 27

II.3   Properties, 30

   Heat Capacity; the Ideal Diatomic Gas, 30
   Diffusivity, 31
   Viscosity, 32
   Thermal Conductivity, 34

II.4   Gas Flow, 34

   Flow Regimes, 34
   Viscous Laminar Flow, 35
   Molecular Flow, 36
   Conductance, 37

II.5   Units of Pressure and Amounts of Gas, 38

   Units of Pressure, 38
   Amounts of Gas, 39

II.6   Summary of Principal Equations, 39
II.7   Appendix, 40

   Arrhenius Plots, 40
   Some Definite Integrals, 41
   Atomic Diameters of the Elements, 42

II.8   Mathematical Symbols, Constants, and Their Units, 43
   References, 44

III Adsorption and Condensation   45

III.1   Adsorption of Gases, 47

   Why Gases Adsorb, 47
   Mean Residence Time, 49
   Langmuir's Adsorption Isotherm, 49
   Atomic Layer Epitaxy, 53

III.2   Vapor Pressure, 57

   The Thermally Activated Vapor Pressure, 57
   Vapor Pressure Data for the Elements, 58
   Vapor Pressures of Alloys and Compounds, 60

III.3   Condensation of Vapors, 62

   Condensation of Pure Elements, 62
   Condensation of Compounds that Produce
      a Stoichiometric Vapor, 64
   Flash Evaporation of Compounds that Dissociate, 65
   Steady-State Techniques for Alloy Films, 65
   Coevaporation with the Three-Temperature Method, 67

CONTENTS ix

     Reactive Evaporation and Sputtering, 70

III.4 Summary of Principal Equations, 71
III.5 Appendix: Thermodynamic Fundamentals, 72

     The Thermodynamic Potentials and the First and
      Second Laws, 72
     The Gibbs Free Energy: The Relevant Potential for
      Equilibria at Fixed Temperature and Pressure, 73
     Standard Reaction and Formation Quantities,
      and the Equilibrium Constant, 74
     Standard Thermochemical Data, 76

III.6 Mathematical Symbols, Constants, and Their Units, 79
     References, 80

IV Principles of High Vacuum   83

IV.1 Basic Vacuum Concepts, 84

     Pumping Speed, 84
     Throughput, 87
     A Throughput Law, 88
     Conductance, 93

IV.2 Behavior of Real Vacuum Systems, 94

     A More Realistic Vacuum System Model, 94
     Desorption, Outgassing, and Permeation, 96

IV.3 Operation Principles of Vacuum Pumps and Gauges, 99

     How Seven Important Pumps Work, 99
     Two Vacuum Gauges in Widespread Use:
      The Thermocouple and Ionization Gauges, 105

IV.4 Summary of Principal Equations, 107
IV.5 Appendix, 107

     How to Draw and Analyze Vacuum Schematic
      Diagrams, 107
     An Electrical Network Analogy, 108
     A Survey of Past Definitions of Throughput, 111

IV.6 Mathematical Symbols, Constants, and Their Units, 112
     References, 112

V Evaporation Sources   115

V.1 The Effusion Cell and Nozzle-Jet Evaporation Sources, 117

     The Ideal Effusion Cell, 117
     The Cosine Law of Emission, 118

		The Nonequilibrium Effusion Cell, 119
		The Near-Ideal Effusion Cell, 121
		The Open-Tube Effusion Cell, 123
		The Conical Effusion Cell, 124
		The Nozzle-Jet Source, 125
	V.2	Free Evaporation Sources, 127
		Free Evaporation, 127
		The Ideal Point Source Model, 129
		How E-Gun Evaporators Work, 129
		Beam Intensity of the E-Gun Evaporator, 131
	V.3	Pulsed Laser Deposition, 133
		Laser-Induced Vaporization, 133
		A Simple Heating Model, 135
		Other Phenomena, 141
	V.4	Materials Aspects of Evaporation Sources, 143
		Evaporation Temperatures of the Elements, 143
		The Problem of Composition Change in the Evaporation of Alloys, 144
		Crucible Interactions, 146
	V.5	Summary of Principal Equations, 147
	V.6	Mathematical Symbols, Constants, and Their Units, 148
		References, 149

VI Principles of Sputtering Discharges   153

    VI.1   Sputtering Arrangements, 155

                 DC Sputtering, 155
                 RF Sputtering, 156
                 The Magnetron, 157
                 Other Sputtering Arrangements, 158

    VI.2   A Practical Sputtering Plasma and its Current Densities and Potentials, 159

                 A Practical Sputtering Plasma, 159
                 The Ideal Langmuir Probe, 161
                 An Experimental Langmuir Probe Characteristic, 165
                 The Enhanced Ion Current Density, 165
                 The Probe Sheath, 168

    VI.3   Gaseous Discharges for Sputtering, 170

                 A DC Discharge Model, 170

The Cathode and Anode Sheaths, 174
The Sputtering Projectiles that Bombard the Cathode, 176
An RF Discharge Model, 178
The RF Sheaths, 182

VI.4 Summary of Principal Equations, 183
VI.5 Appendix, 184

The Voltage–Current Characteristic of a
  DC Discharge, 184
The Voltage–Current Characteristic of an
  RF Discharge, 189
The DC Glow, 190
The RF Glow, 193
Exceptions to the Above, 193

VI.6 Mathematical Symbols, Constants, and Their Units, 195
References, 196

VII Sputtering 199

VII.1 General Characteristics and Background, 199

Definition of Sputtering, 199
The Mechanisms of Sputtering, 201
A Brief History of Sputtering Theory
  and Simulation, 203
Sources of Sputter Yield Data, 205

VII.2 Trends in Sputter Yield Data, 206

Projectile Energy Dependence, 207
Dependence on Surface Binding Energy, 212
Dependence on Choice of Projectile, 214
Effect of Angle of Incidence, 214
Energy Distribution of Sputtered Particles, 219
Angular Distribution of Sputtered Particles, 220
Single-Crystal Targets, 222
Target Conditioning and Dose Effects, 222

VII.3 Basic Concepts for Modeling, 223

The Surface Binding Energy, 223
Energy Transfer in Binary Elastic Collisions
  of Hard Spheres, 225
Threshold Energy for Sputtering at Normal
  Incidence, 227
Nuclear Energy Loss Theory, 229
Linear Cascade Theory, 232

VII.4 A Simplified Collisional Model for Sputter Yield, 238

    A Yield Expression, 238
    Predictions, 241
    Summary, 244

VII.5 An Ideal Sputter Deposition Source, 245

    The Cosine Law of Emission, 245
    The Beam Intensity of a Sputtering Source, 247
    Combined Internal Flux Spectra for the Simplified Collisional Model, 248
    Combined External Spectra Assuming the Spherical Surface Binding Model, 248
    Combined External Spectra Assuming the Planar Surface Binding Model, 249

VII.6 Summary of Principal Equations Not Found in the Sample Calculation of Yield, 250

VII.7 Appendixes, 251

    Appendix A: The Empirical Yield Formula of Matsunami et al. [1984], 251
    Appendix B: A Summary of Target Parameters, 252
    Appendix C: Some Collisional Sputtering Theories, 256
    Appendix D: A Sample Calculation of Yield with the Simplified Collisional Model, 258

VII.8 Mathematical Symbols, Constants, and Their Units, 259
References, 260

VIII Film Deposition   265

VIII.1 Incident Flux and Film Deposition Rate, 267

    The Incident Flux at the Substrate, 267
    Film Deposition Rate, 269
    Associated Substrate Heating Mechanisms, 272

VIII.2 Film Thickness Profiles of the Ideal Small Source, 277

    Three Fundamental Receiving Surfaces, 277
    The Moving-Shutter Technique, 278

VIII.3 Thermalization and Ionization of the Sputtered Beam, 281

    The Thermalization Distance, 283

Reduction of the Incident Flux, 283
Ionized Physical Vapor Deposition, 286

VIII.4 Deposition with Substrate Rotation and with Ideal Large Sources, 289

Off-Axis Substrate Rotation, 290
A Large Disk Source with a Planar Substrate, 291
A Large Ring Source, 293

VIII.5 Deposition Monitors, 295

The Quartz Crystal Microbalance, 295
True Flux Sensors, 298

VIII.6 Summary of Principal Equations, 300
VIII.7 Appendix: Some Definite Integrals, 300
VIII.8 Mathematical Symbols, Constants, and Their Units, 301
References, 302

Index   **305**

PREFACE

Worldwide, there is enormous, accelerating demand for expertise in the technologies of physical vapor deposition. I believe that the most powerful kind of expertise is grounded in a knowledge of the fundamental physical science. The literature is rich but fragmented, and so to gain this expertise, one needs a volume that presents and then integrates the wide range of basic phenomena and principles on which these technologies are based.

In my own teaching and research at Colorado State University, I found no single source adequate as a basic text or reference on physical vapor deposition, so I set out to produce this book. I have tried to develop a unified chain of concepts starting with conditions or events within the vapor source, and concluding with film deposition rate and film thickness profile across a large substrate. The material is relevant to both the delicate art of single-crystal film growth through epitaxy and the production of blanket coatings of polycrystalline or amorphous layers.

This book is more a reference than a text, although there are worked examples that, I hope, make it clear exactly how to calculate things. In most of my treatments of technical subjects, I start with basic experimental data or theories (usually of a classic stature), discuss underlying mechanisms and processes, and finally give the reader the ability to quantitatively model the practical applications. At the end of each chapter, the reader will find what I believe to be the best original references on the topics at hand.

I am grateful to the students of EE671 and EE673, who have helped me learn these subjects, and to Dr. Jack Deeter for our partnership in research, which deepened my knowledge of physical vapor deposition. Also, I would like to thank Professors Jacques Derrien and André Vantomme for their hospitality during two splendid sabbatical years, when much of this book took shape.

Fort Collins, Colorado   JOHN E. MAHAN

I

INTRODUCTION TO PHYSICAL VAPOR DEPOSITION

I.1 PHYSICAL VAPOR DEPOSITION TECHNOLOGIES AND THEIR BASIC PHYSICAL SCIENCE

Physical vapor deposition is the production of a condensible vapor by physical means and subsequent deposition of a thin film from this vapor. The foremost "physical" means of producing a vapor is simple heating of a source material, as with the hot filament source or the evaporation crucible depicted in Figure I.1. These are often called "thermal" sources. More specialized versions of evaporation sources include the precise effusion cell shown in Figure I.2, or the powerful electron-beam evaporator (E-gun) in Figure I.3. But there are a variety of other physical means, such as laser-induced vaporization with an intense photon beam (pulsed laser deposition; see Fig. I.4), and knocking atoms out of a target with energetic ions (sputtering — Fig. I.5). These techniques are performed under vacuum — evaporation typically under high (10^{-6} torr) to ultrahigh (10^{-9} torr) vacuum, and sputtering under moderate to low vacuum (10^{-4} to 10^{-1} torr).

Overview

The purpose of this book is to present the basic physical science underlying physical vapor deposition. To provide an overview of the modeling structure required for a physical vapor deposition source, we portray schematically in Figure I.6 the chain of concepts and principles governing the accumulation of a thin film by thermal evaporation from an effusion cell:

2 INTRODUCTION TO PHYSICAL VAPOR DEPOSITION

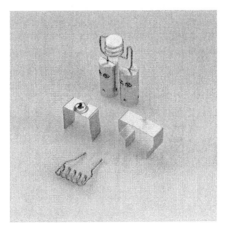

Figure I.1 Material is evaporated from these hot filament, boat, and open-crucible sources, which are resistively heated. Metal wire pieces are hung from the filament, which melts and then evaporates them. (Courtesy of MDC Vacuum Products Corp.)

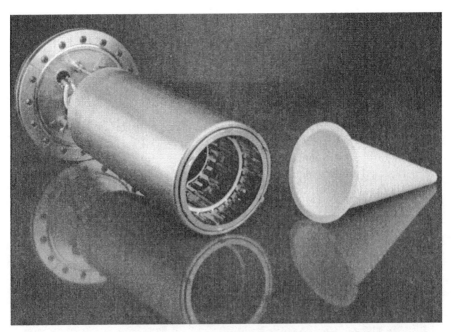

Figure I.2 A beam of evaporant particles is emitted from the end of this effusion cell. The source material is heated with a resistive heating element; the conical BN crucible has been removed for inspection. (Courtesy of VG Semicon).

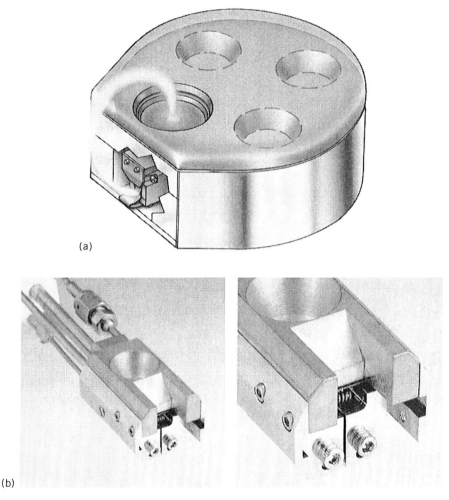

Figure I.3 These E-gun evaporation sources heat the source material through bombardment with energetic electrons, which are thermionically emitted from a hot filament. [(a) Courtesy of Thermionics Vacuum Products, Inc.; (b) Courtesy of MDC Vacuum Products Corp.]

1. The temperature T_{source} within the effusion cell determines.
2. the thermal equilibrium vapor pressure P_{eq} of the source material within the cell.
3. The impingement rate of the vapor (z), the number of particles per unit area per second striking the interior surfaces of the cell, is proportional to P_{eq}.
4. The beam intensity of the source (J_Ω), the number of particles emitted per unit solid angle per second, fundamentally characterizes a vapor

Figure I.4 In a pulsed laser deposition system, an intense laser beam creates a condensible vapor by vaporizing the surface of the target (on the right side of the photo). The visible plume appears when the emitted vapor is ionized by the laser, forming a plasma. The substrate is at left-center in the photo. (Reprinted with permission from D.B. Chrisey and M. A. Savell, *MRS Bulletin*, Vol. XVII, No. 2. Copyright 1992, Materials Research Society.) See color insert.

source. If the orifice size ($\delta A^{1/2}$) is small compared to the distance to the source, the angular distribution may approach the ideal cosine law of emission. This ideal behavior depends on a number of other factors, including whether the emitted particles experience collisions en route to the substrate.

5. The mean free path of the evaporant particles (λ) with respect to collisions with the residual gas in the deposition chamber depends on the level of vacuum attained by the vacuum pump. With evaporation, the mean free path is typically much greater than the distance to the substrate.
6. The incident flux at the substrate (j_i), the number of particles impinging per unit area per second, is a function of emission angle θ, distance R, and deposition angle β (with which the particles arrive at the substrate).
7. The condensation flux j_c is proportional to j_i, but also depends on the condensation coefficient of the film and on any reevaporation flux if the substrate is hot.

PHYSICAL VAPOR DEPOSITION TECHNOLOGIES

(a)

Figure I.5 In a sputter deposition source, ions from a glow discharge bombard a target made of the source material, which is the cathode of the discharge. Atoms ejected from the target condense on a substrate, which is often the anode of the discharge. (*a*) Shown here is a pair of magnetron sources above a substrate (courtesy of Denton Vacuum, LLC). (*b*) Two ion-beam sources on the far left (mostly obscured by the substrate) are bombarding two sputtering targets near the center of the photo. Material accumulates on the substrate on the left. The growing film is being simultaneously bombarded during deposition, with an additional ion source on the far right (out of the photo). (Courtesy of Ion Tech, Inc.) See color insert.

8. The film deposition rate v_n (m/s), the rate at which the film thickens in the direction normal to its surface, is determined by conservation of matter.

We present principles and theories that will allow one to calculate or model each of these steps. It will become evident that the underlying theory of physical vapor deposition is very broad, drawing from many areas of physical science.

Kinetic Theory

Knowing the kinetic theory of gases (Chapter II) allows one to understand and to predict much of the behavior of gases and vapors. The ideal gas law is one

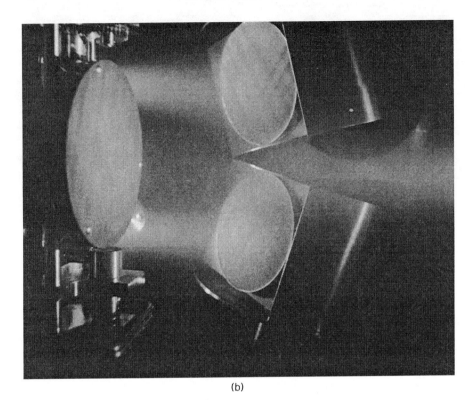

(b)

Figure I.5 *(Continued)*

result of the theory that is of overwhelming importance. For physical vapor deposition, one of the most important concepts of the kinetic theory is the *impingement rate*, which is the number of collisions per unit area per second that a gas makes with a surface, such as a chamber wall or a substrate. The impingement rate as calculated with the kinetic theory is proportional to pressure:

$$z = \frac{P}{\sqrt{2\pi m k T}}, \quad (I.1)$$

where P is the gas pressure, m is the particle mass, k is Boltzmann's constant, and T is the temperature.

Example Calculate the impingement rate of residual gas particles within a vacuum chamber at 10^{-6} torr.

We will assume that the residual gas is composed entirely of nitrogen molecules. At a pressure of 10^{-6} torr, the room temperature impingement rate for particles of 28 amu (atomic mass units) is

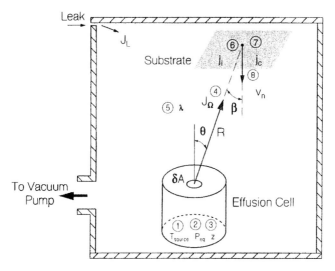

Figure I.6 Vapor deposition of a thin film. An illustration of the chain of concepts and principles leading from the temperature of an evaporation source to the film deposition rate: (1) the source temperature, (2) the thermal equilibrium vapor pressure within the source, (3) the impingement rate of vapor within the source, (4) the source beam intensity, (5) ballistic flight to the substrate (if there is no scattering), (6) the incident flux at the substrate, (7) the condensation flux, and (8) the resulting film deposition rate.

$$z = \frac{10^{-6}\,\text{torr}}{(2\pi \times 28 \times 1.66 \times 10^{-27}\,\text{kg} \times 1.38 \times 10^{-23}\,\text{J/K} \times 300\,\text{K})^{1/2}} = 3.8 \times 10^{18}\,\text{m}^{-2}\,\text{s}^{-1},$$

according to expression (I.1). ■

The impingement rate bears on the question "How long will a clean substrate surface stay clean within the deposition chamber?" With the presence of a residual gas in the chamber, the monolayer adsorption time may be estimated with

$$t_{rg} = \frac{N_s}{\delta \cdot z_{rg}}, \qquad (\text{I.2})$$

where N_s is the surface density of adsorption sites and δ is the trapping probability for an impinging gas particle (presented in Chapter III).

Example Suppose that an atomically clean silicon substrate exists within a deposition chamber having a residual nitrogen gas pressure of 10^{-6} torr. Estimate the time an experimenter might have to deposit a thin film before the surface becomes substantially covered with adsorbed nitrogen atoms.

The number density of atoms in crystalline silicon is $n_{sub} = 5 \times 10^{28}\,\text{m}^{-3}$. As a rough estimate, the surface density of adsorption sites may found from $N_s = n_{sub}^{2/3}$ (although the precise surface site density does vary with crystallographic orientation of the surface). For this example, then, $N_s \approx 1.4 \times 10^{19}\,\text{m}^{-2}$. We will assume that these sites can each accommodate one gas atom and that the gas is nitrogen as in the preceding example.

It is reasonable to assume a trapping probability of unity for a reactive gas such as N_2 striking a reactive surface such as that of silicon. A monolayer formation time may be estimated then as

$$t_{rg} = \frac{N_s}{2z_{rg}} = \frac{1.4 \times 10^{19}\,\text{m}^{-2}}{2 \times 3.8 \times 10^{18}\,\text{m}^{-2}\,\text{s}^{-1}} \approx 1.8\,\text{s}.$$

This is an estimate of the time beyond which an atomically clean surface could *not* be expected to remain clean. (We inserted a 2 into the expression for t_{rg} to account for dissociation of the impinging N_2 molecule.) ∎

The kinetic theory offers useful models for numerous other properties and phenomena that are basic to physical vapor deposition.

Adsorption and Condensation

Adsorption and condensation phenomena (Chapter III) describe, among many things, the accumulation of a thin film on the substrate. The basic science of condensation includes calculation of the thermal equilibrium vapor pressure of substances, and determining whether there is a condition of supersaturation at the substrate.

The thermal equilibrium vapor pressure of substance A is given by

$$P_{A\text{eq}} = P^\circ \exp\left(\frac{\Delta_{\text{vap}} S_A^\circ}{R}\right) \exp\left(\frac{-\Delta_{\text{vap}} H_A^\circ}{RT}\right), \tag{I.3}$$

where P° is the standard pressure (10^5 Pa), $\Delta_{\text{vap}} S_A^\circ$ is the standard entropy of vaporization, $\Delta_{\text{vap}} H_A^\circ$ is the standard enthalpy of vaporization, R is the molar gas constant, and T is the temperature. Vapor pressure is estimated with tabulated values of the thermodynamic quantities. Values for a range of solid elements are given in the Appendix of Chapter III.

As an example of applications of (I.3), we analyze the interesting report by Esposto et al. [1995] that a beverage can will serve as a useful evaporation source of magnesium. They state that the top of a beer can is made of an aluminum alloy that contains 1% magnesium and 1.3% manganese as the major solutes. What sort of a vapor stream might actually be obtained from this alloy when placed in an effusion cell? Could it be pure magnesium and of a useful intensity? What about the fact that it contains more manganese solute than magnesium? Two additional results from Chapters V and VIII are needed — first, the beam intensity of an ideal effusion cell:

$$J_\Omega = \frac{z\,\delta A \cos\theta}{\pi}. \tag{I.4}$$

where δA is the orifice area of the effusion cell and θ is the emission angle.

PHYSICAL VAPOR DEPOSITION TECHNOLOGIES

The second result that we need here is from Chapter VIII — the incident flux at the substrate:

$$j_i = \frac{J_\Omega \cos \beta}{R^2}. \tag{I.5}$$

This relation gives the particle flux at a point on the substrate, in particles per unit area per unit time. Here, β is the deposition angle (see Fig. I.6) and R is the distance from the orifice to the point of interest on the substrate.

Example Calculate the beam intensity when the alloy described above is placed in the effusion cell of Figure I.6. The orifice area is $1\,\text{cm}^2$ and the distance to the substrate is 10 cm. The cell is held at 900 K. Consider the point on the substrate which is directly above the orifice of the cell (i.e., $\theta = 0°$).

First, we will calculate the thermal equilibrium vapor pressures of the three components of the alloy within the effusion cell. We will assume an ideal solution (see Chapter III), so that the vapor pressure of any component is given by the vapor pressure of the pure substance multiplied by its atomic fraction in the alloy: $P_A(T_{\text{source}}) = X_A P_{A\text{eq}}(T_{\text{source}})$. The standard molar entropies and enthalpies of vaporization of the three components (from Chapter III), and their atomic fractions, are

	$\Delta_{\text{vap}} S_A^\circ$ (J/K)	$\Delta_{\text{vap}} H_A^\circ$ (kJ)	X
Magnesium	99	134	0.01
Manganese	106	247	0.013
Aluminum	118	314	0.977

The vapor pressure of magnesium is given by

$$P_{\text{Mg}}(900\,\text{K}) = 0.01 \times 10^5\,\text{Pa} \times \exp\left(\frac{99\,\text{J/K}}{8.31\,\text{J/K}}\right) \times \exp\left(\frac{-134\,\text{kJ}}{8.31\,\text{J/K} \times 900\,\text{K}}\right)$$
$$= 2.47\,\text{Pa}.$$

By similar calculations,

$$P_{\text{Mn}}(900\,\text{K}) = 2.05 \times 10^{-6}\,\text{Pa}$$
$$P_{\text{Al}}(900\,\text{K}) = 8.38 \times 10^{-8}\,\text{Pa}.$$

Note that that $P_{\text{Mg}} \gg P_{\text{Mn}} > P_{\text{Al}}$ in spite of the fact that the magnesium and manganese concentrations are similar, and much less than the aluminum content of the source. This fact is quite important for using the alloy as a magnesium evaporation source.

These vapor pressures correspond to certain impingement rates within the effusion cell. For the magnesium vapor, we obtain

$$z_{Mg}(900\,K) = \frac{2.47\,Pa}{[2\pi \times 24.31 \times 1.66 \times 10^{-27}\,kg \times (1.38 \times 10^{-23}\,J/K) \times 900\,K]^{1/2}}$$
$$= 440\,\text{Å}^{-2} \cdot s^{-1}.$$

By a similar calculation,

$$z_{Mn}(900\,K) = 2.43 \times 10^{-4}\,\text{Å}^{-2} \cdot s^{-1}$$
$$z_{Al}(900\,K) = 1.42 \times 10^{-5}\,\text{Å}^{-2} \cdot s^{-1}.$$

Next, these impingement rates correspond to individual beam intensities for each of the three components. Directly above the orifice,

$$J_{\Omega Mg}(0°) = 440\,\text{Å}^{-2} \cdot s^{-1} \times \cos(0°) \times 1\,cm^2/\pi = 1.40 \times 10^{18}\,s^{-1},$$
$$J_{\Omega Mn}(0°) = 7.72 \times 10^{11}\,s^{-1},$$
$$J_{\Omega Al}(0°) = 4.51 \times 10^{10}\,s^{-1}.$$

Thus the beam intensity for incident is more than six orders of magnitude greater than those of the other two components. The next example will convert these beam intensities to fluxes of atoms onto the substrate. ∎

Example Calculate the incident flux at the substrate for the beam intensities of the previous example. Consider again the point on the substrate that is directly above the orifice of the cell.

Using the previous beam intensity values, the incident fluxes at the point on the substrate under consideration are

$$j_{i\,Mg} = \frac{1.40 \times 10^{18}\,s^{-1} \times \cos(0°)}{(10\,cm)^2} = 1.40\,\text{Å}^{-2} \cdot s^{-1},$$
$$j_{i\,Mn} = 7.72 \times 10^{-7}\,\text{Å}^{-2} \cdot s^{-1},$$
$$j_{i\,Al} = 4.51 \times 10^{-8}\,\text{Å}^{-2} \cdot s^{-1}.$$

The incident flux of magnesium at the substrate is again more than six orders of magnitude greater than the other two, and the magnitude of the flux is on the order of 1 monolayer per second. Thus, this beer can lid can serve as high purity source of magnesium vapor, at least for research-scale evaporations. ∎

Now, if a magnesium thin film is to accumulate on the substrate, the incident magnesium flux must be supersaturated. This fact comes from the Hertz–Knudsen–Langmuir equation, as illustrated in Figure I.7 and discussed further in Chapter V.

This equation expresses the condensation flux at the substrate as proportional to the net difference between the impingement flux from the source,

PHYSICAL VAPOR DEPOSITION TECHNOLOGIES

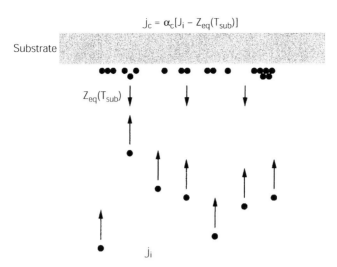

Figure I.7 The net condensation flux at the substrate depends on the difference between the impingement flux j_i from the source and the reevaporation flux $z_{eq}(T_{sub})$ from the substrate. Their difference is multiplied by the condensation coefficient α_c to get the actual condensation flux.

and the reevaporation flux from the substrate:

$$j_c = \alpha_c [j_i - z_{eq}(T_{sub})]. \tag{I.6}$$

where α_c is the condensation coefficient, which gives the fraction of impinging particles that actually condense; it is assumed in the above expression to equal the vaporization coefficient, the probability of success for a particle attempting to vaporize.

The degree of supersaturation and the condition for accumulation of a thin film (Chapter III) are then given by

$$S \equiv \frac{j_i}{z_{eq}(T_{sub})} - 1 \geq 0 \tag{I.7}$$

This inequality means that j_i must be greater than the impingement rate corresponding to the substrate temperature if a film is to condense on the substrate. This does not mean merely that the substrate temperature must be below the source temperature — sometimes it must be *substantially* below the source temperature.

Example Suppose that for the previous example of using a beer can lid as a source of magnesium, the substrate is held at a temperature of 580 K. Will a thin film of magnesium form?

We obtained a value of $1.40 \, \text{Å}^{-2} \cdot \text{s}^{-1}$ for j_{iMg} and must now compare this to the thermal equilibrium impingement rate for magnesium at the substrate temperature. First, the thermal equilibrium vapor pressure of magnesium at the substrate temperature is

$$P_{\text{eqMg}}(T_{\text{sub}}) = 10^5 \text{ Pa} \times \exp\left(\frac{99 \text{ J/K}}{8.31 \text{ J/K}}\right) \times \exp\left(\frac{-134 \text{ kJ}}{8.31 \text{ J/K} \times 580 \text{ K}}\right)$$
$$= 1.26 \times 10^{-2} \text{ Pa}.$$

Next

$$z_{\text{eqMg}}(T_{\text{sub}}) = \frac{1.26 \times 10^{-2} \text{ Pa}}{[2\pi \times 24.31 \times 1.66 \times 10^{-27} \text{ kg} \times (1.38 \times 10^{-23} \text{ J/K}) \times 580 \text{ K}]^{1/2}}$$
$$= 2.79 \text{Å}^{-2} \cdot \text{s}^{-1}.$$

The supersaturation is then $(1.40 \text{Å}^{-2} \cdot \text{s}^{-1}/2.79 \text{Å}^{-2} \cdot \text{s}^{-1}) - 1 < 0$, so no magnesium film will form even though $T_{\text{sub}} \ll T_{\text{source}}$!

Lowering the substrate temperature *would* allow a magnesium film to form — additional calculations show that the degree of supersaturation exceeds zero as T_{sub} falls below ~ 570 K. ∎

High Vacuum

Physical vapor deposition is performed under high vacuum in order to attain a desired level of purity of the thin film. We have already seen that a chamber pressure of 10^{-6} torr allows the film depositor about one second to deposit a film on a clean surface before that surface becomes substantially contaminated with residual gas particles. In a previous example the impingement rate of N_2 at this pressure was calculated to be $3.8 \times 10^{18} \text{ m}^{-2} \cdot \text{s}^{-1}$. If the impingement flux of evaporant particles at the substrate were this same value, the resulting film might well contain 50% residual gas atoms.

Suppose that a film deposition is being performed in a chamber with an inadvertent leak as shown in Figure I.6 and that the leak is equivalent to a cylindrical passage that is only 1.5 μm in radius and 50 μm in length. These are small dimensions, but would this be a serious leak? What is the rate at which air particles enter the chamber? To analyze this situation, we will need several results from Chapter II.

The Poisseuille formula gives the particle flow rate through a cylindrical tube in laminar flow:

$$J = \left(\frac{\pi a^4}{8\eta kT}\right) \frac{P_{\text{av}} \Delta P}{L} \tag{I.8}$$

where a is the radius of the tube, is the viscosity of the gas, P_{av} is the average pressure within the tube, and ΔP is the pressure drop between the ends of the tube. The viscosity according to kinetic theory is

PHYSICAL VAPOR DEPOSITION TECHNOLOGIES

$$\eta = \frac{nv_{av}m\lambda}{4}, \tag{I.9}$$

where n is the gas density, v_{av} is the average particle speed, m is the particle mass, and λ is the mean free path. These quantities are given by

$$v_{av} = \left(\frac{8kT}{\pi m}\right)^{1/2}, \tag{I.10}$$

$$n = \frac{P}{kT}, \tag{I.11}$$

and

$$\lambda = \frac{1}{\sqrt{2}\pi d^2 n}, \tag{I.12}$$

where d is the particle diameter.

Example To calculate the particle flow rate through the leak described above, we will assume that laminar flow occurs in the leak and that the gas is pure nitrogen at 298 K.

The average particle speed is

$$v_{av} = \left(\frac{8kT}{\pi m}\right)^{1/2} = \left(\frac{8 \times 1.38 \times 10^{-23}\,\text{J/K} \times 298\,\text{K}}{\pi \times 28 \times 1.66 \times 10^{-27}\,\text{kg}}\right)^{1/2} = 475\,\text{m/s}.$$

The particle density in the gas is calculated with the ideal gas law:

$$n = \left(\frac{760\,\text{torr}}{1.38 \times 10^{-23}\,\text{J/K} \times 298\,\text{K}}\right) = 2.46 \times 10^{25}\,\text{m}^{-3}.$$

Taking a value of 3.75 Å for the effective diameter of an N_2 molecule from Chapter II, we obtain

$$\lambda = \frac{1}{\sqrt{2}\pi \times (3.75 \times 10^{-10}\,\text{m})^2 \times 2.46 \times 10^{25}\,\text{m}^{-3}} = 6.50 \times 10^{-8}\,\text{m}.$$

Since this value is much less than the dimensions of the leak, our assumption of laminar flow is valid.

The viscosity is now calculated as

$$\eta = \frac{2.46 \times 10^{25}\,\text{m}^{-3} \times 475\,\text{m/s} \times 28 \times 1.66 \times 10^{-27}\,\text{kg} \times 6.50 \times 10^{-8}\,\text{m}}{4} \times \frac{5\pi}{8}$$

$$= 1.73 \times 10^{-4}\,\text{poise(P)}.$$

The particle flow rate then is

$$J = \left[\frac{\pi \times (1.5 \times 10^{-6}\,\text{m})^4}{8 \times 1.73 \times 10^{-4}\,\text{P} \times (1.38 \times 10^{-23}\,\text{J/K}) \times 298\,\text{K}} \right]$$
$$\times \left(\frac{760\,\text{torr}}{2} \right) \times \frac{760\,\text{torr}}{50 \times 10^{-6}\,\text{m}} = 2.87 \times 10^{15}\,\text{s}^{-1}. \quad\blacksquare$$

What would be the steady-state chamber pressure due to such a leak? Before analyzing this, we must preview some results from Chapter IV. Although a vacuum leak fundamentally means that a certain number of air molecules are entering the chamber per second, in order to model vacuum pumping dynamics a leak is characterized with a throughput:

$$Q_L \equiv kTJ_L, \qquad (\text{I.13})$$

where J_L is the particle flow rate of the leak and Q_L is the corresponding throughput. The ultimate pressure of a vacuum system, due to this leak, is given by

$$P_\text{ult} = \frac{Q_L}{S}, \qquad (\text{I.14})$$

where S here is the pumping speed of the vacuum pump. (It is interesting to note that the size of the vacuum chamber has no bearing on the ultimate pressure.)

Example Suppose that a pump of speed 2500 liters/s is used with a deposition chamber having the leak described above. What steady-state (ultimate) pressure will prevail?

The throughput of this leak is

$$Q = (1.38 \times 10^{-23}\,\text{J/K}) \times 298\,\text{K} \times 2.87 \times 10^{15}\,\text{s}^{-1} = 8.85 \times 10^{-5}\,\text{torr}\cdot\text{liters/s}.$$

Then the ultimate pressure of the chamber due to this leak is

$$P_\text{ult} = (8.85 \times 10^{-5}\,\text{torr}\cdot\text{liters/s})/(2500\,\text{liters/s}) = 3.54 \times 10^{-8}\,\text{torr}.$$

The leak could easily be significant for the purity of a deposited thin film. ∎

Sputtering Discharges

The mechanism of collisional sputtering is illustrated in Figure I.8a, where an energetic ion strikes the surface of a sputtering target. An atomic collision cascade ensues, with the ejection of a recoil (a dislodged target atom). Such escaping recoils constitute a condensible vapor that can be utilized to form a thin film of the target material.

PHYSICAL VAPOR DEPOSITION TECHNOLOGIES

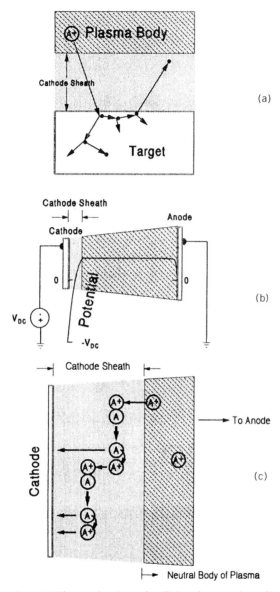

Figure I.8 Sputtering: (*a*) The mechanism of collisional sputtering; (*b*) the electrostatic potential profile of a DC sputtering discharge; (*c*) the charge exchange process within the cathode sheath.

In direct-current (DC) sputtering the energetic ion comes from a plasma that is created by a DC power supply, as suggested in Figure I.8*b*. A substantial pressure of argon gas ($\sim 0.01-1$ torr, more or less) exists within the chamber. We will see in Chapter VI that the enhanced ion current density entering the cathode sheath may be estimated as

$$j_{\text{ion}} \approx qn^{-}\sqrt{\frac{k(T^{+} + T^{-})}{m^{+}}}, \qquad (1.15)$$

where q is the magnitude of the electronic charge, n^{-} is the electron density in the plasma, T^{+} is the ion temperature, T^{-} is the electron temperature, and m^{+} is the mass of an ion. This ion current density is often a substantial fraction of 1 mA/cm^2.

The ion gains kinetic energy as it traverses the cathode sheath, a space-charge region that separates the sputtering target (the cathode of the discharge) from the main body of the plasma. (The anode of the discharge in Fig. I.8 is the substrate.) The kinetic energy with which the ion strikes the target determines the sputter yield, the number of target atoms ejected per incident ion. The maximum possible value of this kinetic energy corresponds to a potential drop at the cathode that is actually greater than the DC applied voltage (as suggested in Fig. I.8b and explained in Chapter VI).

In practice, the kinetic energy of the impinging ions is reduced by the process of symmetric charge exchange in the cathode sheath, as portrayed schematically in Figure I.8c. The ion is neutralized after partly traversing the sheath, and so it acquires only a fraction of the maximum possible kinetic energy. However, with charge exchange a new ion is created, which is then accelerated toward the cathode. The overall result of symmetric charge exchange is that while the sputtering projectile energy is decreased, the total number of projectiles is increased. In Figure I.8c, one ion entering the cathode sheath is converted to three projectiles: one ion and two neutrals. Each of these eventually strikes the cathode with a kinetic energy equal to about one-third of the maximum possible value. These and many other aspects of sputtering discharges are discussed in Chapter VI.

I.2 SUMMARY OF PRINCIPAL EQUATIONS

Impingement rate	$z = P/\sqrt{2\pi m kT}$
Monolayer adsorption time	$t_{\text{rg}} = N_s/(\delta \cdot z_{\text{rg}})$
Thermal equilibrium vapor pressure	$P_{\text{Aeq}} = P° \exp(\Delta_{\text{vap}} S_A°/R)\exp(-\Delta_{\text{vap}} H_A°/RT)$
Beam intensity of an evaporation source	$J_\Omega = z\,\delta A \cos\theta/\pi$
Incident flux	$j_i = J_\Omega \cos\beta/R^2$
Condensation flux	$j_c = \alpha_c (j_i - z_{\text{eq}}(T_{\text{sub}}))$
Supersaturation ratio	$S \equiv j_i/j_{\text{eq}} - 1$
Poisseuille formula	$J = (\pi a^4/8\eta kT)P_{\text{av}}\Delta P/L$
Viscosity	$\eta = nv_{\text{av}}m\lambda/4$
Average particle speed	$v_{\text{av}} = (8kT/\pi m)^{1/2}$

Ideal gas law $\quad P = nkT$
Mean free path $\quad \lambda = 1/\sqrt{2}\pi d^2 n$
Throughput of a leak $\quad Q_L \equiv kTJ_L$
Ultimate pressure $\quad P_{ult} \equiv Q_L/S$
Enhanced ion current density $\quad j_{ion} \approx qn^-\sqrt{k(T^+ + T^-)/m^+}$

I.3 MATHEMATICAL SYMBOLS, CONSTANTS, AND THEIR UNITS

SI (International System of Units) units are given first, followed by other units in widespread use.

a	Radius (m)
d	Particle diameter (m; Å $= 10^{-10}$ m)
j_c	Condensation flux (m$^{-2} \cdot$ s^{-1})
j_i	Incident flux (m$^{-2} \cdot$ s^{-1})
j_{ion}	Enhanced ion current density (A $\cdot$ m^{-2})
k	Boltzmann's constant (1.38×10^{-23} J/K)
m	Mass (kg; amu $= 1.66 \times 10^{-27}$ kg)
n	Number density of atoms (m^{-3}; Å$^{-3} = 10^{-30}$ m^{-3})
q	Magnitude of the electronic charge (1.60×10^{-19} C)
t	Time (s)
v_n	Deposition rate (m/s)
z	Impingement rate (m$^{-2} \cdot$ s^{-1})
J	Particle flow rate (s^{-1})
J_Ω	Beam intensity of source (s$^{-1} \cdot$ steradian^{-1})
L	Length (m)
N_s	Surface density of adsorption sites (m^{-2}; Å$^{-2} = 10^{-20}$ m^{-2})
P	Pressure (Pa; torr $= 133$ Pa; atm $= 1.01 \times 10^5$ Pa)
Q	Throughput (W; torr $\cdot$ liter $\cdot$ s$^{-1} = 0.133$ W)
R	Molar gas constant (8.31 J$\cdot$mol$^{-1}\cdot$K^{-1}); distance from source to substrate (m)
S	Degree of supersaturation (dimensionless)
T	Temperature (K)
α_c	Condensation coefficient (dimensionless)
β	Deposition angle (rad)
δ	Trapping probability (dimensionless)
δA	Area of orifice (m^{-2})
$\Delta_{vap}H$	Enthalpy of vaporization (kJ/mol; kcal/mol $= 4.18$ kJ/mol)
$\Delta_{vap}S$	Entropy of vaporization (kJ $\cdot$ mol$^{-1}\cdot$ K^{-1}; kcal $\cdot$ mol$^{-1}\cdot$ K^{-1} = 4.18 kJ $\cdot$ mol$^{-1}\cdot$ K^{-1})
η	Viscosity (kg $\cdot$ m$^{-1}\cdot$ s^{-1}; poise $= 0.1$ kg $\cdot$ m$^{-1}\cdot$ s^{-1})
θ	Emission angle (rad)

λ   Mean free path (m)
Ω   Solid angle (steradians)

REFERENCE

Esposto, F. J., Cory, C., Griffiths, K., Norton, P. R. and Timsit, R. S., 1995, "Al-Mg Alloy from a Beer Can as a Simple Source of Mg Metal for Evaporators in Ultrahigh Vacuum Applications," *J. Vac. Sci. Technol.* **A13**(6), 3000.

II

THE KINETIC THEORY OF GASES

There is nothing more practical than a good theory.

— Von Karmann

The kinetic theory of gases is of enormous importance to physical vapor deposition. Although it is a classical model, its predictions are remarkably accurate and it affords a physical picture of the behavior of a gas that is easy to grasp and use. It provides estimates of the impingement rate, mean free path, heat capacity, diffusivity, viscosity, and thermal conductivity, parameters that frequently bear on the deposition kinetics and purity of a thin film. It is used in calculating the conductance of gas flow structures and modeling vacuum system pumping dynamics, and predicting the output of physical vapor deposition sources. The kinetic theory enters time and again into models of nearly all aspects of film deposition and growth.

The kinetic theory is a "hard sphere" model. The fundamental assumptions are

- The gas contains an enormous number of classical particles.
- The particles are infinitely hard spheres.
- All collisions are elastic.
- There are no forces on the particles except during collisions with each other or with the container.

If only real billiard balls were so well behaved!

The kinetic theory is a microscopic theory, obtaining properties of gases from the characteristics of the individual particles (atoms or molecules).

We summarize below the parts that are most immediately useful to thin film science; more detailed discussions of the theory may be found in two books of the same name, *Kinetic Theory of Gases* [Present, 1958; Kennard, 1938]. One of this author's favorite references is a short book by Guggenheim [1960].

II.1 STATISTICS

The Boltzmann Distribution

The distribution in energy of the particles (f) is given by the classical Boltzmann distribution function:

$$f(E) = \exp\frac{-E}{kT}, \qquad (II.1)$$

where k is Boltzmann's constant, T is the absolute temperature, and E is the energy. The Boltzmann distribution function is appropriate for identical, distinguishable (and thus classical) particles that do not obey the Pauli exclusion principle. The number of particles having a kinetic energy between E and $E + dE$ is therefore given by $Nf(E)dE/kT$, where N is the total number of particles and $1/kT$ is the necessary normalizing factor.

This distribution function is sometimes described as Maxwellian, sometimes called the Maxwell–Boltzmann distribution, and sometimes called Boltzmann's distribution law. It was first stated by James Clerk Maxwell in 1859 [Guggenheim, 1960] and was directly proved by L. Boltzmann some years later (in 1896; see Present [1958]). By whatever name, this is the only possible energy distribution function for an ideal gas in equilibrium. It is the energy distribution of maximum entropy for a closed system of classical particles.

The *Boltzmann factor* gives the fraction of the total number of particles in the system whose energy is greater than or equal to some specified energy, say, E_a. It is calculated from

$$\int_{E_a}^{\infty} \exp\left[\frac{-E}{kT}\right] \frac{dE}{kT} = \exp\left[\frac{-E_a}{kT}\right]. \qquad (II.2)$$

The Boltzmann factor is important when considering thermally activated rate processes. Such processes exhibit a temperature dependence given by (II.2); when the rate is plotted in the form of an *Arrhenius plot* [log(rate) vs. $1/T$], a straight line is obtained. The slope of the line is $-E_a/k$, and E_a is called the *thermal activation energy*. Physically, such a process is "activated thermally," in the sense that the particles that can participate are those that receive by temperature fluctuations the minimum necessary energy (which is E_a). The fraction of all the particles present that can participate is given by the Boltzmann factor. (See the Appendix for a method for determining the activation energy graphically from an Arrhenius plot.)

STATISTICS

In the kinetic theory of gases, the energy of a particle is entirely translational kinetic energy, given by

$$E = \frac{mv^2}{2} = \frac{m(v_x^2 + v_y^2 + v_z^2)}{2}, \tag{II.3}$$

where v is its speed and v_x and so on are the Cartesian components of its velocity and m is the particle's mass.

It is frequently useful to work with the subgroup of particles that have essentially the same velocity vector. Consider those particles occupying a certain differential volume of "velocity space," as illustrated in Figure II.1. A probability distribution function may be constructed from the Boltzmann distribution. It is based on the probability of having the total energy associated with the velocity $\bar{v} = (v_x, v_y, v_z)$:

$$F(v_x, v_y, v_z) = \frac{\exp[-m(v_x^2 + v_y^2 + v_z^2)/2kT]}{\int_{-\infty}^{\infty} \int_{-\infty}^{\infty} \int_{-\infty}^{\infty} \exp[-m(v_x^2 + v_y^2 + v_z^2)/2kT] dv_x\, dv_y\, dv_z}. \tag{II.4}$$

Expression (II.4) is the probability of occupancy of a differential volume of velocity space $(dv_x\, dv_y\, dv_z)$, per unit volume of that space, by a gas particle. The

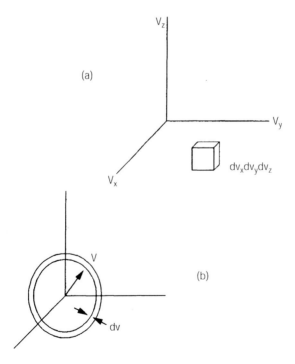

Figure II.1 Velocity space. (*a*) Cartesian coordinates; (*b*) spherical coordinates.

probability itself is $F(v_x, v_y, v_z)dv_x\,dv_y\,dv_z$. The denominator, a normalizing factor, is $(2\pi kT/m)^{3/2}$.

Characteristic Particle Speeds

Speed is the magnitude of the velocity vector. The probability distribution function for speed $[F(v)]$ is obtained from that given above for the Cartesian velocity components after reexpressing it in spherical coordinates. The differential volume in velocity space, instead of an infinitesimal cube, becomes that of a thin spherical shell given by $4\pi v^2 dv$; $v_x^2 + v_y^2 + v_z^2$ becomes simply v^2. The new expression is

$$F(v) = \frac{\exp[-mv^2/2kT]}{\int_0^\infty \int_0^{2\pi} \int_0^\pi \exp[-mv^2/2kT]v^2 \sin\theta\,d\theta\,d\varphi\,dv} \tag{II.5}$$

The probability itself is $4\pi v^2 F(v)dv$. The denominator is, once again, $(2\pi kT/m)^{3/2}$. This function is plotted in Figure II.2.

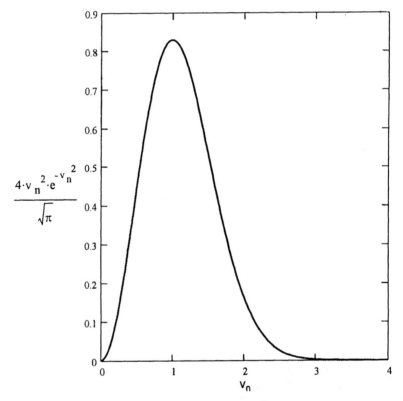

Figure II.2 Probability that a particle's speed is between v_n and $v_n + dv_n$, expressed in units of $(2kT/m)^{1/2}$, where v_n is the particle's speed, normalized to the most probable speed.

There are several characteristic speeds, including the most probable speed, the average speed, and the root mean square(rms) speed.

The *most probable speed* (v_p) is the speed for which $d[4\pi v^2 F(v)]/dv$ is zero (ignoring the $v=0$ case):

$$v_p = \left(\frac{2kT}{m}\right)^{1/2}. \tag{II.6}$$

The *average speed* is equal to $(m/2\pi kT)^{3/2} \int_0^\infty v \cdot \exp[-mv^2/2kT] \cdot 4\pi v^2 dv$:

$$v_{av} = \left(\frac{8kT}{\pi m}\right)^{1/2}. \tag{II.7}$$

The *root-mean-square speed* (v_{rms}) is found from the square root of $(m/2\pi kT)^{3/2} \int_0^\infty v^2 \cdot \exp[-mv^2/2kT] \cdot 4\pi v^2 dv$:

$$v_{rms} = \left(\frac{3kT}{m}\right)^{1/2}. \tag{II.8}$$

Thus, the average kinetic energy (E_{av}) of the particles is

$$E_{av} = \frac{3kT}{2} \tag{II.9}$$

It is always true for the Boltzmann distribution that $v_p < v_{av} < v_{rms}$.

Example Calculate the rms speed, and the average kinetic energy, of a nitrogen molecule at room temperature.

Solution

$$v_{rms} = \left[\frac{3 \times (1.38 \times 10^{-23} \text{ J/K}) \times 298 \text{ K}}{28 \times 1.66 \times 10^{-27} \text{ kg}}\right]^{1/2} = 515 \text{ m/s}$$

and

$$E_{av} = \frac{28 \times 1.66 \times 10^{-27} \text{ kg} \times (515 \text{ m/s})^2}{2} = 6 \times 10^{-21} \text{ J} = 0.04 \text{ eV}. \blacksquare$$

II.2 COLLISIONS

Impingement Rate and Incident Flux Angular Distribution

The *impingement rate* is the number of collisions with a surface to which the gas is exposed, per unit area per unit time. The *cosine law of impingement* specifies the directional distribution of the impinging particles. The former is calculated directly using Cartesian coordinates, while the latter will require polar coordinates.

First, we calculate the basic impingement rate with the aid of Figure II.3a. The number of particles per unit volume having a y-component of velocity between v_y and $v_y + dv_y$ will be denoted $nF_y\,dv_y$, where n is the number of particles per unit volume. Referring to that figure, the total number that will strike the wall, per unit surface area within a time δt is $nF_y\,dv_y \cdot v_y \delta t$. This expression is based on the fact that to strike the wall within time δt, a molecule with velocity component v_y must be somewhere in the surface layer of gas of thickness $v_y\,dt$.

The velocity distribution solely with respect to $v_y (F_y)$ is found by integrating over all possible values of v_x and v_z:

$$F_y = \int_{-\infty}^{\infty} \int_{-\infty}^{\infty} F(v_x, v_y, v_z) dv_x\,dv_z$$
$$= \left(\frac{m}{2\pi kT}\right)^{1/2} \exp\left(\frac{-mv_y^2}{2kT}\right). \tag{II.10}$$

Thus the total number of particles that strike the wall, per unit area, within time δt is

$$N_{\text{total}} = \int_0^{\infty} nF_y v_y \delta t\,dv_y$$
$$= n\delta t \left(\frac{kT}{2\pi m}\right)^{1/2}. \tag{II.11}$$

The impingement rate is $N_{\text{total}}/\delta t$:

$$z = n\left(\frac{kT}{2\pi m}\right)^{1/2}. \tag{II.12}$$

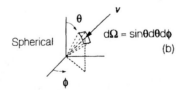

Figure II.3 Impingement: coordinates for deriving (a) the impingement rate and (b) the cosine law of impingement.

There is a law for the angular distribution of the flux of impinging particles. It is necessary to shift to polar coordinates to perform this calculation, proceeding in four steps. The bombarded surface is now assumed to be in the $x-y$ plane:

1. All the particles having speed v that are *capable* of striking the surface must be contained within a thin region near the surface of thickness $v\cos\theta\,\delta t$, where θ is the polar angle as defined in Figure III.3, specifying the orientation of the velocity vector with respect to the surface normal direction. This expression adjusts itself for all directions of travel toward the surface — the region of impingement becomes very thin for large values of θ and is of maximum thickness when the particle is heading directly toward the wall.
2. Next, the number of particles per unit surface area, having speed between v and $v+dv$, which are capable of striking the wall within time δt is calculated from the number density, the thickness of the region of impingement, and the probability distribution function for speed: $n \cdot v\cos\theta\,\delta t \cdot 4\pi v^2 F(v)dv$.
3. Not all of these particles will strike the wall, however, because although they may be contained in the thin layer near the surface, they have a random distribution in direction of travel. The fraction of them that are traveling in a direction defined by the polar angle θ and the azimuthal angle φ, to within a differential solid angle $d\Omega$, is $d\Omega/4\pi$. $d\Omega$ is given by $\sin\theta\,d\theta\,d\varphi$. These quantities are illustrated in Figure III.3b. Thus the flux (per unit area per unit time) of particles onto the wall, having speeds between v and $v+dv$ and coming from directions that are within $d\Omega$ of that specified by θ and φ, is $n \cdot v\cos\theta \cdot v^2 F(v)dv \cdot \sin\theta\,d\theta\,d\varphi$. The total impingement rate is the definite integral of this function over all possible values of v, θ, and φ. However, we are interested here in the directional distribution of impinging particles.
4. Consider a point on the surface where the origin of the polar coordinate system is placed. The total flux of impinging particles per unit solid angle (the *incident flux distribution*, j_Ω) that come from directions within the cone defined by $d\Omega$ is the integral over all possible speeds of the previously derived expression after dividing by $d\Omega$:

$$j_\Omega(\theta,\varphi) = \int_0^\infty n \cdot v\cos\theta \cdot v^2 F(v)dv = \frac{n\cos\theta\, v_{\text{av}}}{4\pi}. \tag{II.13}$$

Thus, this incident flux distribution has a cosine dependence on the polar angle, while it is independent of azimuthal angle. This is the same distribution as that of the emission flux leaving the wall, and the similarity is not coincidental: "by the second law of thermodynamics, the molecules leaving a surface at equilibrium must reflect those incident upon it in all particulars on the average" [Rosenblatt, 1976]. This fact is used later in calculating the flow through a tube in Poisseuille flow.

We return now to a discussion of the total impingement rate. Anticipating the ideal gas law, $PV = NkT$, the impingement rate may be shown to be proportional to pressure:

$$z = P(2\pi mkT)^{-1/2}. \tag{II.14}$$

The impingement rate may be shown from some expressions that were derived above to be proportional to the average speed of the particles:

$$z = \frac{nv_{av}}{4}. \tag{II.15}$$

If there are groups of particles of different chemical species (i.e., different masses and densities), the integrals shown above may be separated into individual integrals — one for each type of particle. The impingement rate for each type is given by expression (II.15) with the appropriate n and m used and where the total impingement rate is the sum of the all the individual impingement rates. This is, actually, the basis for proving Dalton's law with the kinetic theory: "The pressure of a mixture of two or more perfect gases is simply the sum of the pressures which they would exert if each occupied the same volume by itself." The proportionality between partial pressure and an individual specie's impingement rate is the same as that for a pure gas.

The Ideal Gas Law

The ideal gas law follows directly from the assumptions of the kinetic theory of gases. We derive it by calculating the pressure as the rate of transfer of momentum, per unit area, to the wall of the container.

The pressure due solely to particles having velocity component between v_y and $v_y + dv_y$ is equal to the number of particles striking the wall per unit area, times the momentum transferred per collision, per unit time:

$$dP(v_y) = \frac{nF_y dv_y \cdot v_y \delta t \cdot 2mv_y}{\delta t}. \tag{II.16}$$

The total pressure is

$$P = 2mn \int_0^\infty \left(\frac{m}{2\pi kT}\right)^{1/2} \exp\left(\frac{-mv_y^2}{2kT}\right) v_y^2 \, dv_y = nkT. \tag{II.17}$$

The ideal gas law shown above is the equation of state for particles whose behavior is defined by the assumptions of the kinetic theory. It relates the thermodynamic state variables P and T (through n).

COLLISIONS

The partial pressures of a mixture of gases may be calculated by separating the preceding integrals into individual integrals for each gaseous specie. The total pressure is the sum of the partial pressures, as was mentioned in the discussion of impingement rate.

Mean Free Path

The *mean free path* (λ) is the distance a particle travels, on the average, before experiencing a collision. To calculate λ, we must define and calculate first the survival probability $Q(x)$ that a given particle will have traveled a distance x, from the location of the previous collision, before the next collision; $1 - Q(x)$ is the probability of having had a collision after traveling a distance x. Q decreases with x according to the following expression for the probability $(d[1 - Q(x)]/dx)$ of the particle's experiencing a collision while traveling between x and $x + dx$:

$$\frac{-dQ}{dx} = QP_c, \tag{II.18}$$

where P_c is the average number of collisions per unit distance traveled, a constant. From the preceding equation, we see that

$$Q(x) = \exp(-P_c x). \tag{II.19}$$

The mean free path is found by calculating the average distance traveled:

$$\lambda = \int_0^\infty x(-dQ/dx)\,dx = \int_0^\infty xP_c \exp(-P_c x)\,dx = \frac{1}{P_c}. \tag{II.20}$$

Thus, the mean free path is the reciprocal of the probability of a collision per unit distance traveled. We now calculate P_c for "billiard ball" particles.

Referring to Figure II.4, we utilize the assumption that the particles are hard spheres of diameter d, where d is also the *radius* of a sphere surrounding a given particle, into which the center of another particle cannot penetrate. Thus, a particle's collision cross section is πd^2. On the average, a particle sweeps out a volume of space at the rate given by $\pi d^2 v_{av}$ and thus encounters other particles at a rate given by $\pi d^2 v_{av} n$ (v_{av} is the average speed of a particle relative to the container wall).

The last expression was derived assuming the target particles are stationary. However, this is not true, and it may be shown that the average *relative* speed of particles in the kinetic theory is $\sqrt{2} \cdot v_{av}$. Thus, the correct collision rate (r) is

$$r = \sqrt{2}\pi d^2 v_{av} n. \tag{II.21}$$

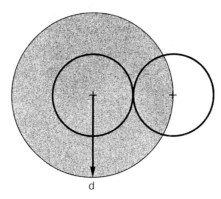

Figure II.4 "Billiard ball" particles. For two hard spheres of the same size, the collision cross section is given by πd^2.

The collision rate is also the average speed multiplied by the collision probability per unit distance traveled:

$$r = v_{av} P_c. \tag{II.22}$$

Thus

$$P_c = \sqrt{2}\pi d^2 n, \tag{II.23}$$

and

$$\lambda = \frac{1}{\sqrt{2}\pi d^2 n}. \tag{II.24}$$

The mean free path is typically determined experimentally from viscosity measurements. Using the preceding relation II.24, the effective particle diameter may be determined. As Table II.1 suggests, it is in the range from 2 to 5 Å for most common gases.

TABLE II.1 Effective Particle Diameter (in Angstroms)

He	2.18
H_2	2.74
O_2	3.61
Ar	3.64
Air	3.72
N_2	3.75
CO_2	4.59
H_2O	4.60

Source: From Kennard [1938].

The end-of-chapter Appendix contains a more complete listing of particle diameters for the elements. Mean free paths of most common gases can be estimated from Figure II.5.

Now consider an impurity molecule at a very low concentration (n_2) within an ideal gas host (of concentration n_1). It may be shown, by evaluating the necessary integrals [Guggenheim, 1960], that the average relative speed (v_{rel}) of the impurity (2) with respect to the host (1) is

$$v_{\text{rel}} = [(v_{2\text{av}})^2 + (v_{1\text{av}})^2]^{1/2}. \tag{II.25}$$

The collision cross section $\pi(d_{21})^2$ is based on the average of their diameters:

$$d_{21} = \frac{d_2 + d_1}{2}. \tag{II.26}$$

Thus, the mean free path of a dilute impurity (λ_2) may be shown to be

$$\lambda_2 = \frac{1}{\pi(d_{21})^2} \left[\frac{1 + (v_{1\text{av}})^2}{(v_{2\text{av}})^2} \right]^{1/2} n_1. \tag{II.27}$$

Example Estimate the mean free path in nitrogen gas at one atmosphere and room temperature.

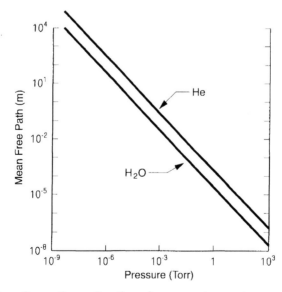

Figure II.5 Mean free path as a function of pressure at room temperature for He and H_2O. The range shown here gives the upper and lower limits for most common gases.

The density under these conditions, from the ideal gas law, is

$$n = \frac{760 \text{ torr}}{(1.38 \times 10^{-23} \text{ J/K}) \times 298 \text{ K}} = 2.46 \times 10^{25} \text{ m}^{-3}.$$

The mean free path is then

$$\lambda = \frac{1}{\sqrt{2}\pi \times (3.75 \text{ Å})^2 \times 2.46 \times 10^{25} \text{ m}^{-3}} = 650 \text{ Å}. \qquad \blacksquare$$

We would like to proceed now with brief and naïve derivations of some fundamental properties of gases: the heat capacity, diffusivity, viscosity, and thermal conductivity. Later, we will calculate the flow through an orifice and a tube for the Poisseuille (viscous) and Knudsen (molecular) flow regimes. It is hoped that the reader will be given mental pictures to which to refer for the basic physical meaning of these properties. More sophisticated treatments are available in the references. The derived quantities are found to disagree with rigorously derived exact expressions only in some values of the numerical coefficients; the correction factors, which are given, are on the order of unity.

II.3 PROPERTIES

Heat Capacity; the Ideal Diatomic Gas

The kinetic theory, when applied to an ideal *mono*atomic gas, predicts a heat capacity at constant pressure (c_p) of $5R/2$ ($20.78 \text{ J} \cdot \text{mol}^{-1} \cdot \text{K}^{-1}$). The heat capacities of real *poly*atomic gases are typically somewhat larger. First, the monatomic gas:

Since the average kinetic energy of a particle is $3kT/2$, the average kinetic energy of a mole of particles is

$$E_{av} = \frac{3RT}{2}. \tag{II.28}$$

This is also the average internal energy of a mole of particles. The heat capacity at constant pressure is defined as

$$c_p \equiv \left. \frac{\delta Q}{\delta T} \right|_P. \tag{II.29}$$

Using the first law of thermodynamics ($dU = \delta Q - dW$, where dU is the change in internal energy per mole and δQ is the heat transferred) and $U = E_{av}$ with only "$P\,dV$" work ($= dW$), we obtain

$$c_p = \left. \frac{\partial U}{\partial T} \right|_P + \left. \frac{P\,\partial V}{\partial T} \right|_P = \frac{3R}{2} + R = \frac{5R}{2}. \tag{II.30}$$

(This is in fact the recommended value for argon in the *JANAF Thermochemical Tables* [Chase et al., 1986].)

The heat capacity c_p for a real diatomic gas, such as H_2, varies roughly from 1.5 to 2 times this value depending on the temperature. The reason is that the diatomic molecule has other degrees of freedom in addition to the three possible translations considered up to this point. The atoms can vibrate along the axis of the molecule, and can rotate about two axes. In addition, electrons may be excited, but these electronic contributions to the internal energy are not of significance except at very high temperatures or high electric fields.

The principle of *equipartition of energy* states that there is a contribution to the average molar kinetic energy of $RT/2$ for each classical degree of freedom. Thus, the three translational degrees of freedom contribute $3RT/2$. The two rotational degrees of freedom contribute an additional RT, making a subtotal for U of $5RT/2$ for a diatomic gas.

The one vibrational degree of freedom of the ideal diatomic molecule contributes a varying amount to E_{av}. This is because the vibrational mode contains potential, as well as kinetic, energy, and the proportion of the latter increases with temperature. The contribution varies from zero at low temperatures to RT at high temperatures.

Thus, E_{av} for the ideal diatomic gas varies with temperature from $5RT/2$ at low temperatures to $7RT/2$ at high temperatures. From this, the molar heat capacity varies from $7R/2$ (29.100 $J \cdot mol^{-1} \cdot K^{-1}$) to $9R/2$ (37.415 $J \cdot mol^{-1} \cdot K^{-1}$) as the temperature varies, which is in reasonable agreement with that of some real diatomic gases. Measured values for H_2 are 28.836 at room temperature and 37.087 at 3000 K [Chase et al., 1986].

Diffusivity

Imagine that it were possible to introduce, into a gas, distinguishable tracer particles of the same mass and cross section as the host particles. A net transport will occur if there is a gradient in the concentration of these tracer particles. This process is called *self-diffusion*, because the tracers have the same properties as the host particles from the point of view of the kinetic theory. The net flux (j) is proportional to the concentration gradient:

$$j = -D\nabla(n_t), \tag{II.31}$$

where D is the self-diffusivity and n_t is the concentration of tracers.

We verify (II.31) and calculate D by considering a one-dimensional case, as shown in Figure II.6. We calculate the net transport of tracer particles across the plane at y as shown in the figure.

The impingement rate onto the plane at y from the left is $n_t(y)v_{av}/4$. The impingement rate onto the plane at $y + dy$ from the right is $n_t(y+dy)v_{av}/4$. Thus, the net flux through the region of width dy is $-(v_{av}/4)[n_t(y+dy) - n_t(y)]$.

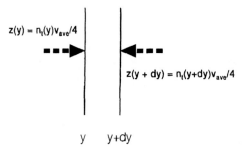

Figure II.6 Diffusion: coordinates for a one-dimensional calculation of diffusivity.

Now we observe that the concentration of tracers cannot vary on a spatial scale smaller than that of the mean free path. In the mathematical limit, $dy \ll \lambda$ but $[n_t(y+dy) - n_t(y)]$ is not given by $(dn_t/dy)dy$; rather, it is given by $(dn_t/dy)\lambda$. We now have, for the flux crossing the plane at y, $j_y = -(v_{av}/4)(dn_t/dy)\lambda$. Thus, we have shown that the net flux is proportional to the concentration gradient and that the self-diffusivity is

$$D = \frac{v_{av}\lambda}{4} = \frac{(kT/\pi^3 m)^{1/2}}{2d^2 n}. \tag{II.32}$$

For a dilute impurity (a more common case in practical applications), the diffusivity is

$$D_2 = \frac{v_{2av}\lambda_2}{4}. \tag{II.33}$$

To make the right-hand side conform to a rigorously derived expression for D, multiply it by $3\pi/4\sqrt{2}$ [Guggenheim, 1960].

Example Estimate the self-diffusivity of nitrogen gas at atmospheric pressure and room temperature.

We will utilize the density value calculated in a previous example.

Solution:

$$D = \frac{[(1.38 \times 10^{-23}\,\text{J/K}) \times 298\,\text{K}/\pi^3 \times 28 \times 1.66 \times 10^{-27}\,\text{kg}]^{1/2}}{2 \times (3.75\,\text{Å})^2 \times 2.46 \times 10^{25}\,\text{m}^{-3}} \times \frac{3\pi}{4\sqrt{2}}$$

$$= 1.28 \times 10^{-5}\,\text{m}^2/\text{s}. \qquad\blacksquare$$

Viscosity

Consider the parallel plates in relative motion as shown in Figure II.7. There is a gas between the plates; its flow velocity $u_z(y)$ varies from zero at $y = 0$ to u_{z0} at $y = h$. There is a shear stress (P_{yz}) within the gas given by

$$P_{yz} = \eta\left(\frac{du}{dy}\right). \tag{II.34}$$

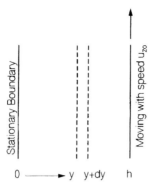

Figure II.7 Viscosity: Coordinates for a one-dimensional calculation of viscosity.

The first subscript (y) gives the orientation of the surface on which the shear stress is felt and the second (z), the direction of the force; η is the *coefficient of viscosity*, defined by (II.34).

Equation (II.34) is verified and the viscosity is calculated by considering the rate of transfer of momentum, in the direction of the flow, across the plane at y in Figure II.7. The rate of transfer of momentum through the plane at y from the left is given by the impingement rate times the momentum per particle: $(nv_{av}/4)mu(y)$; similarly the rate of transfer through the plane at $y+dy$ from the right is $(nv_{av}/4)mu(y+dy)$. The net rate of transfer is (using the same argument for spatial scale as in the derivation of the diffusivity) $P_{yz} = (nv_{av}/4)m(du/dy)\lambda$. Thus, the defining equation for viscosity is demonstrated for this one-dimensional case, and the viscosity is

$$\eta = \frac{nv_{av}m\lambda}{4} = \frac{(mkT/\pi^3)^{1/2}}{2d^2}. \tag{II.35}$$

Since the mean free path is inversely proportional to the density, the viscosity is actually independent of either density or pressure. This result, which is contrary to one's intuition, is confirmed for many gases. To convert the above expression to that obtained via a mathematically rigorous derivation, multiply it by $5\pi/8$ [Guggenheim, 1960].

Example Calculate the viscosity of nitrogen gas at room temperature.

Solution

$$\eta = \frac{[28 \times 1.66 \times 10^{-27}\,\text{kg} \times (1.38 \times 10^{-23}\,\text{J/K}) \times 298\,\text{K}/\pi^3]^{1/2}}{2 \times (3.75\,\text{Å})^2} \times \frac{5\pi}{8}$$

$$= 1.73 \times 10^{-5}\,\text{kg}/(\text{m} \cdot \text{s}) = 1.73 \times 10^{-4}\,\text{poise.} \qquad \blacksquare$$

Thermal Conductivity

The calculation of thermal conductivity proceeds in a fashion exactly analogous to the viscosity. The flux of heat (j_h) is proportional to a temperature gradient according to

$$j_h = -\kappa \nabla T, \qquad (II.36)$$

where κ is the thermal conductivity. The heat flux across a plane is given by the net value of the impingement rate times the energy per particle. The net rate of energy transfer across a plane perpendicular to a temperature gradient, in this one-dimensional analysis, is $j_y = -[z_{\text{right}} E_{\text{right}} - z_{\text{left}} E_{\text{left}}] \approx -z(c_p/N_0) \cdot \frac{3}{5} \cdot (T_{\text{right}} - T_{\text{left}}) = -z(c_p/N_0) \cdot \frac{3}{5} \cdot (dT/dy)\lambda$. Thus, the thermal conductivity is

$$\begin{aligned}
\kappa &= z \cdot \frac{3}{5} \cdot \left(\frac{c_p}{N_0}\right) \lambda \\
&= n\left(\frac{kT}{2\pi m}\right)^{1/2} \cdot \frac{3}{5} \cdot \frac{5k}{2} \cdot \frac{1}{\sqrt{2}\pi d^2 n} \qquad (II.37) \\
&= \left(\frac{k^3 T}{\pi^3 m}\right)^{1/2} \left(\frac{3}{4d^2}\right).
\end{aligned}$$

This result pertains to a monatomic particle. The thermal conductivity of more complex molecules is greater because their heat capacity is greater. To convert expression (II.37) to a mathematically rigorous result, multiply it by $25\pi/16$.

Example Estimate the thermal conductivity of nitrogen gas at room temperature.

Solution

$$\kappa = \left[\frac{(1.38 \times 10^{-23} \text{ J/K})^3 \times 298 \text{ K}}{\pi^3 \times 28 \times 1.66 \times 10^{-27} \text{ kg}}\right]^{1/2} \times \frac{3}{4 \times (3.75 \text{ Å})^2} \times \frac{25\pi}{16}$$

$$= 1.9 \times 10^{-2} \text{ W} \cdot \text{m}^{-1} \cdot \text{K}^{-1}. \qquad \blacksquare$$

II.4 GAS FLOW

Flow Regimes

The nature of gas flow through tubes and orifices varies with the relative size of the structure as compared to the mean free path, and with the flow rate. The size situation is characterized with a *Knudsen number* (K_n), defined as the ratio of the mean free path to a typical dimension (ℓ) of the flow structure

GAS FLOW

(frequently the inside diameter):

$$K_n \equiv \frac{\lambda}{\ell}. \qquad (II.38)$$

The viscous laminar flow regime is observed for $K_n \ll 1$, and for "low" flow rates. Here intermolecular collisions dominate the movement of individual particles. They move in a coordinated fashion, along streamlines. (Two other flow regimes are observed in this size range: turbulent flow for "high" flow rates and choked flow for even higher rates (see later discussion of conductance in viscous flow regime). However, these latter two regimes are rarely relevant to physical vapor deposition situations except perhaps during the initial moments of pumpdown of a deposition chamber from atmosphere.)

Viscous Laminar Flow

Consider the flow within a very long circular tube, as shown in cross section in Figure II.8. The flow velocity $[u(r)]$ is a function only of the radial distance from (r) the tube's axis. It is the greatest in the center of the tube, and equal to zero at the wall $(r=a)$. A hypothetical disk-shaped volume within the fluid is shown in the figure. In dynamic equilibrium, the forces on it must sum to zero. It moves forward by the pressure difference between its ends; this force is $\pi r^2 (dP/dz)dz$. It is restrained by the shearing force on its lateral surface; this force is $\eta(du/dr)2\pi r dz$. In steady state, $\eta(du/dr)2\pi r\, dz + \pi r^2 (dP/dz)dz = 0$. Thus, $du/dr = -(r/2\eta)dP/dz$. The solution is a parabolic velocity profile that satisfies the preceding boundary conditions: $u(r) = ((a^2 - r^2)/4\eta)dP/dz$.

The total flow (J) in particles per second is found from

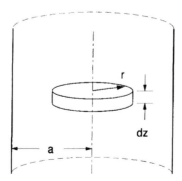

Figure II.8 Laminar Flow: coordinates for deriving the Poisseuille formula with a circular tube of radius a.

$$J = \int_0^a [nu(r)] 2\pi r \, dr$$

$$= \left(\frac{n\pi}{2\eta}\right) \frac{dP}{dz} \int_0^a (a^2 - r^2) r \, dr$$

$$= \left(\frac{n\pi a^4}{8\eta}\right) \frac{dP}{dz} \qquad \qquad \text{(II.39)}$$

$$= \left(\frac{\pi a^4}{8\eta kT}\right) P\left(\frac{dP}{dz}\right).$$

Conservation of matter requires that J be independent of z. If the tube is of length L and if the pressure difference between its ends is small, we may approximate $P(dP/dz)$ with $P_{\text{av}}(\Delta P/L) = (P(L) + P(0))(P(L) - P(0))/2L$.
Thus, the well-known Poiseuille formula is obtained:

$$J = \left(\frac{\pi a^4}{8\eta kT}\right) P_{\text{av}} \frac{\Delta P}{L}, \qquad \text{(II.40)}$$

which describes the flow of a gas through a tube in the viscous flow regime.

For viscous laminar flow through an orifice in a vanishingly thin wall, Roth [1990] gives the following formula for the total flow:

$$J = \pi a^2 \cdot \left[\frac{2\gamma}{\gamma - 1} \frac{kT}{m}\right]^{1/2} \cdot \left[\frac{P_{\text{lo}}}{P_{\text{hi}}}\right]^{1/\gamma} \cdot \left[1 - \left(\frac{P_{\text{lo}}}{P_{\text{hi}}}\right)^{(\gamma-1)/\gamma}\right]^{1/2} \cdot \frac{\Delta P}{kT}, \qquad \text{(II.41)}$$

where γ is the ratio of heat capacities (c_p/c_v), $\frac{5}{3}$ for an ideal monatomic gas and P_{lo} and P_{hi} are static pressures on either side of the orifice. Equation II.41 is valid only for $P_{\text{lo}}/P_{\text{hi}} \geq [2/(\gamma + 1)]^{\gamma/(\gamma-1)}$. (This expression does not seem to be a part of basic kinetic theory, but we include it here anyway, for completeness.)

Molecular Flow

On the other hand, the Knudsen or molecular flow regime is observed when $K_n \gg 1$. Interparticle collisions are rare; flow is controlled by wall collisions. In a wall collision, a particle briefly rests on the wall surface. When it is reemitted, the probability distribution for its new direction of travel is the well-known cosine distribution, which was derived above. Thus, the wall collisions amount to diffuse, rather than specular, reflections of the gas particles.

In this regime, the behavior is found by calculating the net flow through a cross section of the tube, as illustrated in Figure II.9. It is assumed that there is a gradient in the density of gas particles in the y direction. The net reemission flux passing through area element dA' on the cross section is calculated for source area elements dA on both sides of the cross section. A double integration is done over the entire inner area of the tube (A) and over the cross section (A'). The calculation is complex, both geometrically and algebraically. We give one basic result here and refer the reader to the details in Present's

GAS FLOW

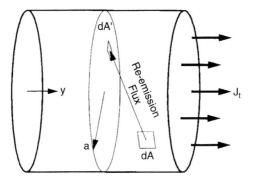

Figure II.9 A depiction of the calculation of molecular flow through a circular tube.

book [1958]. The total flow through a very long tube is $J = -(\pi a^2)(2v_{av}a/3)(dn/dy)$. We may crudely estimate dn/dy as $\Delta n/L$. The total flow becomes

$$J = -(\pi a^2)\frac{8\Delta z a}{3L}, \tag{II.42}$$

where Δz is the difference in impingement rates between the two ends of the tube.

By contrast, the flow through an orifice in the molecular flow regime may be calculated quite simply. It is the difference in impingement rates from either side, multiplied by the area:

$$J = A \cdot \Delta z = A(2\pi mkT)^{-1/2}\Delta p. \tag{II.43}$$

This is known as the *Hertz–Knudsen equation*.

The *Clausing factor* is defined as the ratio of the flow through a tube to that through an orifice of the same cross-sectional area and pressure differential. For the above-mentioned long tube of circular cross section, it is $8a/3L$ in the molecular flow regime.

The transition between molecular and viscous flow is an important regime. However, it is even less tractable to calculations of total flow.

Conductance

The *conductance* (C) of a structure is defined as the total flow through it divided by the difference in gas density causing the flow. For viscous flow in a long circular tube, the conductance is obtained in an obvious manner from the Poisseuille formula:

$$C_{v,\text{tube}} = \left(\frac{\pi a^4 kT}{8\eta L}\right)n_{\text{av}}. \tag{II.44}$$

It is interesting to note that this expression for the conductance of a tube in viscous flow increases without limit as the average gas density in the tube increases. Physically, however, the tube's conductance cannot exceed that of an orifice of the same area (πa^2) in viscous flow. This limitation on the conductances of tubes and other structures leads to the choked flow regime, with its own expressions for conductance [Santeler, 1990].

For viscous flow through an orifice, the conductance is obtained from Equation (II.41):

$$C_{v,\text{orif}} = \pi a^2 \cdot \left[\frac{2\gamma}{\gamma-1}\frac{kT}{m}\right]^{1/2} \cdot \left(\frac{P_{\text{lo}}}{P_{\text{hi}}}\right)^{1/\gamma} \cdot \left[1 - \left(\frac{P_{\text{lo}}}{P_{\text{hi}}}\right)^{(\gamma-1)/\gamma}\right]^{1/2}. \quad (\text{II}.45)$$

(For a more complete discussion of viscous flow, see Santeler [1986].)

The conductance of a long circular tube in the molecular flow regime is given by

$$C_{m,\text{tube}} = (\pi a^2)\left(\frac{2v_{\text{av}}a}{3L}\right) = \frac{2\pi a^3 v_{\text{av}}}{3L}. \quad (\text{II}.46)$$

Finally, that of an orifice in the molecular flow regime is simply

$$C_{m,\text{orif}} = \frac{A v_{\text{av}}}{4}. \quad (\text{II}.47)$$

Example Compare the conductances of an orifice and a tube in molecular flow. Assume a radius of 0.5 cm for both, but a length of 10 cm for the tube, and assume nitrogen gas at room temperature.

The average speed of a nitrogen molecule is 475 m/s. The conductance of the orifice is then

$$C_{m,\text{orif}} = \pi \times (0.5\,\text{cm})^2 \times \frac{(475\,\text{m/s})}{4} = 9.32 \times 10^{-3}\,\text{m}^3/\text{s} = 9.32\,\text{liters/s}.$$

The conductance of the tube is somewhat less:

$$C_{m,\text{tube}} = \frac{\pi \times (0.5\,\text{cm})^2 \times 2 \times (475\,\text{m/s}) \times 0.5\,\text{cm}}{(3 \times 10\,\text{cm})} = 1.24\,\text{liters/s}. \quad \blacksquare$$

II.5 UNITS OF PRESSURE AND AMOUNTS OF GAS

Units of Pressure

The SI unit of pressure is the pascal (1 N/m^2).

To calculate varies quantities in the kinetic theory, it is sometimes necessary to convert alternative pressure unit to pascals. "The pascal is equal to the pressure resulting from 1 N acting uniformly over a $1 - m^2$ area". Some conversions are

$$1 \text{ atm} = 1.01 \times 10^5 \text{ Pa}$$
$$1 \text{ torr} = 133 \text{ Pa}$$
$$1 \text{ bar} = 10^5 \text{ Pa}$$
$$1 \text{ lbf} \cdot \text{in}^{-2} = 6.89 \times 10^3 \text{ Pa}$$

The *standard atmosphere* is a pressure of 1.01325×10^5 pascal. This is equal to exactly 760 torr and 1.01325 bar, and to approximately $14.7 \text{ lbf} \cdot \text{in}^{-2}$ and 29.9 in Hg. The pressure of the *thermodynamic standard state* is 1 bar.

Amounts of Gas

Standard temperature and pressure (STP) are 0°C (273.16 K) and 1 atm. At STP, 1 mole of an ideal gas occupies approximately 22.4 liters and comprises approximately 6.02×10^{23} particles. Its mass is equal to its molecular weight in grams. The atomic mass constant (amu) is approximately 1.66×10^{-27} kg.

II.6 SUMMARY OF PRINCIPAL EQUATIONS

Boltzmann distribution function	$f(E) = \exp(-E/kT)$
Boltzmann factor	$\int_{E_a}^{\infty} \exp[-E/kT] dE/kT = \exp[-E_a/kT]$
Energy of gas particle	$E = mv^2/2 = m(v_x^2 + v_y^2 + v_z^2)/2$
Probability distribution function	$F(v_x, v_y, v_z) = \exp[-m(v_x^2+v_y^2+v_z^2)/2kT]/(2\pi kT/m)^{3/2}$
Most probable speed	$v_p = (2kT/m)^{1/2}$
Average speed	$v_{av} = (8kT/\pi m)^{1/2}$
Root mean square speed	$v_{rms} = (3kT/m)^{1/2}$
Average kinetic energy	$E_{av} = 3kT/2$
Impingement rate	$z = P(2\pi mkT)^{-1/2} = nv_{av}/4$
Incident flux distribution	$j_\Omega(\theta, \Phi) = n \cos\theta v_{av}/4\pi$
Ideal gas law	$P = nkT$
Survival equation	$Q(x) = \exp(-P_c x)$
Collision probability per unit distance traveled	$P_c = \sqrt{2}\pi d^2 n$
Mean free path	$\lambda = 1/P_c$
Average relative speed of like particles	$v_{rel} = \sqrt{2} v_{av}$
Collision rate	$r = v_{av} P_c$
Average relative speed of different particles	$v_{rel} = [(v_{2av})^2 + (v_{1av})^2]^{1/2}$

Effective diameter for different particles	$d_{21} = [(d_2 + d_1)/2]$
Mean free path of a dilute impurity	$\lambda_2 = 1/\{\pi(d_{21})^2 [1 + (v_{1av})^2/(v_{2av})^2]^{1/2} n_1\}$
Average kinetic energy of a mole of particles	$E_{av} = 3RT/2$
Molar heat capacity at constant pressure of a monatomic gas	$c_p = 5R/2$
Diffusive flux	$\boldsymbol{j} = -D\boldsymbol{\nabla}(n_t)$
Self-diffusivity	$D = v_{av}\lambda/4 = [(kT/\pi^3 m)^{1/2}/2d^2 n] \times 3\pi/4\sqrt{2}$
Diffusivity of a dilute impurity	$D_2 = v_{2av}\lambda_2/4$
Shear stress	$P_{yz} = \eta(du/dy)$
Viscosity	$\eta = nv_{av}m\lambda/4 = [(mkT/\pi^3)^{1/2}/2d^2] \times 5\pi/8$
Heat flux	$\boldsymbol{j}_h = -\kappa\boldsymbol{\nabla}T$
Thermal conductivity	$\kappa = [(k^3 T/\pi^3 m)^{1/2}(3/4d^2)] \times 25\pi/16$
Knudsen number	$K_n \equiv \lambda/\ell$
Viscous flow through a very long tube (Poisseuille formula)	$J = (\pi a^4/8\eta kT)P_{av}\Delta P/L$
Viscous flow through an orifice in a vanishingly thin wall	$J = \pi a^2 \cdot \{[2\gamma/(\gamma - 1)](kT/m)\}^{1/2} \times \cdot(P_{lo}/P_{hi})^{1/\gamma} \times \cdot[1 - (P_{lo}/P_{hi})^{(\gamma-1)/\gamma}]^{1/2} \cdot \Delta P/kT$
Ratio of heat capacities	$\gamma = c_p/c_v$
Molecular flow through a very long tube	$J = -(\pi a^2)(8\Delta za/3L)$
Molecular flow through an orifice (Hertz–Knudsen equation)	$J = A(2\pi mkT)^{-1/2}\Delta P$
Conductance of a long circular tube in viscous flow	$C_{v,tube} = (\pi a^4 kT/8\eta L)n_{av}$
Conductance of an orifice in viscous flow	$C_{v,orif} = \pi a^2 \cdot \{[2\gamma/(\gamma - 1)] \times [kT/m]\}^{1/2} \cdot (P_{lo}/P_{hi})^{1/\gamma} \times \cdot[1 - (P_{lo}/P_{hi})^{(\gamma-1)/\gamma}]^{1/2}$
Conductance of a long circular tube in molecular flow	$C_{m,tube} = (\pi a^2)(2v_{av}a/3L)$
Conductance of an orifice in molecular flow	$C_{m,orif} = Av_{av}/4$

II.7 APPENDIX

Arrhenius Plots

As explained in Section II.1, data for thermally activated processes or quantities are frequently given the form of Arrhenius plots. Such a plot is

APPENDIX

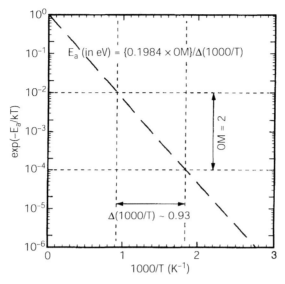

Figure II.10 Arrhenius plot. The calculation of the thermal activation energy from an Arrhenius plot is described in the text. In the example shown, OM is 2, $\Delta(1000/T)$ is ~ 0.93, and E_a is consequently ~ 0.43 eV.

shown schematically in Figure II.10. As suggested in the figure, it is common to have as the abscissa $1000/T$, rather than $1/T$, in order to work with integer labels. To determine the activation energy from such a plot

1. Fit a straight line to the data.
2. Select an order-of-magnitude (OM) change as suggested in Figure II.10.
3. Determine the corresponding temperature interval $\Delta(1000/T)$ as also suggested in the figure.
4. Estimate the activation energy from

$$E_a = \frac{0.1984 \cdot \text{OM}}{\Delta\{1000/T\}} \text{ eV} \quad \text{(per particle)} \tag{II.48}$$

For the example shown in Figure II.10, $E_a = 0.1984 \cdot 2/0.93 \approx 0.43$ eV.

Activation energies are sometimes expressed as molar quantities; 1 eV per particle is equivalent to 96.3 kJ/mol and to 23.0 kcal/mol.

Some Definite Integrals

$$\int_0^\infty x \cdot e^{-ax} dx \equiv 1/a^2$$

$$\int_0^\infty e^{-ax^2} dx \equiv (\pi/4a)^{1/2}$$

$$\int_0^\infty x \cdot e^{-ax^2} dx \equiv 1/(2a)$$

$$\int_0^\infty x^2 \cdot e^{-ax^2} dx \equiv \left(\pi/16a^3\right)^{1/2}$$

$$\int_0^\infty x^3 \cdot e^{-ax^2} dx \equiv 1/(2a^2)$$

$$\int_0^\infty x^4 \cdot e^{-ax^2} dx \equiv (3/8a^2)(\pi/a)^{1/2}$$

Atomic Diameters of the Elements

We present some atomic diameters given by Vályi*:

Element	Diameter (Å)	Element	Diameter (Å)
H	3.0	Ru	2.64
He	2.44	Rh	2.69
Li	3.12	Pd	2.74
Be	2.16	Ag	2.85
B	1.76	Cd	3.0
C	1.54	In	3.08
N	1.26	Sn	3.12
O	1.20	Sb	3.18
F	1.36	Te	2.79
Ne	3.22	I	2.71
Na	3.72	Xe	4.26
Mg	3.20	Cs	5.25
Al	2.85	Ba	4.42
Si	2.32	La	3.75
P	2.16	Ce	3.65
S	2.11	Pr	3.63
Cl	2.04	Nd	3.61
Ar	3.83	Eu	4.08
K	4.68	Gd	3.48
Ca	3.92	Tb	3.54
Sc	3.15	Dy	3.53
Ti	2.96	Ho	3.51
V	2.68	Er	3.50
Cr	2.52	Tm	3.49
Mn	2.54	Yb	3.87
Fe	2.52	Lu	3.43

*Reproduced with permission from L. Vályi, *Atom and Ion Sources*, copyright 1977 by John Wiley & Sons, Ltd.

Element	Diameter (Å)	Element	Diameter (Å)
Co	2.52	Hf	3.24
Ni	2.48	Ta	2.89
Cu	2.54	W	2.76
Zn	2.71	Re	2.73
Ga	2.70	Os	2.68
Ge	2.76	Ir	2.70
As	2.41	Pt	2.76
Se	2.29	Au	2.88
Br	2.34	Hg	3.05
Kr	3.98	Tl	3.68
Rb	5.02	Pb	3.48
Sr	4.26	Bi	3.78
Zr	3.19	Th	3.64
Nb	2.90	U	3.0
Mo	2.78		

II.8 MATHEMATICAL SYMBOLS, CONSTANTS, AND THEIR UNITS

SI units are given first, followed by other units in widespread use.

a	Radius of tube (m)
c_p	Molar heat capacity at constant pressure (J · mol^{-1} · K^{-1})
c_v	Molar heat capacity at constant volume (J · mol^{-1} · K^{-1})
d	Particle diameter (m; Å = 10^{-10} m)
f	Classical energy distribution function (dimensionless)
j	Particle flux density (m^{-2} · s^{-1})
k	Boltzmann's constant (1.38×10^{-23} J/K; 8.62×10^{-5} eV/K)
ℓ	Characteristic dimension (m)
m	Mass (kg; amu = 1.66×10^{-27} kg)
n	Particle number density (m^{-3})
r	Collision rate (s^{-1}); radial coordinate (m)
t	Time (s)
u	Flow velocity (m/s)
v	Speed (m/s)
v_{av}	Average speed (m/s)
v_i	ith component of velocity (m/s)
x, y	Position (m)
z	Impingement rate (m^{-2} · s^{-1}); position (m)
A	Area (m^2)
C	Conductance (m^3/s)
D	Diffusivity (m^2/s)
E	Energy (J; eV = 1.60×10^{-19} J; cal = 4.18 J)

F	Probability distribution function for speed (s^3/m^3)
F_i	Distribution function for velocity component (i) (s/m)
J	Total particle flow (s^{-1})
K_n	Knudsen number (dimensionless)
L	Length (m)
N	Total number of particles
N_0	Avogadro's number (6.02 × 10^{23})
P	Pressure (Pa; torr = 133 Pa; atm = 1.01 × 10^5 Pa)
P_c	Collision probability per unit distance traveled (m^{-1})
Q	Heat content of a system (J; cal = 4.18 J)
$Q(x)$	Probability of traveling a distance x before having a collision (dimensionless)
R	Molar gas constant (8.31 J · mol^{-1} · K^{-1})
T	Temperature (K)
U	Internal energy (J)
V	Volume (m^3; liter = 10^{-3} · m^3)
W	Length of tube (m)
γ	Ratio of heat capacities (dimensionless)
η	Viscosity (kg · m^{-1} · s^{-1}; poise = 0.1 kg · m^{-1} · s^{-1})
θ	Polar angle (rad)
λ	Mean free path (m)
κ	Thermal conductivity (W · m^{-1} · K^{-1})
π	A well-known constant (3.142···)
φ	Azimuthal angle (rad)
Ω	Solid angle (steradian; 1 sphere = 4π steradians)

REFERENCES

Chase, M. W., Davies, C. A., Downey, J. R., Frurip, D. J., McDonald, R. A., and Syverud, A. N., 1986, *JANAF Thermochemical Tables*, 3rd ed., American Institute of Physics, New York.

Guggenheim, E. A., 1960, *Elements of the Kinetic Theory of Gases*, Pergamon Press, Oxford.

Kennard, E. H., 1938, *Kinetic Theory of Gases*, McGraw-Hill, New York.

Present, R. D., 1958, *Kinetic Theory of Gases*, McGraw-Hill, New York.

Rosenblatt, G. M., 1976, "Evaporation from Solids," in N. B. Hannay, ed., *Treatise on Solid State Chemistry*, Vol. 6A, Surfaces I, Plenum Press, New York.

Roth, A., 1990, *Vacuum Technology*, North-Holland, Amsterdam.

Santeler, D. J., 1986, "Exit Loss in Viscous Tube Flow," *J. Vac. Sci. Technol.* **A4**(3), 348.

Santeler, D. J., 1990, "Topics in Vacuum System Gas Flow Applications," *J. Vac. Sci. Technol.* **A8**(3), 2782.

Vályi, L., 1977, *Atom and Ion Sources*, Wiley, New York, p. 388.

III

ADSORPTION AND CONDENSATION

The principal actors in physical vapor deposition are vapors and gases. Indeed, the purpose of physical vapor deposition sources is to produce a vapor that will condense on a substrate. Before or during this condensation, adsorption of residual gases can lead to contamination of the substrate surface. On the other hand, controlled adsorption is the growth mechanism of one physical vapor deposition technique: atomic layer epitaxy. Both adsorption and condensation are fundamental operating mechanisms of certain vacuum pumps.

There are crucial differences between vapors and gases. The pressure of a vapor is *fixed* at the thermal equilibrium vapor pressure by the presence of a condensed phase with which it is in equilibrium. Practically speaking, the sole independent variable for a vapor is the temperature of its source. However, the pressure of a gas is not fixed. A vapor is in equilibrium with its condensed phase, but a gas is just a gas.

Figure III.1 is a "phase diagram" of sorts, showing the pressure-temperature regimes where "gases" and "vapors" exist. Because of the thermally activated character of the vapor pressure, which forms the "phase boundary," it is convenient to portray this relationship as an Arrhenius plot. As a general rule, the pressure of a substance when in the form of a gas is less than P_{eq}, and when a vapor, equal to P_{eq}. Stretching the definition a bit, there are also *supersaturated* vapors. For these, the pressure is greater than P_{eq}. For example, the vapor produced by an ideal effusion cell is in *quasiequilibrium* (infinitely near equilibrium) with the source material within the cell. However, it is normally supersaturated with respect to the substrate, whose temperature is lower than the cell temperature, and where condensation occurs.

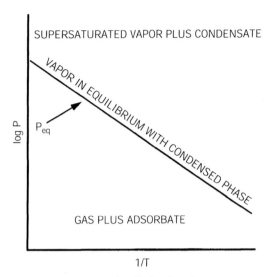

Figure III.1 This diagram portrays the distinction between gases and vapors. The "phase boundary" is the thermal equilibrium vapor pressure, a thermally activated quantity (thus the Arrhenius plot). At low pressures and high temperatures, a gas exists, which is in equilibrium with an adsorbate only. At high pressures and low temperatures, a supersaturated vapor exists and may condense.

A certain kind of equilibrium does exist for gases. The equilibrium is with adsorbed gas particles, which may cover a surface up to a maximum amount of one monolayer, in our ideal model. There is a one-to-one correspondence between the gas pressure and the extent of coverage; the relationship for fixed temperature is called an *adsorption isotherm*.

For vapor creation, if the condensed phase is a liquid, we speak of *evaporation*; if a crystal, of *sublimation*. The two together are referred to as *vaporization* and the reverse, *condensation*. For gases, the *adsorbate*, the assembly of adsorbed particles on a surface, is perhaps the fifth state of matter (the plasma already having been designated fourth)—distinct from solid, liquid, or gas. The adsorbed particles are called *adatoms*; to our knowledge, the term "admolecules" has yet to be coined.

For additional explanations of adsorption phenomena and of the thermodynamics of phase changes, the author would suggest *Physics at Surfaces* by Zangwill [1988], *Surface Physics* by Prutton [1983], *Surfaces I and II* by Hannay [1976], *Surface Physics of Materials*, edited by Blakely [1975], and *Physical Chemistry* by Daniels and Alberty [1966]. We should add that this "gas versus vapor" distinction may not be universal terminology in physics, but one that the author finds particularly useful in the context of thin-film deposition.

III.1 ADSORPTION OF GASES

Why Gases Adsorb

Gas particles adsorb because there are stable positions for them on surfaces, as illustrated schematically in Figure III.2a. The simplest case is represented in this Figure by the diatomic molecule which does *not* dissociate. The desorption energy, E_{dA_2}, is the depth of the potential well with respect to infinite separation of the molecule from the surface.

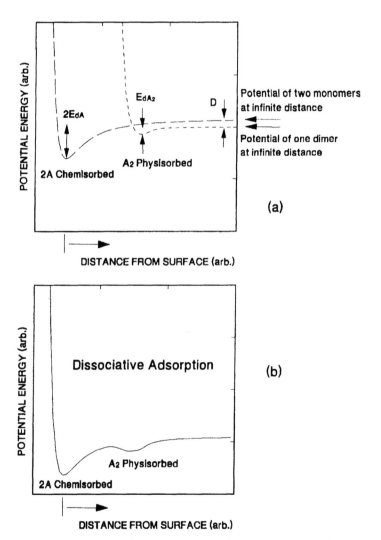

Figure III.2 Absorption energies: (*a*) A schematic representation of the potential energy of a diatomic molecule (A_2) and of its dissociated components (2A) as they approach a surface; (*b*) the resultant potential for dissociative adsorption is portrayed.

This curve for the potential energy of A_2 as it approaches the surface (as well as the curve for 2A) was calculated using the integrated Lennard-Jones potential [Redhead et al., 1993]. It is derived from the well-known Lennard-Jones "6-12" interatomic potential ($4E_d[(r_0/x)^{12}-(r_0/x)^6]$) by integrating over the semiinfinite volume of atoms beneath the surface. The resulting potential energy function is

$$E(x) = 4\pi E_d n r_0^3 \left[\left(\frac{1}{45}\right)\left(\frac{r_0}{x}\right)^9 - \left(\frac{1}{6}\right)\left(\frac{r_0}{x}\right)^3 \right], \tag{III.1}$$

where n is the number density of atoms in the solid and r_0 is a length parameter (the energy minimum occurs at $1.16\,r_0$), x is the distance from the surface, and E_d is the depth of the minimum.

If E_d is less than about 0.4 eV (10 kcal/mol), the particle (whether a diatomic molecule or something else) is said to be in a state of *physisorption*. This is an energy range typical of van der Waals bonding. The noble gases physisorb on all surfaces; a molecule probably also physisorbs if it is able to adsorb without dissociating. If E_d is greater than about 1 eV, the particle is said to be *chemisorbed*. This is a state of relatively strong chemical bonding. Single oxygen atoms chemisorb on virtually all solid surfaces; the enthalpy of formation of solid SiO_2 per oxygen atom, for example, is about 4 eV.

Oxygen is an interesting case, because oxygen gas exists primarily as diatomic molecules, not single oxygen atoms, and yet adsorbed oxygen is often monatomic and in a strongly chemisorbed state. It is also an important case, because of the ubiquitous native oxide that grows on metal surfaces as a result of air exposure. Dissociation occurs simultaneously with adsorption — how can one construct the relevant potential energy diagram for an oxygen molecule approaching a surface?

As illustrated in Figure III.2a, if two oxygen monomers are created in the gas phase by the dissociation of an O_2 molecule, they will have a higher potential energy than the molecule; the difference will be D. If we bring these separated atoms toward the surface (separated from each other but approaching the surface in unison), their potential energy is shown schematically in the curve for 2A, two monomers. Separated, the two oxygen atoms have a much lower potential energy on the surface than a single adsorbed O_2. (The energy required to desorb a *single* monomer would be E_{dA}.)

In reality, a diatomic molecule often initially physisorbs intact, in a precursor state. After some time on the surface, it gives up further energy by dissociating. The dissociation products may then chemisorb. The potential-energy curve experienced in approaching the surface by, for example, the two atoms making up a diatomic molecule such as O_2, is that shown schematically in Figure III.2b. It was Lennard-Jones [1932] who postulated two such minima. Dissociation is exhibited by many gaseous molecules.

The process of adsorption requires *thermal accomodation*, which is the loss of kinetic energy by the gas particle, so that it becomes trapped in the potential well. This excess kinetic energy is typically dispersed as lattice vibrations in the solid.

Mean Residence Time

How long will an individual particle stay adsorbed? The mean residence time may be estimated very simply as the inverse of its desorption rate:

$$\tau = \frac{1}{\nu \cdot \exp(-E_d/kT)}, \tag{III.2}$$

where ν is a frequency of vibration perpendicular to the surface, almost always assumed to be

$$\nu \approx 10^{13} \text{ Hz}. \tag{III.3}$$

It originates with the expression kT/h (1.6×10^{12} Hz at room temperature) for the frequency of a linear harmonic oscillator, where h is Planck's constant. The exponential is a Boltzmann factor giving the probability that the vibrating particle has enough energy to escape from the potential well that it inhabits on the surface.

Example A particle physisorbed with a desorption energy of 0.4 eV has a mean residence time at room temperature of

$$\tau \approx \frac{1}{10^{13} \text{ Hz} \times \exp\{-0.4 \text{ eV}/[(8.62 \times 10^{-5} \text{ eV/K}) \times 298 \text{ K}]\}} = 0.58 \text{ µs}.$$

A desorption energy of 1.2 eV gives a room temperature τ of 1.9×10^7 s; at 500 K this τ is reduced to 0.12 s. The potential utility of baking out ultra-high vacuum systems for eliminating adsorbed gases should be apparent. ∎

Langmuir's Adsorption Isotherm

A solid surface exposed to any gas will become covered with a partial monolayer of the gas particles. If the pressure is less than P_{eq}, the coverage will remain incomplete and no solid or liquid phase of the gas particles will form. However, an equilibrium between the gas phase and this adsorbed layer still occurs, describable with an adsorption isotherm (an expression for the amount of gas adsorbed as a function of pressure, at constant temperature).

Langmuir's adsorption isotherm is one of the fundamental isotherms. The surface coverage in Langmuir's model is expressed using the following variables:

N_s Surface density of adsorption sites
N Adatom surface density
$\theta \equiv N/N_s$ Fractional surface coverage

The isotherm is derived by equating the rate of adsorption to the rate of desorption. The left-hand side (LHS) of the following equation gives the adsorption rate, and the right-hand (RHS), the desorption rate:

$$\delta \cdot z \cdot (1 - \theta) = \frac{N}{\tau}, \tag{III.4}$$

where δ is the trapping probability (and z is the impingement rate of the gas). δ expresses the probability that the impinging particles give up enough of their kinetic energy to become trapped in the surface potential well. It is not at all certain that this thermal accomodation will be accomplished; sometimes impinging particles do not lose the required amount of kinetic energy and are quickly reflected. The value of δ has been found experimentally to vary at least from 0.001 to 0.9 for various gases and substrates [Zangwill, 1988].

There is also a *sticking coefficient* in Langmuir's model, which expresses the probability that an impinging particle is accommodated regardless of where it strikes; the sticking coefficient is given by

$$\alpha_s \equiv \delta(1 - \theta). \tag{III.5}$$

The term α_s contains the probability that the impinging particle strikes a vacant surface site. In the literature, the terminology involving trapping, thermal accomodation, sticking, and condensation coefficients has been confusing. We will adhere religiously to the definitions given above and would emphasize that a "condensation coefficient" plays no role in our discussion of adsorption.

After some substitutions, we have

$$\delta \cdot (1 - \theta) \cdot P(2\pi mkT)^{-1/2} = N_s \theta \cdot \nu \cdot \exp\left(\frac{-E_d}{kT}\right). \tag{III.6}$$

This equation may be rearranged to give

$$\theta = \frac{QP}{1 + QP} \tag{III.7}$$

where

$$Q = \frac{\delta(2\pi mkT)^{-1/2}}{N_s \nu \, \exp(-E_d/kT)}. \tag{III.8}$$

As illustrated in Figure III.3, the predicted surface coverage varies linearly with pressure for $P \ll 1/Q$ and saturates at a value of 1 in the limit of very high pressures. When P reaches P_{eq}, however, the isotherm ceases to hold and the coverage goes to unity as a condensed phase forms. At fixed pressure, the coverage decreases with increasing temperature due to the increasing desorption rate.

Example Suppose that a liquid nitrogen–cooled sorption pump is utilized to evacuate a small vacuum chamber. The chamber volume is 5 liters and the adsorber of the pump is a molecular sieve material having an area of 300 acres (1 acre = 4047 m^3; 300 acres is an actual commercial pump specification).

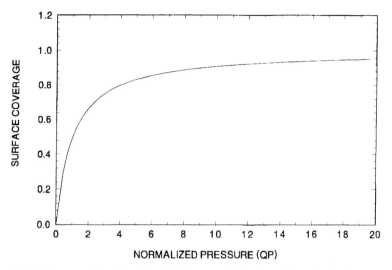

Figure III.3 Langmuir's adsorption isotherm, which gives the extent of coverage as a function of gas pressure. The effect of increasing the temperature is to diminish the coverage, for fixed pressure.

What is the surface coverage of the adsorber after the pumpdown? Assume that the surface is initially clean and that all of the gas molecules go out of the chamber and onto the adsorber, whose adsorption site density is 10^{19} m^{-2}.

The number of gas particles removed from the chamber is calculated with the ideal gas law:

$$\Delta N = \left[\frac{760 \text{ torr}}{1.38 \times 10^{-23} \text{ J/K} \times 298 \text{ K}} \right] \times 5\ell = 1.23 \times 10^{23}.$$

The surface coverage of the adsorber is

$$\Delta \theta = \frac{\Delta N}{N_s \times \text{area}} = \frac{1.23 \times 10^{23}}{10^{19} \text{ m}^{-2} \times 300 \text{ acres} \times 4047 \text{ m}^2/\text{acre}} = 0.010.$$

What is the pressure within the sorption pump after the chamber has been pumped down? Assume a trapping probability of unity, a desorption energy of 0.1 eV, and an average particle mass of 28 amu. The value of Q for the isotherm is

$$Q = \frac{1}{[2\pi \times 28 \times 1.66 \times 10^{-27} \text{ kg} \times (1.38 \times 10^{-23} \text{ J/K}) \times 77 \text{ K}]^{1/2}}$$

$$\times \frac{1}{10^{19} \text{ m}^{-2} \times 10^{13} \text{ Hz} \times \exp[-0.1 \text{ eV}/(8.62 \times 10^{-5} \text{ eV/K}) \times 77 \text{ K}]} = 0.262 \text{ torr}^{-1}.$$

To a very good approximation, the isotherm is in the linear range, so the pressure within the pump may be estimated from

$$P \approx \frac{\Delta \theta}{Q} = 0.010/0.262 \text{ torr}^{-1} = 0.039 \text{ torr}. \qquad \blacksquare$$

There is also a classical Langmuir isotherm for dissociative adsorption. We will assume one specific set of conditions here and derive the isotherm. First, to adsorb, the impinging dimer must strike a vacant surface site that also has a vacant neighboring site so that it can dissociate. Suppose that the crystal surface is a close-packed plane of atoms. There would be six possible neighboring sites, so the rate of adsorption becomes $\delta \cdot z \cdot (1-\theta) \cdot 6(1-\theta)$. Second, we will assume that an adatom must have an occupied site next to it so that it can desorb as a dimer. The desorption rate is then $N_s \cdot \theta \cdot 6\theta/\tau$. After some algebra, the adsorption isotherm for this particular form of dissociative adsorption becomes

$$\theta = \frac{\sqrt{QP}}{1 + \sqrt{QP}}, \qquad (\text{III.9})$$

where z for dissociative adsorption must be the impingement rate of the dimers and τ is the mean residence time of a *pair* of adatoms, calculated with the desorption energy $2E_{dA}$ from Figure III.2. Thus

$$Q = \frac{\delta(2\pi mkT)^{-1/2}}{N_s \nu \exp(-2E_{dA}/kT)}. \qquad (\text{III.10})$$

Other factors (not considered here) could influence the dissociative adsorption–desorption kinetics and alter the expressions for θ and Q. It might be necessary for adatoms to move together by surface diffusion before desorption. The desorption energy might vary with number of neighboring adatoms, if there is a strong lateral interaction. Desorption of single atoms, rather than the dimers just considered, can dominate if D (in Fig. III.2) is less than E_{dA}. A more detailed discussion can be found in Menzel [1975].

High-vacuum practitioners are well acquainted with the adsorption of water vapor onto the inner surfaces of high-vacuum chambers as a result of exposure to the atmosphere. This occurs to a significant degree even when the partial pressure is well below P_{eq} for water, and is difficult to reverse after the chamber is again under vacuum. Other atmospheric gases also adsorb onto vacuum chamber surfaces and onto substrates. These include N_2, CO, CO_2, and O_2. Desorption may be hastened by baking the vacuum chamber or by heating the substrates under vacuum.

In the modeling just presented, the maximum coverage by an adsorbing gas is a single monolayer; however, there are cases of multilayer adsorption where a small number of monolayers adsorb before true condensation commences

Atomic Layer Epitaxy

Atomic layer epitaxy (ALE) is a technique for growing compound films with thickness control of monolayer precision. The technique first appeared in the literature in a Finnish patent dealing with the preparation of ZnS layers and was applied to the manufacture of electroluminescent displays using this semiconductor:

> In the earliest and perhaps simplest example of ALE, viz., the growth of ZnS by evaporation in a vacuum, Zn vapor was allowed to impinge on the heated glass.... If the initial substrate surface is heated sufficiently one can achieve a condition such that only the chemisorbed layer remains attached.... Evaporation was then stopped and if physisorbed Zn were present because of the impinging flux, it would reevaporate. The process was then repeated with sulfur, the first layer of which would chemisorb on the initial Zn layer; any subsequent physisorbed sulfur would gradually come away from the heated substrate when the sulfur flux was cut off, leaving one (double) layer of ZnS. This complete cycle could be repeated indefinitely.

This concept is illustrated in Figure III. 4.

The method has been applied mainly to II–VI compound semiconductor growth. Other examples of the successful application of such *additive reactions* (meaning that pure elements are used as the reactants) in ALE include ZnSe, ZnTe, CdS, CdSe, and CdTe [Suntola, 1994].

The *film growth mechanism* was described in a review by Goodman and Pessa [1986]:

> A particularly simple way to describe the ALE approach is to say that it makes use of the difference between chemical adsorption and physical adsorption.... When the first layer of atoms or molecules of a reactive species reaches a solid surface there is usually a strong interaction (chemisorption); subsequent layers tend to interact much less strongly (physisorption). If the initial substrate surface is heated sufficiently one can achieve a condition such that only the chemisorbed layer remains attached.
>
> The formation of a 'layer per cycle' is the specific feature that conceptually distinguishes the ALE mode from other modes of vapor deposition; the latter all give a growth rate, ALE gives a growth per cycle.

However, we wonder whether the technique might be better explained as exploiting the difference between chemisorption and *condensation*. (Can a vapor species A physisorb on solid A? How would this differ from condensation of A?)

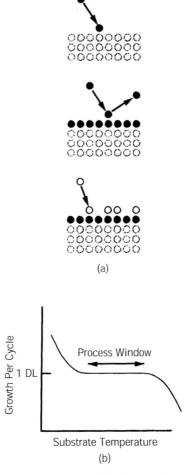

Figure III.4 Atomic layer epitaxy. (*a*) in ALE the substrate is exposed alternately to fluxes of components A and B. During each exposure one monolayer adsorbs, and no more. Thus, each full growth cycle results in one additional double (AB) layer. (*b*) The substrate temperature must be held within a process window in order to achieve the selective adsorption.

As a general statement, we believe that ALE may be applied to the growth of compound AB if the following is true: A on B and B on A chemisorb, while A vapor over A solid and B vapor over B solid are undersaturated. Furthermore, the substrate temperature must be practically chosen such that the mean residence times of the chemisorbed layers are $\gg 1$ second, because it is difficult to start and stop an exposure at times much below this.

A *typical growth procedure* is as follows:

ADSORPTION OF GASES

- Using shuttered evaporation sources, expose the substrate alternately to fluxes of the components A and B.
- Allow sufficient time during each exposure for a complete monolayer to chemisorb.
- After closing one shutter and before opening the other, allow sufficient time for any physisorbed particles to desorb, but open the shutter before any significant desorption of chemisorbed particles occurs.

The growth rate is precisely one double (AB) layer per shutter cycle. There is a process window for substrate temperature. If it is too low, there will be a failure to desorb the physisorbed reactants or perhaps condensation will occur; if the temperature is too high, there will be a failure to retain the chemisorbed species.

Example Select a substrate temperature for the preparation of ZnS by atomic layer epitaxy. Assume that there are effusion cells that create the effective vapor pressures at the substrate (sulfur vapor is predominantly diatomic and exhibits dissociative adsorption): $P_{Zn} = P_{S_2} = 10^{-5}$ torr. (The effusion cells actually create an incident flux j_i at the substrate. This j_i is equivalent to an impingement rate z, from which the effective vapor pressure may be calculated.) Assume $\delta = 1$ and the following desorption energies:

Zinc on a sulfur surface of ZnS $\qquad E_{dZn} = 2.27$ eV

Two sulfurs on a zinc surface of ZnS $\qquad 2E_{dS} = 3.31$ eV

Assume a surface adsorption site density of 10^{19} cm^{-2}.

We repeat here the conditions which the substrate temperature must satisfy: (1) the equilibrium coverages of zinc on ZnS and of sulfur on ZnS must be essentially 100%; (2) the zinc and sulfur vapors must be undersaturated with respect to pure zinc and pure sulfur, respectively; and (3) the monolayer formation time for each vapor is on the order of 0.1 s at the pressures specified above; consequently, the mean residence times on ZnS must be at least a few seconds to allow for reasonable shutter manipulation.

We now show that a substrate temperature of 550 K is workable:

1. First, we will calculate the Q parameters of the Langmuir adsorption isotherms for zinc and sulfur vapor on ZnS:

$$Q_{Zn} = \frac{1}{[2\pi \times 65.4 \times 1.66 \times 10^{-27}\,\text{kg} \times (1.38 \times 10^{-23}\,\text{J/K}) \times 550\,\text{K}]^{1/2}}$$
$$\times \frac{1}{10^{19}\,\text{m}^{-2} \times 10^{13}\,\text{Hz} \times \exp[-2.27\,\text{eV}/(8.62 \times 10^{-5}\,\text{eV/K}) \times 550\,\text{K}]}$$
$$= 1.15 \times 10^{13}\,\text{torr}^{-1}.$$

$$Q_{S_2} = \frac{1}{[2\pi \times 2 \times 32.1 \times 1.66 \times 10^{-27}\,\text{kg} \times (1.38 \times 10^{-23}\,\text{J/K}) \times 550\,\text{K}]^{1/2}}$$

$$\times \frac{1}{10^{19}\,\text{m}^{-2} \times 10^{13}\,\text{Hz} \times \exp[-3.31\,\text{eV}/(8.62 \times 10^{-5}\,\text{eV/K}) \times 550\,\text{K}]}$$

$$= 3.91 \times 10^{22}\,\text{torr}^{-1}.$$

The surface coverages of zinc and sulfur monomers on ZnS at the given pressure and substrate temperature are

$$\theta_{Zn} = \frac{1.15 \times 10^{13}\,\text{torr}^{-1} \times 10^{-5}\,\text{torr}}{1 + 1.09 \times 10^{13}\,\text{torr}^{-1} \times 10^{-5}\,\text{torr}} = 0.9999\cdots$$

$$\theta_S = \frac{(3.91 \times 10^{22}\,\text{torr}^{-1} \times 10^{-5}\,\text{torr})^{1/2}}{1 + (3.91 \times 10^{22} \times 10^{-5}\,\text{torr})^{1/2}} = 0.9999\cdots$$

Thus, each of the two steady-state coverages is well above 99%.

2. The thermal equilibrium vapor pressures of zinc and sulfur at the proposed substrate temperature are $P_{Zn} = 5.8 \times 10^{-4}$ torr and $P_{S_2} = 6.4 \times 10^{-2}$ torr. (These values were taken from Chase et al. [1986].) Thus, condensation of neither pure zinc nor pure sulfur will occur with the given vapor pressures.

3. The mean residence time for zinc on ZnS at the substrate temperature is

$$\tau_{Zn} = \frac{1}{10^{13}\,\text{Hz} \times \exp[-2.27\,\text{eV}/(8.62 \times 10^{-5}\,\text{eV/K}) \times 550\,\text{K}]}$$

$$= 6.20 \times 10^{7}\,\text{s}.$$

The mean residence time for sulfur on ZnS is estimated by assuming that the sulfur adatoms desorb as dimers:

$$\tau_{S_2} = \frac{1}{10^{13}\,\text{Hz} \times \exp[-3.31\,\text{eV}/(8.62 \times 10^{-5}\,\text{eV/K}) \times 550\,\text{K}]}$$

$$= 2.09 \times 10^{17}\,\text{s}.$$

These are sufficient to avoid any desorption during exposures of just a few seconds. ∎

A variant of ALE utilizes pulsed flows of source gases in chemical vapor deposition (CVD). The mechanism with CVD-ALE is more complicated, because an exchange reaction, instead of a simple additive reaction, is required. The mechanism is

- Adsorption of a monolayer of AB molecules
- Adsorption of CD molecules and simultaneous surface exchange reaction:

AB(adsorbed) + CD(adsorbed) → AD(crystal) + BC(gas).

Again, one monolayer of the desired compound forms per full cycle. This approach has produced thin films of oxides, nitrides, and III–V compounds, as well as II–VI compounds [Leskelä and Niinistö, 1990].

III.2 VAPOR PRESSURE

In this section we show how to estimate the thermal equilibrium vapor pressure from thermochemical data. (Thermodynamics fundamentals are presented in the end-of-chapter Appendix.) Next, we present vapor pressure data for silicon as an example, which was taken from a widely used reference for thermochemical data, the JANAF tables [Chase et al., 1986]. Finally, we present a theoretical model for the vapor pressures over alloys and compounds and then discuss the nonideal behavior of GaAs when heated.

The Thermally Activated Vapor Pressure

Consider the evaporation of element A: $A_\ell \leftrightarrow A_v$. In equilibrium, since the molar Gibbs free energies of liquid (ℓ) and vapor (v) are equal, it must be true that

$$\Delta_f G_\ell^\circ = \Delta_f G_v^\circ + RT \ln\left(\frac{P_{Aeq}}{P^\circ}\right), \tag{III.11}$$

where $\Delta_f G^\circ$ represents the standard molar Gibbs free energy of formation of vapor or liquid as explained in the Appendix. By rearranging the above equation, we find that the vapor pressure of A in equilibrium with the condensed phase is a function of the standard Gibbs free energy change for the reaction:

$$P_{Aeq} = P^\circ \exp\frac{-\Delta_{evap} G_A^\circ}{RT}, \tag{III.12}$$

where

$$\Delta_{evap} G_A^\circ = \Delta_f G_{A,v}^\circ - \Delta_f G_{A,\ell}^\circ. \tag{III.13}$$

This particular free energy change is called the *standard Gibbs free energy of evaporation*. It may be expanded in terms of the standard enthalpy and entropy of evaporation:

$$\begin{aligned} P_{Aeq} &= P^\circ \exp\frac{(-\Delta_{evap} H_A^\circ + T\Delta_{evap} S_A^\circ)}{RT} \\ &= P^\circ \exp\frac{\Delta_{evap} S_A^\circ}{R} \exp\frac{-\Delta_{evap} H_A^\circ}{RT}. \end{aligned} \tag{III.14}$$

Thus, P_{Aeq} is a thermally activated quantity. The thermal activation energy is the standard enthalpy of evaporation and the prefactor is $P^\circ \exp(\Delta_{evap} S_A^\circ)$.

58 ADSORPTION AND CONDENSATION

For many substances, a plot of $\log(P_{Aeq})$ versus $1/T$ is linear over a wide temperature range. It is continuous through the melting point, although there is a slope change in the Arrhenius plot due to the difference between the enthalpies of evaporation and sublimation.

Vapor Pressure Data for the Elements

Some vapor pressure data for silicon taken from the JANAF tables [Chase, 1986] are shown in Figure III.5. It is an interesting fact that while the silicon vapor phase is predominantly monatomic, there are substantial concentrations of Si_2 and Si_3 molecules present in equilibrium [Tanaka and Kanayama, 1997].

Example Set up the expressions from which the Si_1 vapor pressure curve of Figure III.5 was calculated.

Two expressions were used: one for sublimation and one for evaporation. In the JANAF tables, the entropies of vaporization are not given explicitly, but must be calculated as $S^\circ_{vap} - S^\circ_{con}$. However, the enthalpies of vaporization *are* given, as the enthalpies of formation of the vapor. The thermodynamic quantities vary modestly with temperature; for sublimation we selected values just below the melting point of 1685 K and assumed them to be constant:

	S°	$\Delta_{sub}S^\circ$	$\Delta_{sub}H^\circ$
Si_1	204 J·mol^{-1}·K^{-1}	142 J·mol^{-1}·K^{-1}	444 kJ/mol
Crystal	61.8		

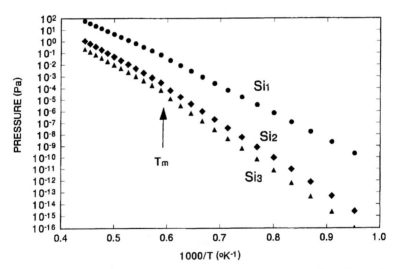

Figure III.5 Thermal equilibrium vapor pressures. Although the silicon thermal equilibrium vapor is predominantly monatomic, there are measurable densities of diatomic and triatomic molecules. One may note a distinct change in slope of these Arrhenius plots near the melting point. (Data from Chase et al. [1986].)

The expression for the vapor pressure of Si_1 below the melting point is then

$$P_{sub}(T) = 10^5 \text{ Pa} \times \exp\frac{142 \text{ J}\cdot\text{mol}^{-1}\cdot\text{K}^{-1}}{8.31 \text{ J}\cdot\text{mol}^{-1}\cdot\text{K}^{-1}} \times \exp\frac{-444 \text{ kJ/mol}}{8.31 \cdot T \cdot \text{J}\cdot\text{mol}^{-1}\cdot\text{K}^{-1}}.$$

For evaporation we selected values just above the melting point and assumed these to be constant:

	$S°$	$\Delta_{evap}S°$	$\Delta_{sub}H°$
Si_1	205 J·mol^{-1}·K^{-1}	113 J·mol^{-1}·K^{-1}	393 kJ/mol
Liquid	91.6		

The expression for the vapor pressure of Si_1 above the melting point is then

$$P_{sub}(T) = 10^5 \text{ Pa} \times \exp\frac{113 \text{ J}\cdot\text{mol}^{-1}\cdot\text{K}^{-1}}{8.31 \text{ J}\cdot\text{mol}^{-1}\cdot\text{K}^{-1}} \times \exp\frac{-393 \text{ kJ/mol}}{8.31 \cdot T \cdot \text{J}\cdot\text{mol}^{-1}\cdot\text{K}^{-1}}.$$

Note: To obtain the entropies of vaporization for the molecular species, one must take into consideration the number of condensed phase atoms forming the vapor molecule. For example, $\Delta_{evap} S°_{Si_3} = S°_{vap} - 3S°_{liq}$. ∎

The equilibrium constant at fixed pressure for the formation reaction of the vapor (K_p, reviewed in the Appendix) has a special significance; K_p itself is equal to the thermal equilibrium vapor pressure normalized to the standard pressure:

$$K_{p,evap} = \frac{P_{Aeq}}{P°}. \tag{III.15}$$

Values of the logarithm of K_p for the formation reaction of vapors are tabulated as "log K_f" in the JANAF tables. Thus, this column of data actually gives the logarithm of the thermal equilibrium vapor pressure, referenced to standard pressure. Some data for silicon monatomic vapor is presented in the Appendix (Table III.1).

A recent reference gives an empirical expression for the vapor pressures of 65 metallic elements [Alcock, 1995]—it is stated that the equations reproduce the observed vapor pressures to within ±5%. The empirical equations are of the form

$$\log[P_{eq}(\text{in Pa})] = 5.006 + A + \frac{B}{T} + C \log T + \frac{D}{T^3}. \tag{III.16}$$

Example Estimate the vapor pressure of aluminum at 1500 K using the equilibrium constant found in the JANAF tables and using the above empirical expression of Alcock [1995].

In the JANAF tables one finds log $K_f = -4.872$. Thus

$$P_{A\ell}(1500 \text{ K}) = 10^{-4.872} \text{ atm} = 1.343 \times 10^{-5} \text{ atm} = 1.361 \text{ Pa}.$$

In the Alcock reference, one finds $A = 5.911$, $B = -16211$, $C = 0$, and $D = 0$. Thus

$$P_{A\ell}(1500 \text{ K}) = 10^{5.006+5.911-16211/1500} \text{ Pa} = 1.287 \text{ Pa}. \qquad \blacksquare$$

Vapor Pressures of Alloys and Compounds

The equilibrium vapor pressure of an element, when the condensed phase is an alloy or compound, naturally differs from the thermal equilibrium vapor pressure as just calculated. Often it is reduced, because there is a lower concentration of atoms present in the condensed phase. An example of this would be the vapor pressures of silicon and germanium over a Si-Ge alloy. Other behavior is possible, however; the vapor pressure may be increased if the chemical bonding within the condensed phase is particularly weak. We will see below an example of an enhancement of vapor pressure—that of gallium over GaAs as compared to gallium over pure gallium metal. Predicting the vapor pressures of multicomponent phases is usually an empirical science, but there is an idealized model.

The *ideal solution* is a good starting place for understanding the vapor pressure of alloys and compounds. For an ideal solution, the activity of component A is equal to its numerical fraction (X_A). The Gibbs free energy of the component in the ideal solution is reduced from its standard value, according to

$$G_{A,\text{ideal}} = G_A^{\circ} + RT \ln X_A. \qquad (\text{III.17})$$

The vapor pressure of component A will be established over the solution according to the following equilibrium relation:

$$A(\text{alloy}) \Longleftrightarrow A(\text{vapor}).$$

The activity product is $(P_A/P^{\circ})/X_A$, while the equilibrium constant is $\exp(-\Delta_{\text{vap}}G^{\circ}/RT)$. These quantities must be equal to each other. Thus, the vapor pressure,

$$P_{A,\text{ideal}}(T) = X_A P_0 \exp\frac{-\Delta_{\text{vap}}G^{\circ}}{RT} = X_A P_{A\text{eq}}, \qquad (\text{III.18})$$

is reduced by the factor X_A as compared to the equilibrium value it would have (P_{aeq}) if pure A were present. This is called *Raoult's law* behavior.

Since the composition of the thermal equilibrium vapor phase ($X_A P_{A\text{eq}}/X_B P_{B\text{eq}}$) is not in general equal to that of the condensed phase (X_A/X_B), one may take advantage of the vaporization of an alloy to either purify the alloy, or perhaps concentrate a solute in it, depending on the ratio of vapor pressures. On the other hand, this noncongruent evaporation can be destructive to a compound condensed phase. For example, the sublimation of GaAs generally leads to disproportionate loss of arsenic, often along with the formation of liquid droplets of gallium on the sample surface (Fig. III.6).

*Non*ideal behavior occurs—it is modeled by replacing the concentration with the activity, which is itself some function of the concentration:

$$P_{A,\text{nonideal}}(T) \equiv a_A P_{A\text{eq}}. \tag{III.19}$$

The nonideal behavior is put into the function $a_A(X_A)$.

Gallium arsenide is an interesting and technologically significant semiconductor that exhibits nonideal behavior. Gallium vapor may be assumed to be purely monatomic, but arsenic vapor consists of molecules, predominantly As_2 and As_4. As shown in Figure III.7, the vapor pressure of gallium over GaAs is slightly *increased* compared to that over pure liquid gallium. By contrast, the vapor pressures of the two arsenic species are reduced by orders of magnitude compared to those over pure solid arsenic. These nonidealities are due to huge differences in bond strength between GaAs and the pure elements. Furthermore, the predominant vapor species for arsenic switches from As_4 to As_2. For more details, see Arthur [1967] and Foxon et al. [1972].

These particular relationships among the vapor pressures of gallium and arsenic permit the application of the three-temperature method to the vapor deposition of GaAs thin films, a technique whereby stoichiometric films may be deposited while using an inaccurate gallium/arsenic flux ratio. The principles of this technique are explained in the following section.

Ga/GaAs (100) 5 min 660°C

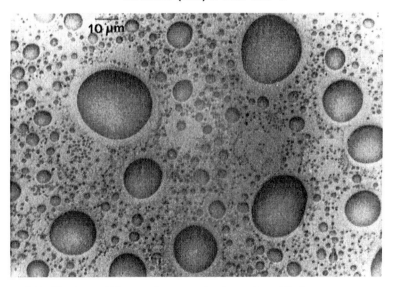

Figure III.6 The large difference between the thermal equilibrium vapor pressures of gallium and arsenic cause a disproportionate loss of arsenic from a heated GaAs sample. (Courtesy of Prof. Leonard C. Feldman)

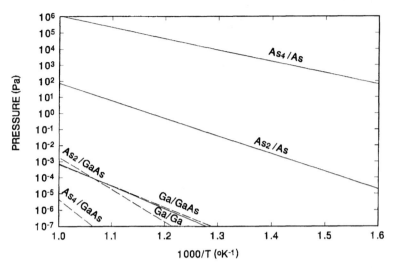

Figure III.7 Thermal equilibrium vapor pressures of gallium and arsenic over GaAs and over the pure elements. Extremely nonideal behavior is exhibited. The vapor pressure of gallium is scarcely different for the two sources. For arsenic, however, the predominant vapor species changes from As_4 to As_2 in going from the elemental arsenic source to GaAs. The data for pure element sources came from Barin [1995]; GaAs data came from Borisenko and Hesketh [1997]. (The lines for the GaAs source were fitted to data over a more limited temperature range than is shown. Consequently, this plot is not intended to be a vapor pressure data reference for GaAs, but merely to illustrate some of the behaviors of nonideal systems.)

III.3 CONDENSATION OF VAPORS

We present here the principles governing the condensation of a thin film from the vapor phase. Here, we are concerned only with the *conditions* that foster the condensation of vapors — not with the actual atomic processes of condensation and crystal growth, or with the film microstructure that results.

A difficult challenge is to get alloys and compounds to condense in the correct composition. We present some practical techniques for obtaining stoichiometric compound films when it is difficult or impossible to obtain stoichiometric vapor molecules or codeposition fluxes in exact proportion.

Condensation of Pure Elements

A film will condense on a substrate when there exists a supersaturated vapor above the substrate. The idealized behavior is portrayed schematically in Figure III.8, which shows the *condensation flux* (the actual amount that condenses) as a function of *incident flux* (that which merely impinges on the

CONDENSATION OF VAPORS

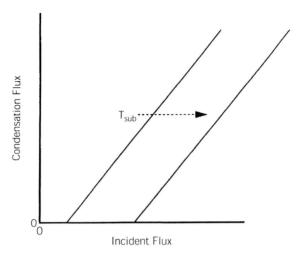

Figure III.8 Film condensation. For film deposition by condensation from a vapor, there is a critical incident flux at the substrate below which no film accumulates. The value of this critical flux increases with substrate temperature.

substrate). The following principles are illustrated:

- For a given substrate temperature, there is a critical incident flux above which a film will form, but below which no deposit is obtained.
- The greater the substrate temperature, the greater the critical incident flux.

The condensation flux is given by

$$j_c = \alpha_c j_i - \alpha_v z(T_{sub}) \equiv \alpha_c j_i - \alpha_v \frac{P_{eq}(T_{sub})}{\sqrt{2\pi m k T_{sub}}}, \tag{III.20}$$

which is close in concept to, if not a precise expression of, the historical Hertz–Knudsen equation. j_i is the actual incident flux of particles impinging on the substrate.

In (III.20) α_c is the *condensation coefficient*, expressing the fraction of incident particles that actually condense; α_c embodies myriad effects that can hinder film condensation. These include initial reflection (δ) and possible desorption of single adatoms before they attach to stable structures. In order to become part of the growing film, the impinging atom must first be adsorbed. Then it must then migrate across the surface and bond to another atom or group. During this surface diffusion it may instead desorb (the probability of such a desorption could be estimated from the mean diffusion time required and the mean residence time of adatoms). Finally, starting with a bare substrate, there can also the requirement of nucleation, which can lead to an "incubation time" for condensation, giving rise to a condensation coefficient that varies with the amount of material that has been deposited. In spite of all these difficulties, α_c is often practically equal to one in common film deposition conditions for many elements.

The effective condensation rate, however, is even less than $\alpha_c j_i$ because of reevaporation of condensed particles. The pressure term is the impingement rate expression for these same particles, from the kinetic theory of gases, evaluated at the substrate temperature. By the principle of microscopic reversibility, this impingement rate is also the theoretical reemission flux from the film or substrate surface.

The term $z(T_{\text{sub}})$ is multiplied by a vaporization coefficient, α_v, because vaporization fluxes less than the theoretical value occur [Glang, 1970]. A material exhibiting a low α_v is said to exhibit "retarded vaporization." It is hard at first to imagine how $\alpha_v \neq 1$ — it is much easier to see how condensation might be difficult. Rosenblatt [1976] writes that

> Solids whose structure is such that considerable electronic and/or atomic rearrangement is required to form gaseous molecules from the crystal may have vaporization coefficients very much less than unity... (examples include arsenic, alkali halides, the claudetite modification of arsenic trioxide, gallium nitride, and ammonium chloride).... It appears that the primary factors which cause low vaporization coefficients are *chemical* ones associated with the molecular nature of the evaporation reaction.

For a surface in equilibrium with its vapor, $\alpha_v = \alpha_c$, but in general, these coefficients differ because the atomistic pathways for condensation and vaporization are not identical.

Supersaturation means that the vapor pressure exceeds the thermal equilibrium value corresponding to the substrate temperature. The degree of supersaturation is calculated with the incident and reevaporation fluxes:

$$S \equiv \frac{\alpha_c j_i}{\alpha_v z(T_{\text{sub}})} - 1. \tag{III.21}$$

The Hertz–Knudsen equation shows that deposition occurs only with supersaturation ($S > 0$).

Condensation of Compounds that Produce a Stoichiometric Vapor

There is a class of compounds, molecular solids, whose vapors consist of particles having the stoichiometric composition, or are at least composed primarily of such molecules [Klabunde, 1985]. Stoichiometric thin films may be obtained by *direct* vaporization of these compounds, with varying degrees of success according to the specific requirements for purity and microstructure. The ones known to the author are

- Oxides (including SiO, MoO$_3$, WO$_3$),
- Chalcogenides (ZnS, PbS)
- Halides (NaCl, KCl, AgCl, MgF$_2$, CaF$_2$)

The predominant vapor particles are molecules, and thermochemical data are usually available for calculating their vapor pressures. The net condensation rate expression is the same as that given above for pure elements.

An impractical but important idea for obtaining a stoichiometric vapor: One may deduce from Figure III.7 that there is a temperature (around 900 K) where the composition of the vapor over GaAs is precisely Ga/As = 1.0. This is the *congruent evaporation temperature*. However, it is not practical to use this idea as an approach for the deposition of GaAs films from a GaAs compound evaporation source, because the requirement for precise control of source temperature is greater than that which is usually possible. There exists a much more forgiving experimental method for obtaining compositional exactitude, the three-temperature method to be discussed below.

A stoichiometric vapor may also be obtained during the sputtering of certain materials, whose ejected particles are molecules [Urbassek and Hofer, 1993]. These include oxides, molecular crystals again, bioorganic materials, and polymers.

Flash Evaporation of Compounds that Dissociate

Compounds which dissociate strongly have been deposited as stoichiometric thin films by *flash evaporation*, a technique that utilizes very rapid vaporization, typically by dropping powders or grains of the source material onto a hot surface [Richards et al., 1964], as shown schematically in Figure III.9. The vapor condenses rapidly onto a relatively cold substrate, usually with the same gross composition as that of the source material.

Flash evaporation has not worked well for the deposition of high-quality semiconductor films. On the microscopic scale the film may be compositionally stratified or layered over distances of a few monolayers, due to different arrival times for the various elements, and the frequent necessity of a low substrate temperature often makes the crystallinity quite poor.

Flash evaporation had fallen into disuse until the advent of pulsed laser deposition, a relatively new version (see Chapter V). Instead of bringing the source material to the heat, the heat is instead deposited on the surface of the source material in the form of an intense laser pulse. Pulsed laser deposition has been intensively and quite successfully applied to the deposition of thin films of high-temperature superconductors.

Steady-State Techniques for Alloy Films

During the evaporation of a liquid alloy, the composition of the source material must change. This is due to the practical impossibility of maintaining

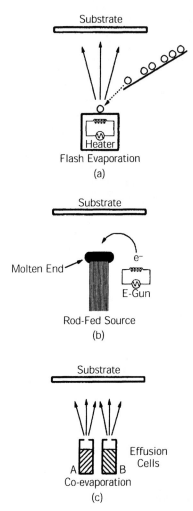

Figure III.9 Deposition of alloy and compound films. Techniques for obtaining alloys and compound films by condensation of supersaturated vapors include (a) transient flash evaporation, (b) nonequilibrium evaporation from a rod-fed source, and (c) coevaporation.

the composition of the vapor stream the same as that of the source. The proportion of the two monatomic components in free evaporation of a binary alloy would be, for example, $X_A P_{Aeq} m_B^{1/2} / X_B P_{Beq} m_A^{1/2}$, assuming Raoult's law behavior. The proportion of the components in the liquid source material would be, on the other hand, simply X_A/X_B. Whenever $P_{Aeq} m_B^{1/2} \neq P_{Beq} m_A^{1/2}$, the source will become more and more depleted in the component with the higher vapor pressure (the mass difference is usually negligible by comparison). The unfortunate result is that the composition of any films deposited from such a source will vary continuously.

Rod-fed evaporation sources solve this problem, through enforcement of the law of conservation of matter. In a rod-fed source, typically an electron-beam-heated evaporator such as is shown schematically in Figure III.9, the source material evaporates from the molten end of the rod. The rod advances as material is lost from the molten end. In steady state, the composition of the vapor stream must equal that of the rod. This requires that the molten end be enriched in the less volatile component. The adjustment is automatic, since diffusion in the liquid state is rapid. The composition of the molten region, which develops automatically, is found from

$$\frac{X'_A P_{Aeq} m_B^{1/2}}{X'_B P_{Beq} m_A^{1/2}} = \frac{X_A}{X_B} \quad (III.22)$$

with $X'_A + X'_B = 1$. The ($'$) signifies the molten region. The rod-fed source has found commercial application in the deposition of metal alloy thin films.

Similarly, during sputtering from an alloy or compound target, the composition of the surface must change. This is due to the fact that the sputter yields of the constituent elements are never in exact proportion to their concentrations. The proportion of the two elements in the sputtered flux of a binary alloy target may be expressed as $Y_A X_A / Y_B X_B$, where the Y_i values are *component sputter yields* (the actual sputter yield of a component is $Y_i X_i$) [Betz and Wehner, 1983]; Y_i is often approximated with the sputter yield of pure element i. Since $Y_A / Y_B \neq 1$ in general, the sputtered flux will be enriched in the component having the higher component sputter yield.

An *altered surface layer* thus forms, which is enriched in the other component. Its depth in steady state is a measure of the escape depth of recoils within the target. In steady state, and neglecting diffusion within the target, conservation of matter sets the following constraint on the concentrations X'_i within the altered layer:

$$\frac{Y_A X'_A}{Y_B X'_B} = \frac{X_A}{X_B} \quad (III.23)$$

Altered layer formation is automatic; its development is called "conditioning" of the target. It is accomplished by presputtering a target for a time typically on the order of one hour.

Thus, rod-fed evaporation sources and multicomponent sputtering targets share a common principle. By conservation of matter, the composition of the source flux is equal to that of the underlying solid state deposition source.

Coevaporation with the Three-Temperature Method

The three-temperature method has been an effective technique for the compositionally accurate deposition of compound semiconductor films whose components' vapor pressures differ greatly. It was the forerunner of molecular beam epitaxy [Freller and Günther, 1982] and is sometimes called the *Günther technique*.

The vapor pressures of the components of a suitable binary compound are shown schematically in Figure III.10. The method exploits a fact responsible for a main feature of this figure—the standard Gibbs free energy of reaction (see Appendix) for the dissociation of AB is greater than that for the vaporization of pure B. This fact creates a wide temperature interval within which B vapor is supersaturated with respect to AB but undersaturated with respect to pure B.

The method works as follows:

1. Choose a vapor pressure at the substrate corresponding to a reasonable deposition flux of the less volatile component, A. This will ultimately determine the deposition rate of the compound AB.
2. Choose a vapor pressure at the substrate giving a deposition flux of the more volatile component, B, somewhat greater (say, 10%) than that corresponding to the stoichiometric ratio. The actual flux ratio need not be accurately controlled or even known, but an excess of B is needed to make sure that A is fully reacted. (The range of practical flux ratio has been termed the *stoichiometric interval*. Unlike the condensation window for substrate temperature, the stoichiometric interval does not lend itself to precise calculation.) The sources for A and B are typically effusion cells. Thus, choosing the two vapor pressures *at the substrate* really amounts to choosing two cell

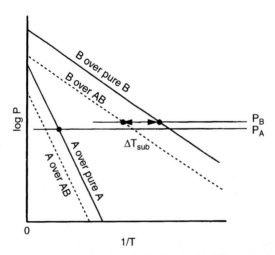

Figure III.10 The three-temperature method. To apply this method, first choose the vapor pressure at the substrate of the less volatile component (P_A). Then choose a modestly excess vapor pressure of the other component (P_B). Finally, choose a substrate temperature with the condensation window (ΔT_{sub}).

temperatures. These are two of the three temperatures of the method. It now remains to choose the third temperature, that of the substrate.

3. The substrate temperature must be chosen within a "condensation window," ΔT_{sub}, as shown in Figure III.10. The upper limit corresponds to the vapor pressure of B over AB; with a higher temperature B would not condense, and no AB film would form. The lower limit corresponds to the vapor pressure of B over pure B. At and below this substrate temperature, B will condense as pure B in addition to AB, resulting in a two-phase film.

The condensation coefficient for B automatically assumes the following value:

$$\alpha_{cB} = \frac{j_{cA}}{j_{iB}}, \qquad (III.24)$$

ignoring the reevaporation of B. The amount of B that condenses is determined by the amount of A that condenses. Whatever excess incident B there may be desorbs.

Example Apply the three-temperature method to the growth of GaAs.

The starting place is to choose a reasonable gallium incident flux, say, 10^{15} cm^{-2} s^{-1} (about one monolayer per second). This corresponds to a vapor pressure at the substrate of approximately 10^{-6} torr (10^{-4} Pa). Figure III.7 shows that pure gallium has this vapor pressure at a $1000/T$ value of ~ 1.06, or $T = 943$ K. The substrate must be colder than this, so that the gallium is sure to condense.

We will assume that the arsenic source produces As$_4$. The stoichiometric flux ratio would be As$_4$/Ga $= \frac{1}{4}$. Choosing an excess arsenic incident flux of 10% (and ignoring the different masses that go into the impingement rate expression), a workable arsenic vapor pressure at the substrate is

$$P_{As_4} \approx (1.1/4) \times 10^{-4}\,\text{Pa} = 2.8 \times 10^{-5}\,\text{Pa}.$$

Now the vapor pressure of As$_4$ over elemental arsenic is 2.8×10^{-5} Pa at a temperature of ~ 420 K [Barin, 1995]. This sets the lower temperature limit of the condensation window. The substrate must be hotter than this in order to avoid condensation of pure arsenic.

The vapor pressure of As$_4$ over GaAs is 2.8×10^{-5} Pa at a $1000/T$ value of ~ 0.98, corresponding to $T = 1020$ K (see Fig. III.7). The substrate must certainly be colder than this, but the upper limit of the condensation window is actually set by the gallium, rather than by the arsenic.

To sum up, the substrate temperature must be within the range 420 K $\leq T_{sub} \leq$ 943 K in order to satisfy the requirements of the three-temperature method.

The other two temperatures of the three-temperature method are, of course, those of the gallium and arsenic evaporation sources. They must provide the incident fluxes that we assumed above; these temperatures could be predicted when the geometries of the sources are known, with the models of Chapter VIII. ∎

The three-temperature method can be flexible, and forgiving; the condensation window for various compound semiconductors has been reported to be hundreds of degrees wide. It *not* necessary to precisely control the flux ratio since the condensation coefficient of B compensates automatically. Freller and Günther [1982] observed that "the three-temperature method is almost universal and can be used for epitaxial deposition over broad ranges of substrate temperatures and growth rates."

Of course, the effects of crystal growth processes on film microstructure usually lead to an optimum substrate temperature within the window, and there may be other requirements on thickness control. In practice, slight deviations from stoichiometry occur with variations in flux ratio and substrate temperature. The principles of the three-temperature method should work with cosputtering, also, but we are unaware of any attempt to extend them to sputter deposition.

Reactive Evaporation and Sputtering

Reactive evaporation and reactive sputtering give compound films by vapor deposition of a substance in the presence of a reactive gas, such as $Al + O_2$ and $SiO + O_2$ [Glang, 1970]. Thus, they are the integrations of condensation and adsorption in a film growth technique.

One of the best examples is TiN, which is obtained by evaporating a titanium film within a certain background pressure of nitrogen. A practical incident flux of titanium is chosen, and then a nitrogen partial pressure is established in the deposition chamber, such that the impingement rate of N_2 molecules is greater than or equal to half the titanium flux. The N_2 molecules adsorb, dissociate, and then combine with free titanium atoms to form TiN. Incorporating the various coefficients (which are typically unity anyway), the actual relationship determining the minimum required nitrogen partial pressure is found from $\alpha_s Z_{N_2} > \alpha_c j_{iTi}/2$.

Reactive evaporation resembles the three-temperature method. First, the incident flux of the less volatile component (the metal, rather than the reactive gas, of course) is chosen. Then, the impingement rate of the reactive gas is selected to be slightly higher than j_i (taking into account any stoichiometric coefficients from the chemical reaction). For the substrate temperature, the condensation window is in principle enormous, extending from a cryogenic temperature at which the reactive gas condenses, to a very high temperature at which the gas would not properly chemisorb. As with the three-temperature method, other factors may limit the choice of substrate temperature, such as optimizing the crystallinity of the film or controlling reactive gas incorporation or stress.

As a final comment, it should be remembered that all evaporation and sputtering is reactive, because there will always be residual gases present in the deposition chamber. The concentration of such impurities trapped in a thin film is given by

$$f = \frac{\alpha_s z}{\alpha_s z + \alpha_c j_i}. \tag{III.25}$$

The way to minimize unwanted reactive deposition is to have a deposition flux that is much higher than the impingement rate of the residual gases.

III.4 SUMMARY OF PRINCIPAL EQUATIONS

Ideal gas law	$n = P/kT$
Integrated Lennard-Jones potential	$E(x) = 4\pi E_d N r_0^3 [\frac{1}{45}(r_0/x)^9 - \frac{1}{6}(r_0/x)^3]$
Mean residence time	$\tau = 1/[\nu \cdot \exp(-E_d/kT)]$
Attempt frequency	$\nu \approx 10^{13}\,\text{Hz}$
Sticking coefficient	$\alpha_s = \delta(1-\theta)$
Langmuir's adsorption isotherm	$\theta = QP/(1+QP)$ $Q = [\delta(2\pi mkT)^{-1/2}]/[N_s \nu \exp(-E_d/kT)]$
Isotherm for dissociative adsorption	$\theta = \sqrt{QP}/(1+\sqrt{QP})$
Thermal equilibrium vapor pressure	$P_{\text{Aeq}} = P° \exp(-\Delta_{\text{vap}} G_A^\circ/RT)$
Free energy of vaporization	$\Delta_{\text{vap}} G_A^\circ = \Delta_f G_{A,v}^\circ - \Delta_f G_{A,c}^\circ$ $= \Delta_{\text{vap}} H_A^\circ - T\Delta_{\text{vap}} S_A^\circ$
Equilibrium constant for vaporization of monatomic vapors	$K_{p,\text{evap}} = P_{\text{Aeq}}/P°$
Gibbs free energy of a component in an ideal solution	$G_{A,\text{ideal}} = G_A^\circ + RT \ln X_A$
Partial pressure of a component of an ideal solution	$P_{A,\text{ideal}}(T) = X_A P_{\text{Aeq}}$
Partial pressure of a component of a nonideal solution	$P_{A,\text{nonideal}}(T) \equiv a_A P_{\text{Aeq}}$
Net deposition flux	$j_c = \alpha_c j_i - \alpha_v z(T_{\text{sub}})$
Supersaturation ratio	$S \equiv j_i/z(T_{\text{sub}}) - 1$

Composition of molten end of rod-fed source	$(X'_A P_{Aeq} m_B^{1/2})/(X'_B P_{Beq} m_A^{1/2}) = X_A/X_B$
Composition of altered layer	$Y_A X'_A / Y_B X'_B = X_A/X_B$
Condensation coefficient of more volatile component in three-temperature method	$\alpha_{cB} = j_{cA}/j_{iB}$
Concentration of residual gas	$f = \alpha_s z/(\alpha_s z + \alpha_c j_i)$
First law of thermodynamics	$dU = \delta Q - dW$
Second law of thermodynamics	$dS - \delta Q/T \geq 0$
Thermodynamic potentials	$H = U + PV$
	$F = U - TS$
	$G = H - TS$
Combined first and second laws	$dG - V\,dP + S\,dT = 0$
Criterion for equilibrium at constant temperature and pressure	$dG\|_{T,P} = 0,\ G_c = G_v$
Pressure dependence of G for a vapor	$\partial G/\partial P\|_T = RT/P$
	$G_v(T, P) = G_v^\circ(T) + RT \ln(P/P_0)$
Pressure independence of G for a condensed phase	$\Delta_f G_\ell(T, P) \approx \Delta_f G_\ell^\circ(T)$
Standard Gibbs free energy of reaction	$\Delta_r G^\circ(T) = c\,\Delta_f G_C^\circ(T) + d\,\Delta_f G_D^\circ(T)$ $\quad\quad -a\,\Delta_f G_A^\circ(T) - b\,\Delta_f G_B^\circ(T)$ $\quad\quad \equiv \sum i\,\Delta_f G_I^\circ(T)$
Equilibrium constant	$K_p = a_C^c a_D^d / a_A^a a_B^b \equiv \Pi\, a_I^i;$ $\ln K_p = -\Delta_r G^\circ / RT$
Activity	$a(\text{pure condensed phase}) \equiv 1$ $a(\text{vapor}) \equiv P/P_0$

III.5 APPENDIX: THERMODYNAMIC FUNDAMENTALS

The Thermodynamic Potentials and the First and Second Laws

The *internal energy* (U) is a basic state function of a system. The *first law of thermodynamics* expresses the principle of *conservation of energy*:

$$dU = \delta Q - dW, \tag{III.26}$$

where δQ is the amount of heat added to a system and dW is the amount of work done by the system.

APPENDIX: THERMODYNAMICS FUNDAMENTALS

The *second law of thermodynamics* says that the *entropy* (S) of an isolated system will increase spontaneously until, in equilibrium, S is maximized:

$$dS \geq 0, \tag{III.27}$$

while if heat transfer is allowed, then

$$dS - \delta Q/T \geq 0. \tag{III.28}$$

The enthalpy (H), Helmholtz free energy (F), and Gibbs free energy (G) are defined as follows:

$$\begin{aligned} H &= U + PV, \\ F &= U - TS, \\ G &= H - TS, \end{aligned} \tag{III.29}$$

where V is the volume. These five state functions (U, S, H, F, and G) are referred to as the *thermodynamic potentials*. Henceforth, they are assumed to be molar quantities.

The Gibbs Free Energy: The Relevant Potential for Equilibria at Fixed Temperature and Pressure

Combining the first and second laws and the preceding definitions of thermodynamic potentials, the criterion for equilibrium is

$$dG - V\,dP + S\,dT = 0, \tag{III.30}$$

For a system maintained by external influences at constant temperature and pressure, the criterion for equilibrium becomes

$$dG|_{T,P} = 0. \tag{III.31}$$

This criterion based on the Gibbs free energy sets it apart in practical importance, as compared to the other state functions, because it applies to the practical situation of fixed temperature and pressure. It is applicable to a system of liquid and vapor in the following way: If the system is in equilibrium, then during the transfer of an infinitesimal amount of material (dn moles) from one phase to the other, we must have $dG = dn_\ell G_\ell + dn_v G_v = 0$. Since $dn_\ell = -dn_v$, we must also have

$$G_\ell = G_v, \tag{III.32}$$

which says that the molar Gibbs free energy of the liquid phase is equal to that of the vapor for equilibrium at constant temperature and pressure. It was this result that we used in Section III.2 to calculate the thermal equilibrium vapor pressure.

Now, thermodynamic state functions such as the Gibbs free energy depend on state variables. For example, G varies with temperature according to $\partial G/\partial T|_P = -S$ [according to Eq. (III.30)], and with pressure according to $\partial G/\partial P|_T = V$. If the substance is a gas or vapor that obeys the ideal gas law, then

$$\left.\frac{\partial G}{\partial P}\right|_T = RT/P. \quad \text{(III.33)}$$

The value of G at one pressure may be related to that at another pressure by integrating:

$$\begin{aligned} G_v(T,P) &= G_v^\circ(T) + \int_{P^\circ}^{P} \left(\frac{RT}{P'}\right) dP' \\ &= G_v^\circ(T) + RT \ln\left(\frac{P}{P^\circ}\right). \end{aligned} \quad \text{(III.34)}$$

The degree superscript (°) refers to the *standard pressure*, which we discuss below as part of the definition of the standard state. The term $G^\circ(T)$ will be the standard molar Gibbs free energy (i.e., the value under standard state conditions, at temperature T).

If the substance is a pure condensed phase (either liquid or solid), the molar volume is typically negligible compared to that of a gas. Thus, for condensed phases it is frequently assumed that $\partial G/\partial P = 0$, so that, for example

$$G_\ell(T,P) \approx G_\ell^\circ(T). \quad \text{(III.35)}$$

Unfortunately, it is impossible to give absolute values for thermodynamic state functions involving the internal energy, such as G. Equilibrium is expressed, instead, in terms of *formation* quantities (to be explained in the next section). Thus, for a liquid phase, for example, we must really write Equation (III.35) as

$$\Delta_f G_\ell(T,P) \approx \Delta_f G_\ell^\circ(T) \quad \text{(III.36)}$$

to express the pressure invariance of G.

Standard Reaction and Formation Quantities, and the Equilibrium Constant

Suppose that a product is formed from some reactants at a fixed temperature:

$$a(A) + b(B) \rightarrow c(C) + d(D),$$

where A, B, C, and D are chemical species and a, b, c, and d are the stoichiometric coefficients of the reaction equation. This could be either a chemical reaction, or perhaps a phase change such as evaporation or sublimation. There is a *standard Gibbs free energy of reaction* for the preceding reaction equal to

$$\Delta_r G°(T) = c\,\Delta_f G°_C(T) + d\,\Delta_f G°_D(T) - a\,\Delta_f G°_A(T) - b\,\Delta_f G°_B(T), \quad \text{(III.37)}$$

where $\Delta_f G°_I(T)$ is the *standard Gibbs free energy of formation* of species I. For any reaction, $\Delta_r G°(T)$ is the sum of the Gibbs free energies of formation of all the products and reactants in the chemical equation, with each term multiplied by the corresponding *stoichiometric coefficient* in the equation (the stoichiometric coefficients of reactants are preceeded by a negative sign in this sum). The general expression is

$$\Delta_r G°(T) \equiv \sum i\,\Delta_f G°_I(T) \quad \text{(III.38)}$$

(and the same formulation also pertains to the standard enthalpy and entropy of reaction).

Standard "formation" thermochemical quantities for a substance are reaction quantities pertaining to its *formation* reaction from the elements (in their reference states, the subject of the next section). For example, the standard enthalpy of formation of vapor A is the standard enthalpy of reaction for the formation reaction by evaporation of the liquid (if the temperature is between the melting and the boiling points, because within this temperature range the liquid is the reference phase):

$$A_\ell \to A_v,$$

with, for example

$$\Delta_f H°_{A,v} = \Delta_{evap} H°_A. \quad \text{(III.39)}$$

In general, the enthalpy of formation of a substance is the heat input required, at constant pressure and temperature, to form the substance from the elements in their reference states. When, as is often the case, chemical compounds have a lower potential energy than the free elements, their enthalpies of formation are negative. (Typically, the enthalpies of formation of vapors are positive.)

The standard Gibbs free energy of reaction is not necessarily the Gibbs free energy change of the system if the reaction goes forward to some extent; it is simply the algebraic sum of the formation quantities as defined above. The extent to which a reaction does progress is described with an *equilibrium constant*. Because of our current interest in vapor pressure, we consider the situation of constant pressure, which is described with the equilibrium constant at constant pressure, K_p. For the previous reaction

$$K_p = \frac{a_C^c\,a_D^d}{a_A^a\,a_B^b} \quad \text{(III.40)}$$

where a_I is the *activity* of species *I*. While the activity of a pure condensed phase is defined to be unity, that of a vapor or gas is its pressure divided by $P°$, the standard pressure:

$$a(\text{condensed phase}) \equiv 1 \tag{III.41}$$

and

$$a(\text{vapor}) \equiv \frac{P}{P^\circ}. \tag{III.42}$$

The general formulation for K_p is

$$K_p \equiv \Pi a_I^i. \tag{III.43}$$

The physical significance of the equilibrium constant is that it is an indicator of how far the reaction must proceed in order to attain thermodynamic equilibrium. The larger is K, the farther the reaction will proceed to the right as written.

The relationship between the equilibrium constant and the standard Gibbs free energy of reaction is

$$\ln K_p = \frac{-\Delta_r G^\circ}{RT}. \tag{III.44}$$

We are now ready to explain more fully the concept of the standard state and the conventions of standard thermochemical data:

Standard Thermochemical Data

Certain standard states and reference states have been adopted to facilitate the tabulation of thermochemical data and the calculation of thermochemical quantities from those data. This practice is necessary because absolute values of internal energies, and thus enthalpies and Gibbs free energies, are unknown. The values of these thermodynamic state functions for a substance may be calculated, each referenced to an assumed (but arbitrary) value of that function for when the substance is in a well-defined standard state. This is suggested by the relation derived previously for the variation with pressure of the Gibbs free energy of a vapor.

"In the *standard* state" means that the substance in question is perfectly pure and at the standard pressure (P°). The standard state does *not* refer to the microscopic structure of a substance or to its temperature—only to the independent state variables of composition and pressure. Thus, a diamond may exist in the standard state when worn on one's finger under normal circumstances. Because it is a metastable phase of carbon at normal pressures, diamond is *not* the reference state. The value of the standard pressure is $P^\circ \equiv 10^5 \, \text{Pa} \approx 0.9869 \, \text{atm} \approx 750 \, \text{torr}$.

On the other hand, the "*reference* state" at a specified temperature does refer to the microscopic structure of the substance. An element may have a multitude of reference states, each of which is the reference phase for that element within a specified temperature range. The reference state for an element below the melting point is a specified crystal structure. (Manganese is a peculiar example, as it has four distinct reference states below the melting point, due to its ability to exhibit four distinct equilibrium crystal structures. As another example, amorphous silicon is never a reference state for silicon because it is not an equilibrium structure; however, amorphous silicon may exist in the standard state.) From the melting point to the boiling point, the reference state is a liquid; above the boiling point, it is a gas assumed to obey the ideal gas law. (It might have been better if the reference state had been named the "reference phase," to lessen the opportunity for confusion with the standard state.)

It is assumed that the enthalpy of formation, the entropy of formation, and thus the Gibbs free energy of formation of a chemical element, when it is in both a reference state and the standard state, are zero.

The JANAF tables [Chase et al., 1986] are an excellent source of standard thermochemical data. The introduction to the tables discusses the standard and reference state conventions; the notation of this chapter conforms with these tables. In Table III.1 we show some data from the tables that illustrate basic points about formation quantities.

There are several points to be noted in the data in Table III.1. Since the liquid is the reference state above the melting point, its standard enthalpy and Gibbs free energy of formation are zero above 1685 K. However, below the melting point the crystal is the reference state, and since the heat content of the liquid is larger, the standard enthalpy of formation of the liquid is positive below 1685 K. By comparison, the standard enthalpy of formation of the vapor is enormous at all temperatures shown. $\log K_f$ is always $\Delta_f G°/2.303RT$; for the monatomic vapor it is also the logarithm of the vapor pressure, normalized to standard pressure.

In Table III.2 we present a selection of thermochemical data that may be used to estimate vapor pressures. These quantities were taken from, or calculated from data in, the JANAF tables. Values at room temperature and at

TABLE III.1 Thermochemical Data for Silicon from the JANAF Tables

	Liquid			Monatomic Vapor		
T(K)	$\Delta_f H°$(kJ/mol)	$\Delta_f G°$(kJ/mol)	$\log K_f$	$\Delta_f H°$(kJ/mol)	$\Delta_f G°$(kJ/mol)	$\log K_f$
1500	50.512	5.531	−0.193	444.803	228.879	−7.970
1600	50.365	2.537	−0.083	444.070	214.508	−7.003
1685		Crystal ↕ Liquid				
1700	0	0	0	393.131	200.631	−6.165
1800	0	0	0	392.569	189.323	−5.494

Source: Chase et al. [1986].

TABLE III.2 Standard Enthalpies and Entropies of Vaporization for Pure Elements

Element	Melting Point (K)	$\Delta_{vap}H°(298)$ (kJ/mol)	$\Delta_{vap}H°(T_m)$ (kJ/mol)	$\Delta_{vap}S°(298)$ (J·mol^{-1}·K^{-1})	$\Delta_{vap}S°(T_m)$ (J·mol^{-1}·K^{-1})
Lithium	454	159	158	109.8	97.6
Beryllium	1560	324	303	126.9	111.4
Boron	2350	560	500	147.6	122.3
Carbon	M	717	M	152.4	M
Sodium	371	107	104	148.0	93.3
Magnesium	923	147	134	115.9	98.8
Aluminum	933	330	314	136.2	117.8
Silicon	1685	450	393	149.2	113.2
Phosphorus P_2	317	144	142	135.9	129.5
Sulfur S_2	388	129	123	164.0	147.5
Potassium	336	89	86	95.6	86.4
Calcium	1115	178	158	113.3	90.6
Titanium	1939	474	438	149.5	124.6
Vanadium	2190	516	477	153.4	131.5
Chromium	2130	398	349	150.7	118.8
Manganese	1519	283	247	141.7	106.2
Iron	1809	416	377	153.2	123.4
Cobalt	1768	427	395	149.4	125.8
Nickel	1728	430	401	152.3	129.2
Copper	1358	338	317	133.2	113.9
Zinc	693	130	120	119.3	103.3
Gallium	303	272	266	128.2	109.7
Rubidium	313	81	78	93.3	85.2
Strontium	1050	164	149	108.9	90.2
Zirconium	2125	610	580	144.1	125.8
Niobium	2750	733	693	157.7	135.4
Molybdenum	2896	659	589	153.4	119.7
Cesium	302	77	74	90.5	83.4
Barium	1000	179	166	107.7	78.8
Hafnium	2500	618	599	143.3	115.7
Tantalum	3258	782	744	143.7	129.1
Tungsten	3680	851	807	141.3	129.7
Mercury	234	61	62	99.0	99.0
Lead	601	195	188	110.6	97.6

Source: Chase et al. [1986].

the melting point of each element are given; it is clear that these quantities are quite temperature-dependent. Values at 100° intervals may be found in the original reference [Chase et al., 1986]. All pure elements represented in the JANAF tables are given in Table III.2. For additional elements, the reader might consult Barin [1995] or recent editions of the CRC Press' *Handbook of Chemistry and Physics*. There are two elements for which the assumed vapor is not monatomic, but diatomic; M denotes a missing value.

III.6 MATHEMATICAL SYMBOLS, CONSTANTS, AND THEIR UNITS

SI units are given first, followed by other units in widespread use.

a	Activity; stoichiometric coefficient
b	Stoichiometric coefficient
c	Stoichiometric coefficient
f	Concentration of impurities (dimensionless)
h	Planck's constant (6.63×10^{-34} J·s)
j_s	Source flux (m^{-2}·s^{-1})
j_d	Deposition flux (m^{-2}·s^{-1})
j_{eq}	Reevaporation flux (m^{-2}·s^{-1})
k	Boltzmann's constant (1.38×10^{-23} J/K; 8.62×10^{-5} eV/K)
m	Particle mass (kg; amu = 1.66×10^{-27} kg)
n	Particle density (m^{-3})
r_0	Length parameter (nm)
x	Distance (m)
z	Impingement rate (m^{-2}·s^{-1})
A	Chemical species
B	Chemical species
C	Chemical species
E	Potential energy (J; eV = 1.602×10^{-19} J)
F	Helmholtz free energy (kJ/mol)
G	Gibbs free energy (kJ/mol)
H	Enthalpy (kJ/mol; kcal/mol = 4.18 kJ/mol)
K_f	Equilibrium constant for the formation reaction
K_p	Equilibrium constant at constant pressure
N	Surface density of adatoms
N_s	Surface density of adsorption sites
P	Pressure (Pa; torr = 133 Pa; atm = 1.01×10^5 Pa)
Q	Heat (J; cal = 4.18 J); desorption parameter (Pa^{-1})
R	Molar gas constant (8.31 J·mol^{-1}·K^{-1})
S	Entropy (J·mol^{-1}·K^{-1}); supersaturation ratio (dimensionless)
T	Temperature (K)
U	Internal energy (J/mol)
V	Volume (m^{-3}; liter = 10^{-3} m^3)
W	Work (J)
X_i	Atomic fraction of element I
Y_i	Component sputter yield of I (dimensionless)
α_c	Condensation coefficient (dimensionless)
α_s	Sticking coefficient (dimensionless)
α_v	Vaporization coefficient (dimensionless)
δ	An inexact differential; trapping probability (dimensionless)
ΔT_{sub}	Condensation window (K)

θ Fractional surface coverage (dimensionless)
ν Vibration frequency (s^{-1})
τ Mean residence time (s)

REFERENCES

Alcock, C. B., 1995, "Vapor Pressure of the Metallic Elements," in D.R. Lide, ed., *CRC Handbook of Chemistry and Physics*, CRC Press, Boca Raton, FL.

Arthur, J. R., 1967, "Vapor Pressures and Phase Equilibria in the Ga-As System," *J. Phys. Chem. Solids* **28**, 2257.

Barin, I., 1995, *Thermochemical Data of Pure Substances*, 3rd ed., VCH, Weinheim.

Betz, G., and Wehner, G. K., 1983, "Sputtering of Multicomponent Materials," in R. Behrisch, ed., *Sputtering by Particle Bombardment II*, Springer-Verlag, Berlin.

Blakely, J. M., ed., 1975, *Surface Physics of Materials*, Vol. I, Academic Press, New York.

Borisenko, V. E., and Hesketh, P. J., 1997, *Rapid Thermal Processing of Semiconductors*, Plenum Press, New York, p.116.

Chase, M. W., Davies, C. A., Downey, J. R., Frurip, D. J., McDonald, R. A., and Syverud, A. N., 1986, *JANAF Thermochemical Tables*, 3rd ed., American Institute of Physics, New York.

Daniels, F., and Alberty, R. A., 1966, *Physical Chemistry*, 3rd ed., Wiley, New York.

Foxon, C. T., Harvey, J. A. and Joyce, B. A., 1973, "The Evaporation of GaAs Under Equilibrium and Nonequilibrium Conditions Using a Modulated Beam Technique," *J. Phys. Chem. Solids* **34**, 1693.

Freller, H., and Günther, K. G., 1982, "Three-Temperature Method as an Origin of Molecular Beam Epitaxy," *Thin Solid Films* **88**, 291.

Glang, R., 1970, "Vacuum Evaporation," in L. I. Maissel and R. Glang, eds., *Handbook of Thin Film Technology*, McGraw-Hill, New York.

Goodman, C. H. L., and Pessa, M. V., 1986, "Atomic Layer Epitaxy," *J. Appl. Phys.* **60**(3), R65.

Hannay, N. B., ed., 1976, *Surfaces I and II*, Vols. 6A and 6B of Treatise on Solid State Chemistry, Plenum Press, New York.

Klabunde, K. J., 1985, "Introduction to Free Atoms and Particles," in K. J. Klabund, ed., *Thin Films from Free Atoms and Particles*, Academic Press, 1985.

Lennard-Jones, J. E., 1932, "Processes of Adsorption and Diffusion on Solid Surfaces," *Trans. Faraday Soc.* **28**, 334.

Leskelä, M., and Niinistö, L., 1990, "Chemical Aspects of the Atomic Layer Epitaxy Process," in T. Suntola and M. Simpson, eds., *Atomic Layer Epitaxy*, Blackie and Son, Ltd., Glasgow.

Menzel, D., 1975, "Desorption Phenomena," in R. Gomer, ed., *Interactions on Metal Surfaces*, Springer-Verlag, New York, p.101.

Prutton, M., 1983, *Surface Physics*, 2nd ed., Oxford Univ. Press, Oxford.

REFERENCES

Redhead, P. A., Hobson, J. P., and Kornelsen, E. V., 1993, *The Physical Basis of Ultrahigh Vacuum*, American Institute of Physics, New York.

Richards, J. L., Hart, P. B., and Müller, E. K., 1964, "Epitaxy by Flash Evaporation," in M. H. Francombe and H. Sato, eds., *Single-Crystal Films*, Pergamon Press, Oxford.

Rosenblatt, G. M., 1976, "Evaporation from Solids," in N. B. Hannay, ed., *Surfaces I and II*, Vols. 6A and 6B of Treatise on Solid State Chemistry, Plenum Press, New York.

Steele, W. A., 1974, *The Interaction of Gases with Solid Surfaces*, Pergamon Press, Oxford, Chapter 5.

Suntola, T., 1994, "Atomic Layer Epitaxy," in D. T. J. Hurle, ed., *Handbook of Crystal Growth*, **3** Elsevier, Amsterdam, Chapter 14.

Tanaka, H., and Kanayama, T., 1997, "Thermal Desorption of Si Clusters from Si and Si-deposited Ta Surfaces," *J. Vac. Sci. Technol.* **B15**(5), 1613.

Tu, K. N., Mayer, J. W., and Feldman, L. C., 1992, *Electronic Thin Film Science: For Electrical Engineers and Materials Scientists*, Macmillan, New York.

Urbassek, H. M., and Hofer, W. O., 1993, "Sputtering of Molecules and Clusters: Basic Experiments and Theory," in P. Sigmund, ed., *Fundamental Processes in Sputtering of Atoms and Molecules*, Symp. on Occasion of the 250th Anniversary of the Royal Danish Academy of Sciences and Letters, Copenhagen, Aug. 30–Sept. 4, 1992, The Royal Danish Academy of Sciences and Letters, Copenhagen.

Zangwill, A., 1988, *Physics at Surfaces*, Cambridge Univ. Press, Cambridge, UK.

IV

PRINCIPLES OF HIGH VACUUM

Nature abhors a vacuum.

—Otto von Guericke

Vacuum integrity is critical to the quality of vapor deposited thin films. Hashim et al. [1997] write that "Currently, the most stringent vacuum requirements of all semiconductor [manufacturing] processes are for physical vapor deposition processes." Thus, while nature abhors a vacuum, the thin-film scientist and engineer abhor vacuum leaks or other limitations to the highest attainable vacuum. A thorough understanding of the behavior of vacuum systems is fundamental to designing and controlling the physical vapor deposition environment.

The *throughput law* is the basic principle governing the behavior of vacuum systems. It is simply a statement of conservation of matter assuming that the ideal gas law holds and that the system is of uniform and constant temperature. Throughput is calculated in pressure–volume units probably because the degree of vacuum has historically been expressed in pressure rather than, say, gas density (which seems a more rational choice).

We focus in this chapter on the fundamental physical principles and mechanisms underlying high-vacuum components and systems. There are several reference books with more of a technology-oriented emphasis [Dushman, 1949; Dennis and Heppell, 1968; Holland et al., 1974; Weissler and Carlson, 1979; O'Hanlon, 1980; Chambers et al., 1989].

IV.1 BASIC VACUUM CONCEPTS

Pumping Speed

Consider the vacuum chamber shown in Figure IV.1. A schematic diagram using symbols adopted by the American Vacuum Society [AVS, 1967] is also shown in the figure, together with an analogous electrical network that will be discussed in the end-of-chapter Appendix. An "appendage" pump may be attached to the pumping port of the chamber. The effect of the pump is not to *pull* the gas particles out of the chamber, but rather to prevent a return flux from the port into the chamber, making the port a sink for gas particles.

The chamber becomes evacuated because gas molecules pass through the port at a rate given by the port area (A) multiplied by the net impingement rate ($z_{\text{out}} - z_{\text{in}}$). Thus, the *particle flow* (J, positive if there is a net flux *into* the chamber) is

$$J \equiv \frac{dN}{dt} = -(z_{\text{out}} - z_{\text{in}})A. \tag{IV.1}$$

The maximum possible flow rate out of the chamber occurs when z_{in} is zero:

$$J_{\max} = -n v_{\text{av}} \frac{A}{4}, \tag{IV.2}$$

using an expression for z from the kinetic theory of gases. Here, N is the total number of gas particles, t the time, n the particle volume density, and v_{av} the average thermal speed.

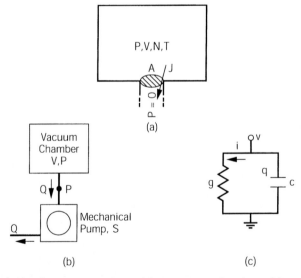

Figure IV.1 A simple vacuum system: (*a*) a vacuum chamber with pump, (*b*) their schematic diagram, and (*c*) an analogous electrical network.

BASIC VACUUM CONCEPTS 85

The port's pumping ability may be characterized with a pumping speed (S, occasionally called the *volumetric flow rate*). To understand pumping speed, imagine the port to be a butterfly net and the gas particles to be butterflies, though in reality the port is stationary. The net sweeps out spatial volume at a rate given by S and captures butterflies at a rate given by nS. This effective "rate of volume sweeping out" is the definition of pumping speed:

$$S \equiv -\frac{J}{n}. \tag{IV.3}$$

The maximum possible pumping speed is

$$S_{\max} \equiv v_{av}\frac{A}{4}. \tag{IV.4}$$

It was a deliberate choice to expound on the pumping ability of the port, rather than the pump itself. Ideally, the pump plays a very passive role since its job is to prevent any return flux. While the maximum possible pumping speed of any pump is given by expression (IV.4), the actual pumping speeds of real pumps are always less than this ideal limit, because some return flux is inevitable. A "black hole" might represent the ideal.

S is usually quoted in liters per second when characterizing a particular vacuum pump. The concept is easy to visualize with a rotary or piston-type mechanical pump. Assume that whenever the intake valve is open, the pump's bore fills with gas from the vacuum chamber until equilibrium is attained. The pumping speed of such a pump is given by the filled volume times the rate of revolution of the rotor (multiplied by the number of times the bore is filled per revolution). The concept probably originated with piston-type pumps, which are rare nowadays, and it has since been applied to all vacuum pumps. The concept really means that a pump is capable of removing from a vacuum chamber the number of gas particles contained in S liters per second. As the pressure drops, the rate of removal of particles obviously decreases, even if S remains constant.

Example A commercial diffusion pump has a pumping speed for "air" of 5600 liters/s. The diameter of its inlet is 31 cm. How does the actual pumping speed compare to the theoretical maximum value?

In order to make a quick estimate of the pumping speed for air, we will assume air to be entirely N_2, of mass 28 amu. Then

$$v_{av} = \left[\frac{8 \times (1.38 \times 10^{-23}\,\text{J/K}) \times 298\,\text{K}}{\pi \times 28 \times 1.66 \times 10^{-27}\,\text{kg}}\right]^{1/2} = 475\,\text{m/s}.$$

The inlet area is $7.55 \times 10^{-2}\,\text{m}^2$, so

$$S_{\max} = \frac{(475\,\text{m/s}) \times 7.55 \times 10^{-2}\,\text{m}^2}{4} = 8{,}956\,\text{liters/s}.$$

The theoretical pumping speed is substantially greater than the commercial specification; there must be some nonidealities in the real pump. ∎

The vacuum chamber port in Figure IV.1 may be said to have a *conductance* (C) that relates the particle flow rate through the port to the difference in gas density on either side. The defining equation for conductance is

$$J \equiv -C \Delta n. \tag{IV.5}$$

Thus, the conductance of the port is equal to its ideal pumping speed since if there is no return flux, $\Delta n = n$. Other structures, such as tubes, baffles, and valves, are characterized with conductance as well.

The preceding expressions for S give the theoretical maximum pumping speed of any pump for a gas of a certain particle mass at a specified temperature. Thus, *the maximum possible pumping speed of any real pump is equal to the conductance of its inlet port*. (Real high-vacuum pumps such as diffusion pumps may approach the theoretical pumping speed under ideal conditions, but mechanical displacement pumps generally do not.)

How does the pressure vary with time for the chamber of Figure IV.1? Using the ideal gas law, $dP/dt = (dN/dt)kT/V = (-nS)kT/V = -SP/V$. The solution of this simple differential equation in P is

$$P = P(0) \exp\left(-\frac{St}{V}\right). \tag{IV.6}$$

Thus, the pressure of this simple system decays exponentially with a characteristic time given by V/S.

What if there were a minor leak in the chamber? A leak may be modeled with a conductance value, and if the chamber pressure is far below that of the atmosphere, the rate at which particles enter the chamber through the leak is a constant, $dN/dt|_L = C_L n$, since $\Delta n \to n$ (the particle density of the air outside the chamber). Eventually during the pumpdown the chamber pressure will reach a steady-state value. In steady state, both dP/dt and dN/dt are zero. The value of dN/dt is given by $dN/dt|_L - nS = dN/dt|_L - PS/kT$. Thus, the ultimate pressure in the presence of a leak is

$$P_{\text{ult}} = \frac{kT(dN/dt|_L)}{S}. \tag{IV.7}$$

Example A particular leak in a vacuum system is known to have a conductance of 10^{-4} liters/s. What is the rate at which gas particles enter through this leak?

The gas density at room temperature and atmospheric pressure is

$$n = \frac{760 \text{ torr}}{(1.38 \times 10^{-23} \text{ J/K}) \times 298 \text{ K}} = 2.46 \times 10^{25} \text{ m}^{-3}.$$

Then

$$\frac{dN}{dt} = (10^{-4}\,\text{liter/s}) \times 2.46 \times 10^{25}\,\text{m}^{-3} = 2.46 \times 10^{18}\,\text{s}^{-1}.$$

Assuming that the pump of the previous example is used, what is the ultimate pressure of the system in the presence of this leak?

$$P_{\text{ult}} = \frac{1.38 \times 10^{-23}\,\text{J/K} \times 298\,\text{K} \times 2.46 \times 10^{18}/\text{s}}{5600\,\text{liters/s}} = 1.36 \times 10^{-5}\,\text{torr}. \quad \blacksquare$$

Throughput

The throughput concept is employed with pumping speed and conductance expressions to calculate gas flow and to model the pumping dynamics of vacuum systems. It is not difficult to *use* the concept, but *explaining* its meaning and why there is a conservation law for throughput has been neglected in the literature. In essence, the concept is used to express the principle of conservation of matter in terms of pressure changes.

It should be easy to accept that one is interested in relating the rate of change of the number of gas particles in a vacuum chamber to variations in the chamber pressure. We can use the ideal gas law to do this, assuming quasiequilibrium (spatially uniform P and T, even near a pumping port or leak):

$$\frac{d(PV)}{dt} = \frac{d(NkT)}{dt}. \tag{IV.8}$$

With both the chamber volume and the temperature fixed, we have

$$V\frac{dP}{dt} = kT\frac{dN}{dt} = kTJ. \tag{IV.9}$$

The left-hand side is the throughput of a vacuum chamber, a positive quantity if the pressure is increasing:

$$Q_c \equiv V\frac{dP}{dt}. \tag{IV.10}$$

Thus, the American Vacuum Society officially defines throughput [Kaminsky and Lafferty, 1980] as "The quantity of gas in pressure-volume units, at a specified temperature, flowing per unit time across a specified open cross-section." We will see next that throughput is the particle flow rate (dN/dt) multiplied by kT.

The right-hand side might be, correspondingly, the throughput of a leak, positive if particles are entering the chamber:

$$Q_L \equiv kTJ_L. \tag{IV.11}$$

The rate at which gas particles pass through the leak, J_L, fundamentally characterizes the leak, but the throughput of the leak is given by J_L multiplied by kT.

The throughput removed from a vacuum system by a pump is

$$Q_p \equiv kTJ_p = SP, \tag{IV.12}$$

where Q_p, the throughput of a pump, is defined as positive if the particles are entering the pump.

Finally, the throughput through some pathway such as a tube, which is characterized with a conductance, is given by

$$Q_t \equiv kTJ = kTC_t \Delta n = C_t \Delta P. \tag{IV.13}$$

Example Calculate the throughput of the leak in the previous example.

Solution

$$\begin{aligned} Q_L &= (1.38 \times 10^{-23} \text{ J/K}) \times 298 \text{ K} \times 2.46 \times 10^{18}/\text{s} \\ &= 0.076 \text{ torr} \cdot \text{liter} \cdot \text{s}^{-1} (\text{or } 1.01 \times 10^{-2} \text{ Pa} \cdot \text{m}^3/\text{s}). \end{aligned}$$ ∎

The leak is, perhaps, the starting place for *explaining* throughput. A leak of J particles per second has a throughput of J multiplied by kT. (Thus, the throughput of a leak will change if the temperature of the gas changes, even though the number of particles entering remains the same.) A National Bureau of Standards writer objects to this usage, preferring to represent a leak simply as so many moles of particles per second [Ehrlich, 1986], and it is easy to sympathize with this view.

A Throughput Law

If we consider a vacuum system to be a lumped-parameter system, in analogy to a lumped-parameter electrical circuit, there are nodes where system elements (e.g., chamber and pump) are joined. These should be obvious in the schematic vacuum system diagrams of Figures IV.1 and IV.2. We would like to propose a throughput law: At any node, it must be true that

$$\Sigma Q_i = 0 \text{ (node version)}, \tag{IV.14}$$

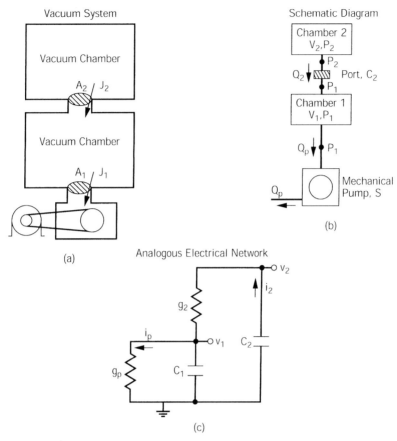

Figure IV.2 (*a*) A two-chamber vacuum system with (*b*) its schematic diagram and (*c*) its analogous electrical network.

where Q_i represents each individual throughput contribution *entering* the node under consideration. This is the *node version*, that is, the throughput law applied to a *node*, which has in principle no volume to consider.

On the other hand, if we focus on a chamber or a pump instead of a node, we have

$$\Sigma Q_i = Q_c \quad \text{(chamber version)}, \tag{IV.15}$$

$$\Sigma Q_i = Q_p \quad \text{(pump version)}. \tag{IV.16}$$

Here, Q_i represents each individual throughput contribution *entering* a chamber or pump. The term Q_c, when positive, represents the accumulation of gas within a chamber.

Equations (IV.14)–(IV.16) are the three forms of the throughput law. We would like to illustrate the application of this law by analyzing the two-chamber system shown in Figure IV.2. The chambers have their respective pressures and volumes, are at the same temperature T, and are connected by a port of area A_2. The area of the pump inlet is A_1, which is contained in its pumping speed, S.

There are four variables—Q_p, P_1, Q_2, and P_2—and four equations are needed. First, the throughput of chamber 2 is $V_2\, dP_2/dt$, which equals the throughput *entering* it, $-Q_2$:

$$Q_2 = -V_2 \frac{dP_2}{dt}. \tag{IV.17}$$

Second, the throughput of chamber 1 is $V_1\, dP_1/dt$, which equals the throughput entering it, $Q_2 - Q_p$:

$$Q_2 - Q_p = V_1 \frac{dP_1}{dt}. \tag{IV.18}$$

Third, the throughput of the pump, SP_1, equals the throughput entering the pump, Q_p:

$$Q_p = SP_1. \tag{IV.19}$$

Fourth, the relation between the two chamber pressures is determined by the conductance of the orifice A_2:

$$Q_2 = C_2(P_2 - P_1). \tag{IV.20}$$

Equations (IV.18) and (IV.19) yield

$$V_1 \frac{dP_1}{dt} = Q_2 - SP_1, \tag{IV.21}$$

and Equation (IV.17) makes this

$$V_1 \frac{dP_1}{dt} + V_2 \frac{dP_2}{dt} = -SP_1. \tag{IV.22}$$

Substituting Q_2 from Equation (IV.20) into Equation (IV.21), differentiating the result, and then using Equation (IV.22) to replace dP_2/dt yields

$$V_1 \frac{d^2 P_1}{dt^2} + [S + C_2(1 + V_1/V_2)] \frac{dP_1}{dt} + (SC_2/V_2)P_1 = 0. \tag{IV.23}$$

This equation represents a second-order system. To solve for the pumpdown from atmosphere, we require its step response. The two initial conditions for

BASIC VACUUM CONCEPTS

P_1 are $P_1(0) = 1$ atm and $dP_1(0)/dt = -SP_1(0)/V_1$. The solution is surprisingly complex; $P_1(t)$ is the sum of two exponentials having different time constants [Fastenau, 1993]:

where
$$P_1(t) = K_1 e^{-\lambda_1 t} + K_2 e^{-\lambda_2 t}, \qquad \text{(IV.24)}$$

with
$$\begin{aligned}\lambda_1 &= a + \sqrt{a^2 - b^2} \\ \lambda_2 &= a - \sqrt{a^2 - b^2}\end{aligned} \qquad \text{(IV.25)}$$

$$\begin{aligned}a &= \frac{1}{2}\left[\frac{C_2}{V_2} + \frac{C_2}{V_1} + \frac{S}{V_1}\right] \\ b &= \sqrt{\frac{SC_2}{V_1 V_2}}.\end{aligned} \qquad \text{(IV.26)}$$

The coefficients of the two exponentials are given by

$$\begin{aligned}K_1 &= 1\,\text{atm} \cdot \frac{(S/V_1) - \lambda_2}{\lambda_1 - \lambda_2}, \\ K_2 &= 1\,\text{atm} \cdot \frac{\lambda_1 - (S/V_1)}{\lambda_1 - \lambda_2}.\end{aligned} \qquad \text{(IV.27)}$$

We may go so far as to provide a continuity equation for throughput. The well-known continuity equation expressing the principle of conservation of matter is

$$\vec{\nabla} \cdot \vec{j} = -\frac{\partial n}{\partial t}, \qquad \text{(IV.28)}$$

where $\vec{j}$ is the vector flux of particles and t is the time. If we multiply through by kT, assuming the ideal gas law and that T is constant and uniform, we obtain

$$\vec{\nabla} \cdot (kT\vec{j}) = \frac{-\partial P}{\partial t}. \qquad \text{(IV.29)}$$

This is the *throughput continuity equation*. Applying the divergence theorem of vector calculus, we immediately find that

$$\oint_A kT\vec{j} \cdot d\vec{A}' = \int_V \frac{-\partial P}{\partial t} dV', \qquad \text{(IV.30)}$$

where V is any volume bounded by a closed surface A (such as a bell jar or the body of a pump). The left-hand side (LHS), recognizing $kT\vec{j}$ to be the throughput density, is simply a totalling up of all the throughputs *leaving* through the surface A (such as $-kT\,dN/dt|_L$ or SP, since the pumped particles adsorb, condense, or actually depart, depending on the type of pump). Assuming uniform pressure, the right-hand side (RHS) is simply $-V\,dP/dt$.

The throughput law follows from Equation (IV.30). For a vacuum chamber, A is the chamber wall area and V is the chamber volume. The RHS of Equation (IV.30) is $-V\,dP/dt$ and the LHS is $-\Sigma Q_i$. We thus arrive at the chamber version of the throughput law. For a pump, A is the pump wall area and V is the volume of the pump. The RHS of Equation (IV.30) is 0 (dP/dt is zero within the pump), and the LHS is $-\Sigma Q_i + SP$. (The term $-\Sigma Q_i$ is the sum of the throughputs leaving the pump through its mouth. The pump throughput, SP, also leaves the pump. It literally leaves the pump in the case of throughput pumps and effectively leaves in the case of entrainment pumps.) We thus arrive at the pump version of the throughput law. For a node, we get $0 = \Sigma Q_i$ (since $V = 0$). We have thus arrived at the node version of the throughput law. In summary, *the three forms of the throughput law are expressions of the principle of conservation of matter under conditions of uniform and constant temperature, uniform pressure, and assuming the ideal gas law for the equation of state.*

There is an excellent electrical network analogy to vacuum systems, which is summarized in the Appendix; Kirchhoff's current law may be derived in the same fashion as the throughput law from the current continuity equation, $\nabla \cdot j_e = -\partial\rho/\partial t$, where ρ is the volume charge density and j_e is the current density.

It is worthwhile to reflect further on previously attempted definitions of throughput (see the Appendix for a sampling), and some alternative suggested physical meanings of throughput continuity. One is tempted to try to develop a thermodynamic interpretation. First, some authors have suggested a more generalized definition of throughput as $Q_{\text{bogus}} \equiv d(PV)/dt$. We believe that this came from a desire to tie the necessity of throughput continuity to conservation of energy; Q_{bogus} does *not* represent throughput at all. The term $d(PV)/dt$ is the power required to do flow work: the power required to move a gas adiabatically (without exchange of heat with the environment) from one region to another, with both pressure and volume able to change as a unit of gas moves. However, in vacuum system pumpdowns the flow of gas is nearly isothermal, not adiabatic (until the point when it might enter a cryogenic pump), and furthermore the throughput concept requires that the temperature be both constant and spatially uniform. It is simply the conservation of matter that is the controlling principle. The term Q_{bogus} is misleading.

Second, Q is *not* generally equal to the "flow of power across an open cross section," as has been suggested by others [O'Hanlon, 1980] (and neither is Q_{bogus}). (If Q were defined as $\tfrac{3}{2}kT\,dN/dt$, then it would be equal to the flow of kinetic energy of the moving gas particles, but this is not the case.)

Throughput is dimensionally equivalent, but only dimensionally and not in meaning, to power. The natural meaning of throughput as we have developed it to represent particle flow is "particles per second at temperature T." "Mole" is the recognized unit for "amount of substance" in the SI system; thus "moles per second at temperature T" would be the logical SI meaning for throughput.

The AVS is correct in its official definition as "quantity of gas in pressure–volume units per unit time." Of course, when one uses Q in a mathematical expression, dN/dt multiplied by kT is effectively watts (1 torr·liter·s^{-1} = 0.133 W).

What really counts in a physical vapor deposition system is the impingement rate, the rate of capture of particles by the pump, the rate at which particles enter through a leak or desorb from the walls. The throughput concept was invented to relate these fluxes to pressure — and does so successfully, although it is harder to understand than it needs to be, because the particle flow rate is tangled up with the temperature. Simplicity and transparency are desirable; one might conceive an alternative vacuum science, modeling framework, and measurement technology that is based not on pressure measurement at all, but rather on particle density and impingement rate, which are the gas parameters that are directly relevant to thin film deposition.

Conductance

In the analysis of vacuum system dynamics it is necessary to represent structures, and combinations of them, with conductances. The effective conductance of a pair of vacuum elements connected in series, such as two valves, is calculated by summing their reciprocals just as one does for the electrical conductance of two series-connected electrical resistors. Another instance:

In Chapter II we obtained expressions for the conductances of orifices and tubes of idealized geometries (different expressions are needed under different conditions). For example, the conductances of a long circular tube are $C_{m,tube} = \pi a^2 (2 v_{av} a / 3L)$ and $C_{v,tube} = (\pi a^4 / 8 \eta k T L) P_{av}$ for the molecular and viscous flow regimes, respectively. While the former expression suggests that a tube's conductance approaches infinity as its length goes to zero, this cannot be true. The difficulty may be overcome by expressing its total conductance as the series combination of an orifice and a tube, since the tube, after all, has at least one end:

$$1/C_{total} = 1/C_{tube} + 1/C_{orif}. \qquad (IV.31)$$

It has been found that the above equation works well for very long and very short tubes, but in the intermediate range it has not been quantitatively acceptable. Dushman [1949] analyzed the problem in some depth, and obtained expressions for the intermediate range as well as for numerous other structure shapes. Holland et al. [1974] review in some detail the conductance expressions for pipes of various cross sections in various flow regimes, and for combinations of various structures. Santeler [1987] states that "In the process of pumping down from atmospheric pressure to high vacuum, we necessarily pass through all pressure regions — from the turbulent choked flow range, through laminar viscous and slip flow, and on to free molecular flow."

IV.2 BEHAVIOR OF REAL VACUUM SYSTEMS

A More Realistic Vacuum System Model

A more realistic model of a vacuum system is shown schematically in Figure IV.3. There is an air leak in the chamber (represented by a throttle valve, since the American Vacuum Society has not approved a symbol for an inadvertent leak). Its throughput is Q_L, which will remain practically constant so long as $P_3 \ll$ one atmosphere. There are a baffle and an isolation valve interposed between the diffusion pump and the chamber. These components are characterized with conductances C_b and C_v, respectively. The pump has an intrinsic pumping speed, S, but will have an *effective pumping speed*, due to the baffle and isolation valve, of $S_{eff} < S$. This real pump has its own intrinsic *ultimate pressure*, P_{ult}, that prevails within the pump when the inlet is closed. Furthermore, there is gas desorption from the interior surfaces of the chamber. *Desorption* is simply the release of adsorbed gas particles from a surface. The various pressure and throughput variables are indicated in the figure, and an analogous electrical network is also shown.

The leak throughput is a constant, while the desorption process gives a time-varying throughput that depends on the history of the vacuum. Ultimately the desorption throughput approaches zero. (The desorption process is not represented with any symbol in the vacuum system schematic, which is limited to hardware.) The switch in the electrical circuit, which is closed after $t = 0$, represents the opening of the main isolation valve to start the pumpdown

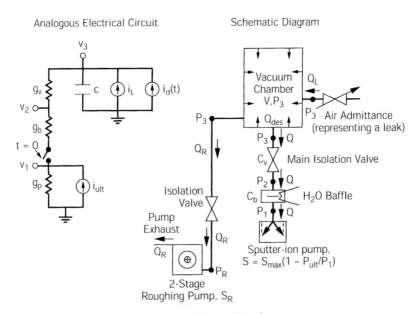

Figure IV.3 A more realistic model of a vacuum system.

BEHAVIOR OF REAL VACUUM SYSTEMS

(assuming that the chamber has already been roughed down by the two-stage pump).

The four variables are P_1, P_2, P_3, and Q. The differential equations describing the pumpdown are

$$V \frac{dP_3}{dt} = -kT \frac{dN_a}{dt} + Q_L - Q, \qquad (IV.32)$$

$$Q = C_v(P_3 - P_2), \qquad (IV.33)$$

$$Q = C_b(P_2 - P_1), \qquad (IV.34)$$

and

$$Q = SP_1 = S_{max}\left(1 - \frac{P_{ult}}{P_1}\right)P_1, \qquad (IV.35)$$

$N_a(t)$ is the total number of adsorbed particles within the vacuum chamber. Therefore, the desorption throughput contribution to the chamber is $-kT\,dN_a/dt$. As discussed in Chapter III, the desorption rate can be modeled as the total number of adsorbed particles, divided by the mean residence time (τ). Desorption is a thermally activated process; the mean residence time is the average time that an adsorbed particle stays on the surface before it acquires enough vibrational energy, by random fluctuations, to desorb. Therefore, the desorption rate is

$$\frac{-dN_a}{dt} = \frac{N_a}{\tau}, \qquad (IV.36)$$

neglecting any adsorption process (which would contribute a term $-\alpha_s \bar{z} A_c$ to the right-hand side). The number of adsorbed particles will decay with time as

$$N_a(t) = N_a(0) \exp\left(\frac{-t}{\tau}\right). \qquad (IV.37)$$

Ultimately N_a becomes negligibly small.

The last of the four governing equations shows how the ultimate pressure of the pump is incorporated. The pressure P_{ult} exists because the pump itself has its own nonidealities, such as a small return flux to the chamber that is intrinsic to the pump's design or materials. As the inlet pressure of the pump approaches its ultimate pressure, the pump throughput approaches zero. The term SP_{ult} is represented in the electrical circuit by i_{ult}; in the vacuum schematic, it is incorporated into the expression for S.

The *effective pumping speed* is obtained in the following manner: First, we obtain the *effective conductance* (C_{eff}) of the baffle–isolation valve combination. This combination is like electrical resistors in series; the reciprocals of their conductances must be added together to get the effective total conductance.

Thus

$$\frac{1}{C_{\text{eff}}} = \frac{1}{C_b} + \frac{1}{C_v}. \qquad \text{(IV.38)}$$

or

$$C_{\text{eff}} = \frac{C_b C_v}{C_b + C_v}. \qquad \text{(IV.39)}$$

Now, to find the effective *initial* pumping speed (before the point in the pumpdown when P_1 may approach P_{ult}), we recognize that the throughput of the baffle–gate valve combination must equal the throughput of the pump: $C_{\text{eff}}(P_3 - P_1) = S(P_1 - P_{\text{ult}})$. We solve for the ratio of the two pressures after dropping P_{ult}: $P_1/P_3 = C_{\text{eff}}/(S + C_{\text{eff}})$. The effective pumping speed of the pump–gate valve–baffle combination is defined by

$$S_{\text{eff}} P_3 \equiv SP_1. \qquad \text{(IV.40)}$$

Thus

$$S_{\text{eff}} = \frac{SC_{\text{eff}}}{S + C_{\text{eff}}}. \qquad \text{(IV.41)}$$

In words, *the effective pumping speed of a pump is calculated as the series combination of the pump with any conductance-limiting devices which are interposed between the pump and the chamber.*

The *ultimate pressure* ($P_{3\text{ult}}$) of the chamber may be controlled by the leak if Q_L is large compared to SP_{ult}. To calculate this quantity, we assume that in steady state, the leak throughput is equal to the pump throughput: $Q_L = S_{\text{eff}} P_{3\text{ult}}$. Thus

$$P_{3\text{ult}} = \frac{Q_L}{S_{\text{eff}}}. \qquad \text{(IV.42)}$$

Although the chamber volume is a factor in the pumpdown *rate*, it has no bearing on the ultimately obtainable pressure. The value of Q_L may be assumed to be independent of P_3 if P_3 is small, and thus equals $C_L \cdot 1$ atm. C_L is typically in either molecular or transition flow, depending on the area of the hole [Ehrlich and Basford, 1992].

Desorption, Outgassing, and Permeation

In most real high-vacuum systems, desorption-limited pumpdown kinetics are observed. This generally occurs a few seconds after opening the main isolation valve and may continue for several days, or even weeks. The time variation of the vacuum chamber pressure is controlled, in this regime, by the desorption of gases from the chamber walls. The transient behavior may be predicted by equating the desorption throughput to the pump throughput:

$$-kT\frac{dN_a}{dt} = S_{\text{eff}} P_3. \quad \text{(IV.43)}$$

Thus

$$\begin{aligned} P_3(t) &= \left(\frac{-kT}{S_{\text{eff}}}\right)\frac{dN_a}{dt} \\ &= \left(\frac{kT}{S_{\text{eff}}}\right)\left(\frac{N_a(0)}{\tau}\right)\exp\left(\frac{-t}{\tau}\right). \end{aligned} \quad \text{(IV.44)}$$

The desorption process frequently causes the pumpdown of high-vacuum systems to be much, much slower than one would expect for the given chamber volume and pumping speed. The characteristic time is τ, rather than V/S_{eff}, during desorption-limited pumpdown; it is interesting that in both instances it is an exponential decay.

The initial value of N_a (at the start of a pumpdown) depends on the environmental exposure history of the vacuum system. Sometimes, high-vacuum systems are vented with dry nitrogen gas to load the surfaces with adsorbed nitrogen, which has a relatively brief residence time. Water molecules, in particular, have a very large τ compared to most other gases. Sample introduction chambers, or "loadlocks," are used to allow exchange of samples without exposing the interior of the chamber to the atmosphere. Since desorption is a thermally activated process, pumpdown can be accelerated greatly by baking a high-vacuum system at typically 200°C for from several hours to several days. *Bakeout* refers to this process when applied to whole vacuum systems; *degassing* popularly refers to a "local bakeout" of an ionization vacuum guage using its internal heater.

There are other fundamental physical processes that can play a significant role in the performance of real vacuum systems. *Outgassing* is the diffusion of gas from within the material of a chamber component (where it was *dissolved*) to the surface where it is then desorbed. Li and Dylla [1993] state that "The outgassing of unbaked [steel or aluminum vacuum] systems is dominated by H_2O. The remaining outgassing species for clean systems are H_2, CH_4, CO, and CO_2." O-rings and other elastomers commonly outgas hydrocarbons.

Li and Dylla developed a model for the outgassing of water from metal surfaces. Metals are typically covered with a native oxide, which is porous. During air exposure, water vapor diffuses into this oxide layer. The resulting concentration profile within the oxide results in a theoretical outgassing rate that varies as $1/t^{3/2}$, where t represents time. There is also an amount of water vapor or hydrogen dissolved within the bulk of a metal and having a uniform concentration; it diffuses through the native oxide and then desorbs. This second outgassing source has a theoretical outgassing rate that varies as $1/t^{1/2}$.

Li and Dylla [1993] fitted experimental measurements of outgassing rate with an expression of the form

$$Q_{\text{out}}(t) = \frac{q_1}{t^{3/2}} + \frac{q_0}{t^{1/2}}, \qquad \text{(IV.45)}$$

finding regimes in which each outgassing mechanism was important. For lengthy air exposure, q_1 was found to be on the order of 4×10^{-3} torr·liter·cm^{-2}·s^{-1} and q_0 was found to be on the order of 4×10^{-8} torr·liter·cm^{-2}·s^{-1} for t expressed in seconds.

Permeation refers to the passage of gas entirely through solid bodies. Permeation of helium through glass vessels and elastomers is observed frequently. It really amounts to a distributed leak on the atomic scale and is modeled by a constant throughput.

Adsorption can sometimes work to one's advantage. This is the process by which gas particles are attached (by physi- or chemisorption) to a surface and is the principle of operation of *sorption* pumps and cryopumps, which are discussed below. The term *sorption* itself generally describes the uptake of gas particles, referring to both adsorption and absorption (dissolution within the bulk of a solid material).

The preceding definitions were taken from *Dictionary of Terms for Vacuum Science and Technology, Surface Science, Thin Film Technology, and Vacuum Metallurgy* [Kaminsky and Lafferty, 1980].

Materials cleaning is a very important part of high vacuum technology. The purpose is to minimize gross contamination of material surfaces (such as grease), as well as desorption and outgassing. Sasaki [1991] prepared an interesting survey of cleaning, handling, and packaging procedures that were in current use by research labs and commercial equipment manufacturers. The procedures are concerned first and foremost with stainless steel since it is the most widely used material in ultrahigh vacuum hardware. The report included insightful commentary on how the procedures work. Some interesting observations include the following:

> Although no two organizations practiced identical procedures, there was sufficient commonality among those canvassed to suggest that there is a sound basic procedure around which many variations have evolved. ... one should be cautious about using grinding wheels. The abrasive particles are normally bonded with epoxy which will rub off onto the work surface.
>
> Chromium enrichment at the [stainless steel] surface will provide a stable barrier of chromium oxide against gas sorption, and some of the cleaning procedures ... contribute to chromium enrichment. ... its [the surface chromium oxide layer on stainless steel] effect on outgassing of hydrogen from within the bulk is not clearly established. ... stainless steel accumulates a large amount of atomic hydrogen in solution interstitially. Under vacuum the hydrogen diffuses out, accounting for about 99% of the total outgas from clean stainless steel.
>
> The most commonly used method of reducing hydrogen outgas is baking ... [glow discharge cleaning] is superior to vacuum bakeout in removing nonvolatile or

tightly bonded surface contaminants such as carbon and carbon monoxide. It produces near atomically clean surfaces. ... Aluminum forms a very tenacious oxide which could be very thick and porous, acting as a gas source.

Acetone dissolves heavy organics effectively [on copper surfaces] but tends to leave residues and is chemically unstable on surfaces in vacuum. Methanol as well as propanol has the advantage of good stability in vacuum plus the ability to remove any small amounts of acetone that may remain from the previous acetone rinse even though these alcohols alone are not effective in removing heavy organic contamination.

It is widely believed that "finger oil" is a stubborn contaminant.

Latex gloves, vinyl gloves, and nylon gloves coated with urethane showed up to 30 µg of dry transfer per fingering and up to 300 µg of wet transfer per fingering as well as visible stains.

As for packaging ... PVDC [a copolymer of vinylidene chloride and vinyl chloride] transferred its plasticizers most [among three types of widely used polymer wrap] to the vacuum component.

—Reprinted by permission from Y. T. Sasaki, 1991, *J. Vac. Sci. Technol.* **A9**(3), 2025. Copyright 1991 by the American Vacuum Society.

IV.3 OPERATION PRINCIPLES OF VACUUM PUMPS AND GAUGES

How Seven Important Pumps Work

As was emphasized in the beginning of this chapter, the purpose of a vacuum pump is to prevent a return flux into the chamber. Both *throughput pumps* (they ultimately exhaust the pumped gases) and *entrainment pumps* (they immobilize and retain the pumped gases) have been developed. We present the operating principles of two common roughing pumps and five ultra-high-vacuum pumps that include both of these types. The reader is referred to technology-oriented sources for actual specifications and operating parameters [O'Hanlon, 1980].

A *rotary mechanical pump* is shown schematically in Figure IV.4a. As the rotor turns, it creates three volumes:

V_1 Between the junction of rotor and bore, and vane on the intake side. It is connected to the pump inlet and is being filled with gas.
V_2 Between the two vanes. Gas is being moved and compressed.
V_3 Between the vane on the exhaust side and the junction. Gas is exhausted if there is enough pressure to open the flapper valve. Some gas is trapped after the vane passes the exhaust port.

The pumping speed of the rotary mechanical pump is given by

$$S = V_1 \cdot \text{rotation frequency} \cdot 2, \quad \text{(IV.46)}$$

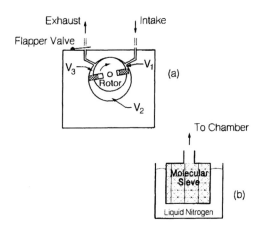

Figure IV.4 Two common roughing pumps are shown schematically in cross section: (a) a rotary mechanical pump and (b) a sorption pump.

assuming that V_1 fills with gas to equilibrium with the chamber during each revolution. The value of S is far below the conductance of the inlet port.

A *sorption pump* is shown schematically in Figure IV.4b. As a roughing pump, this device possesses one enormous advantage over mechanical pumps — the complete absence of oil that could backstream into the chamber.

The pump body is filled with a molecular sieve material of enormous surface area, typically a "zeolite" (an alkali metal aluminosilicate). The pump is cooled with liquid nitrogen. The pumping action occurs through the conventional *condensation* of some gases, whose partial pressures are above their thermal equilibrium vapor pressures at the boiling point of N_2 (as is typically true for H_2O and CO_2), and through the *adsorption* of other atmospheric species, for which this condition is not met (helium and hydrogen as examples), onto the porous molecular sieve. As discussed in Chapter III, adsorption is not the formation of the customary liquid or solid phase, but rather a unique state of matter in the form of one or a few monolayers on a surface. There is an equilibrium relationship between the partial pressure and the state of coverage of the surface that supplants the normal thermal equilibrium vapor pressure. In fact, this adsorption process is capable of pulling the vapor pressure far below the thermal equilibrium vapor pressure at 77 K if only modest amounts of gas need be adsorbed.

The pumping speed and ultimate pressure for a selected gas specie may be calculated by considering the net particle flux into the sorption pump:

$$\frac{dN_c}{dt} = -(z_{\text{out}} - z_{\text{in}})A$$
$$= \frac{-A}{4}(n_c v_{\text{av,c}} - n_p v_{\text{av,p}}), \qquad (\text{IV.47})$$

OPERATION PRINCIPLES OF VACUUM PUMPS AND GAUGES

where the subscripts c and p refer to chamber and pump, respectively. Taking into account the different temperatures with the chamber and pump, we obtain

$$\frac{dN_c}{dt} = \left(\frac{AP_c v_{av,c}}{4kT_c}\right)\left[1 - \left(\frac{P_p}{P_c}\right)\left(\frac{T_c}{T_p}\right)^{1/2}\right]. \qquad (IV.48)$$

Since the throughput leaving the chamber is $kT_c \, dN_c/dt$, the pumping speed is

$$S = \left(\frac{A v_{av,c}}{4}\right)\left[1 - \left(\frac{P_p}{P_c}\right)\left(\frac{T_c}{T_p}\right)^{1/2}\right]. \qquad (IV.49)$$

Thus S is reduced below the conductance of the inlet port by the factor in the brackets, which is a function of the pressure and temperature ratios. Assuming quasiequilibrium within the pump, P_p is given by the thermal equilibrium vapor pressure of the selected gas at T_p, if the gas condenses. On the other hand, if the gas adsorbs, P_p is determined by the adsorption isotherm. The ultimate chamber pressure is that for which $S = 0$:

$$P_{ult,c} = P_p \left(\frac{T_c}{T_p}\right)^{1/2}. \qquad (IV.50)$$

It is interesting to note that the ultimate pressure in the chamber is greater than that in the pump by the factor derived above. This occurs through the requirement of having zero net rate of transfer of particles. The tendency to transfer particles from a cold region to a hotter region is called *thermal transpiration*. The above-mentioned temperature factor is called the thermal transpiration ratio.

The sorption pump is regenerated by removing the liquid nitrogen and allowing the zeolite to warm to room temperature. The trapped gases desorb — and exit the pump through a safety valve as the internal pressure exceeds that of the atmosphere.

The *diffusion pump* has been the most widely used of all high-vacuum pumps. This pump is portrayed schematically in Figure IV.5a. The pumping action is achieved by the entrainment of gas by a supersonic jet of pumping fluid vapor, which creates an active pumping surface in the shape of a truncated cone.

It is the pumping fluid that forms the entraining jet. The fluid is vaporized within the reservoir by the heater. Small nozzles form high-velocity jets directed downward that form the pumping cone. A gas particle that by chance enters the throat of the pump, if it impinges on the jet, is entrained and imparted a large downward velocity. The pumping fluid itself impinges upon the cooled interior wall of the pump, condenses, and drains back down into the reservoir to be recycled. Noncondensable gas from the vacuum chamber

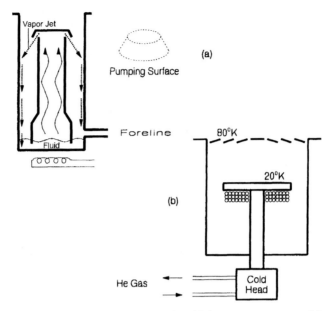

Figure IV.5 Schematic diagrams of two ultra-high-vacuum pumps: (*a*) a diffusion pump and (*b*) a cryopump.

accumulates in the volume below the jets and is exhausted by a foreline (or backing) mechanical pump. Thus, in spite of its name, diffusion does not play a role in the mechanism of pumping.

Ideally, the pumping speed is some fraction of the conductance of the inlet port, due to the reduction of the active pumping surface by the pumping stack's cap, as shown in Figure IV.5:

$$S = \sigma \cdot v_{av} \frac{A}{4}, \qquad (IV.51)$$

where σ is a "capture coefficient" for the pump, which includes the above area effect and any other nonidealities.

A cryopump is shown schematically in Figure IV.5*b*. It functions, as does the sorption pump, via condensation and/or adsorption. It is a much more powerful pump, though, as it is cooled by a closed-cycle helium refrigerator instead of liquid nitrogen, and its capacity is much greater than that of the sorption pump. Water vapor is condensed on the "80 K" condensing array in the throat of the pump. A very cold (20 K) stage is capable of effectively adsorbing all common gases except hydrogen and helium. These latter two gases are pumped by an activated charcoal charge also held at 20 K. The pumping speed and ultimate pressure are calculated as for the sorption pump.

The titanium sublimation and sputter-ion pumps are shown in Figure IV.6. They are often used in combination in ultra-high-vacuum systems. The

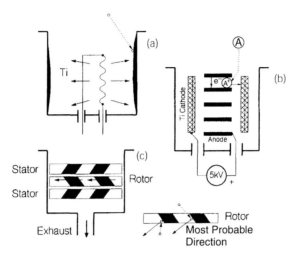

Figure IV.6 Schematic diagrams of three ultra-high-vacuum pumps: (*a*) a sublimation pump, (*b*) a sputter-ion pump, and (*c*) a turbomolecular pump. In each case, the basic pumping mechanisms are suggested.

pumping principle of the sublimation pump is gettering, via chemisorption. A chemically reactive coating is deposited on the interior walls of the pump by the sublimation of titanium from a heated filament. This coating may be renewed, when it is saturated, by reheating the filament. At high chamber pressures, the sublimator may be run continually.

The titanium film has a pumping speed per unit area (say, liters $\cdot$ s^{-1} $\cdot$ cm^{-2}) for each gaseous specie of

$$S_a = \frac{\alpha_s v_{av}}{4} \qquad \text{(IV.52)}$$

Both the sticking coefficient and the average speed are specific to each species. The value of α_s on a fresh titanium film is near unity for many gases, but is essentially zero for the inert gases and certain molecules that do not decompose. The pumping speed of the entire titanium coating (of area A_{Ti}) within the pump is given by $A_{Ti}S_a$. (We do not introduce the conductance of the inlet port because the sublimation pump is typically quite an open structure, and often the pump body is simply the lower portion of the vacuum chamber.)

This maximum pumping speed pertains only as long as the titanium surface remains mostly clean — that is, when the titanium deposition flux is much greater than $\alpha_s z$ for the gaseous specie of interest. As the titanium surface becomes loaded with chemisorbed gas particles, the value of α_s falls. In this high-pressure "titanium-limited" regime, the condensation coefficient becomes equal to j/z, and therefore S varies inversely with the pressure. It is a sobering experience for those engaged in the deposition of reactive metal films to consider the mechanism of the sublimation pump. Their films may be performing this same pumping function in removing residual gas from the chamber during deposition.

There are three pumping mechanisms operative in the sputter-ion pump, which are based on either chemisorption or implantation. Residual gas particles that enter the throat of the pump are ionized and accelerated to the cathode by a potential of typically 5 kV. Pumping of inert gases proceeds by implantation into the cathode and then burial (which occurs on certain regions of the cathode where sputtered titanium accumulates). The light gases (He and H_2) are also implanted but they are permanently trapped by then diffusing into the bulk of the cathode. Chemically active gases are pumped by gettering onto the sputtered titanium coating which accumulates on the anode.

We also show a *turbomolecular pump* in schematic cross section in Figure IV.6. This pump is a high-speed turbine, with a rate of rotation typically in the range 10,000–90,000 rpm (rev/min), giving a blade velocity greater than v_{av} for the residual gas. The reason for the enormous speed is to ensure that any gas particle attempting to return to the chamber experiences a collision with the leading edge of the rotor blade, and never with the trailing edge. The gas particle actually physisorbs on the blade surface and then desorbs; thus, collisions are diffuse, not specular, and desorbing particles are re-emitted with a cosine distribution about the surface normal direction of the blade. The tilt of the blade causes the most probable direction of travel following a collision to be toward the exhaust side of the pump. And they have a substantially greater velocity upon re-emission than they did on striking the blade.

Finally, we have not meant to imply from our simple estimates of pumping speed for the ultra-high-vacuum pumps that the pumping speed is a constant. First, for a given pump, there is a unique pumping speed for each gas specie. Second, Figure IV.7 portrays approximately the kind of pressure dependence that is observed for all the types. The speed is roughly constant over a range of several orders of magnitude of pressure in the high- to ultra-high-vacuum range. At extremely low pressures the pumping speed falls to zero as the ultimate pressure is approached. A simple mathematical model of pumping speed near ultimate pressure is

$$S = S_{\max}\left(1 - \frac{P_{\text{ult}}}{P_{\text{p}}}\right). \tag{IV.53}$$

This ultimate limitation occurs through some sort of nonideality in the pumping mechanism or the technology, such as desorption in the sublimation pump, sputter release of previously implanted gas particles in the ion pump, or permeation of gases through the pump body. At the high-pressure end, the pumping speed falls to zero as a result of some technical limitation giving a nonzero z_{out}. With diffusion pumps, this is a breakup of the pumping fluid jet.

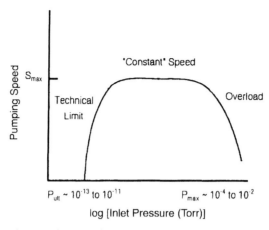

Figure IV.7 Generic pumping speed curve: pumping speed versus inlet pressure for an ultrahigh vacuum pump.

With cryopumps, it is a warming of the cryosurfaces. With ion pumps, it is a loading down of the high-voltage supply.

We turn now to consider the principles of operation of perhaps the two most widely used of all vacuum gauges.

Two Vacuum Gauges in Widespread Use: The Thermocouple and Ionization Gauges

The *thermocouple gauge* is a "thermal conductivity-type" gauge. It measures the cooling effect that the residual gas has on a heated thermocouple within the molecular flow regime. The pressure range of utility is typically $\sim 10^{-3} - 1$ torr.

As shown schematically in Figure IV.8, a single thermocouple is heated by a wire heater internal to the gauge tube. (Actually, the temperature difference between the junction and the voltmeter terminals is measured by the thermocouple.) By a tricky analysis based on our derivation of the thermal conductivity of the ideal gas in Chapter II, this temperature difference is proportional to the gas density or the pressure. In Chapter II, we calculated the heat flux across a plane to be $-z(C_p/N_0)(dT/dy)\lambda$. This leads to a thermal conductivity expression that is independent of pressure because of cancellation between z and λ — not very useful for a pressure gauge. However, for the thermocouple gauge in the molecular flow regime ΔT is given not by $(dT/dy)\lambda$ but rather by $(dT/dy)L$, where L is the internal dimension of the gauge tube. Thus the only pressure-dependent term in the heat flux is z, which is $P/(2\pi mkT)^{1/2}$. As long as the gauge remains in the molecular flow regime, the heat flux through the gas is proportional, or at least sensitive, to the pressure. This sets the upper limit of about 1 torr.

As the pressure decreases, the heat flux via conduction through the gas within the gauge also falls in proportion. The lower limit is fixed by the

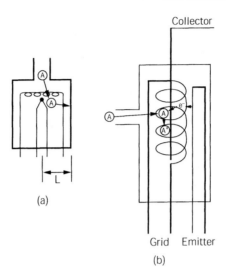

Figure IV.8 Two common vacuum gauges are shown schematically: (*a*) the thermocouple gauge ($L \ll \lambda$ as discussed in the text) and (*b*) the Bayard–Alpert ionization gauge.

occurrence of parallel heat conduction paths, such as radiation and conduction through metal parts, which again make the thermocouple temperature insensitive to pressure.

The thermocouple gauge is often used in conjunction with the ionization gauge to cover the pressure range from 1 to $\sim 10^{-11}$ torr. The principle of operation of the latter is that residual gas particles are ionized and an ion current is measured. The ion current is proportional to the ion density, which is proportional to the neutral particle density, which, in turn, is proportional to the pressure.

The component parts of the Bayard–Alpert ionization gauge tube are shown schematically in Figure IV.8 [Bayard and Alpert, 1950]. A hot filament thermionically emits electrons, and the total emission current is carefully calibrated at typically 5 mA to allow for quantitative pressure measurements. A large helical grid accelerates the electrons, but is inefficient at capturing them because of its very open design. The electrons ionize residual gas particles within the grid by impact ionization at a rate per unit volume given by $(j_e/q)A_i n$, where j_e/q is the electron particle flux and A_i is the ionization cross section, a property of the particle identity and the electron's fixed kinetic energy. A collector in the center of the tube is biased negatively to collect the ions. The ion current collected is typically a fixed fraction of the ionization rate, and thus proportional to pressure.

The upper pressure limit of the ion gauge is fixed by the necessity to avoid oxidizing the filament and by nonlinearity of current. The lower limit is determined by the processes of photoemission of electrons at the collector, as a result of X rays created by that small percentage of electrons that strike the grid. The point of the Bayard–Alpert design is to have the filament outside, and the collector inside, the grid; it was this design that proved the "X-ray hypothesis" for the origin of the residual current.

IV.4 SUMMARY OF PRINCIPAL EQUATIONS

Maximum possible pumping speed	$S_{max} \equiv v_{av} A/4$
Definition of conductance	$J \equiv -C \Delta n$
Throughput of a vacuum chamber	$Q_c \equiv V\, dP/dt$
Of a leak	$Q_L \equiv kTJ_L$
Of a pump	$Q_p \equiv kTJ_p = SP$
Of a tube or other structure	$Q_t = C_t \Delta P$
Throughput law	
Node version	$\Sigma Q_i = 0$
Chamber version	$\Sigma Q_i = Q_c$
Pump version	$\Sigma Q_i = Q_p$
Throughput continuity equation	$\vec{\nabla} \bullet (kT j) = -\partial P/\partial t$
Series combination of conductances	$1/C_{total} = 1/C_{tube} + 1/C_{orif}$
Effective pumping speed	$S_{eff} = SC_{eff}/(S + C_{eff})$
Ultimate pressure	$P_{3ult} = Q_L/S_{eff}$
Desorption-limited pumpdown	$P_3(t) = (kT/S_{eff})(N_a(0)/\tau)\exp(-t/\tau)$
Outgassing rate	$Q_{out}(t) = (q_1/t^{3/2}) + (q_0/t^{1/2})$
Pumping speed	
Of rotary mechanical pump	$S = V_1 \cdot$ rotation frequency $\cdot 2$
Of sorption pump	$S = (Av_{av,c}/4)[1-(P_p/P_c)(T_c/T_p)^{1/2}]$
Of diffusion pump	$S = \sigma \cdot v_{av} A/4$
Of titanium sublimation and sputter ion pumps	$S_a = \alpha_s v_{av}/4$ (per unit area of pumping element)
Pumping speed near ultimate pressure	$S = S_{max} - (Q_{ult}/P_p)$

IV.5 APPENDIX

How to Draw and Analyze Vacuum Schematic Diagrams

A vacuum schematic diagram, like an electrical circuit diagram, is a lumped-parameter model of a system. To prepare a diagram

- Draw the symbols of the individual elements, using those adopted by the American Vacuum Society [AVS, 1967]. (See Fig. IV.9.) It is desirable that the relative positions, orientations, and sizes of the symbols suggest the spatial layout of the real vacuum system.
- Connect the elements with lines. (These lines do not represent physical components of vacuum systems, because the elements are directly

connected to each other—e.g., a pump and its gate valve. Thus, the lines have no pressure difference from one end to the other.)
- To represent physical tubulation that is lossy (not available among the AVS symbols), use Mahan's special symbol:

C

- Indicate a node between adjoining elements with a dot:

- Where two or more lines join to form a node, use a dot to indicate their connection:

- Label the diagram: Elements have *names* (e.g., type of valve, chamber, type of pump) and *parameters* (e.g., conductance, volume, pumping speed)

(See Figs. IV.1–IV.3, which were prepared in this manner.)

To analyze a diagram, label it with pressure and throughput variables: Give every node and chamber a pressure (a node next to a chamber has the chamber pressure). Give every line a throughput, with an arrow showing your defined direction of positive flow. Lines in series and not separated by a chamber should have the same throughput variable. If only two lines enter a node, they have the same throughput variable. Apply the throughput law and element relations to obtain a number of independent equations equal to the number of variables. Solve them as best you can.

An Electrical Network Analogy

As suggested in Figures IV.1 and IV.2, there is an electrical network analogy to vacuum systems. A tube or orifice is like an electrical resistor, with

$$Q = C\,\Delta P \quad \text{and} \quad i = g\,\Delta v, \tag{IV.54}$$

where i is the current, g the conductance of the resistor, and Δv the electrostatic potential drop across it. Throughput is analogous to electrical current. A leak is like a current source. Pressure is analogous to voltage.

A vacuum chamber is like a charged capacitor having one side grounded, with

$$NkT = PV \quad \text{and} \quad q = cv, \tag{IV.55}$$

APPENDIX

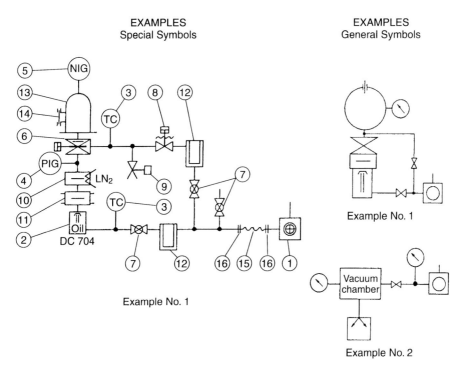

Figure IV.9 Standard vacuum symbols of the American Vacuum Society (reprinted with permission from The American Vacuum Society, 1967, "Graphic Symbols in Vacuum Technology," *J. Vac. Sci. Technol.* **4**, 139. Copyright 1967 by the American Vacuum Society.)

where q is the electrical charge and c is the capacitance. The total number of particles, multiplied by kT, is analogous to total charge (number of electrons multiplied by the electronic charge). The chamber volume is analogous to the capacitance value. More precisely, the chamber is represented by the ungrounded plate of the capacitor. Maintaining high vacuum in a chamber is like keeping a capacitor discharged when one side of it is grounded and it is surrounded by bodies of very high electrostatic potential.

An ideal pump, one with infinite pumping speed, is like a perfect electrical ground. A complete list of analogous variables and laws is given in Table IV.1.

Let's consider the vacuum system and electrical circuit presented in Figure IV.1. Analysis of the circuit shows that $i = -c\,dv/dt$ and also that $i = gv$. Thus, for the node at which v is measured

$$-c\frac{dv}{dt} - gv = 0. \tag{IV.56}$$

The solution is

TABLE IV.1 Analogous Behavior of Vacuum Systems and Electrical Circuits

Vacuum variable	Electrical variable
Throughput Q	Current i
Pressure P	Voltage v
NkT	Charge q
Vacuum parameter	Electrical parameter
Conductance C	Conductance g
Volume V	Capacitance c
Vacuum law	Electrical law
Throughput of a component: $Q = C\Delta P$	Ohm's law: $i = g\Delta v$
Ideal gas law: $NkT = PV$	Capacitor characteristic: $q = cv$
Throughput law: $\Sigma Q_j = 0$	Kirchhoff's current law: $\Sigma i_j = 0$

$$v = v(0)\exp\left(\frac{-gt}{c}\right). \qquad (IV.57)$$

Thus, the voltage decays exponentially as does the pressure of the chamber.

Now reconsider the two-chamber system and its electrical analog in Figure IV.2. Kirchhoff's current law applied to node one yields

$$\frac{-dq_1}{dt} - i_p + i_2 = 0, \qquad (IV.58)$$

where q_i is the charge on capacitor i. This is analogous to the throughput equation, $V_1 \, dP_1/dt + Q_p - Q_2 = 0$. We also have

$$i_p = g_p v_1, \qquad (IV.59)$$

and

$$\frac{dq_2}{dt} = -i_2. \qquad (IV.60)$$

in analogy with $Q_p = SP_1$ and $P_2 \, dV_2/dt = -Q_2$.

The general procedures for constructing an analogous electrical network are as follows:

> Make a drawing, representing each vacuum chamber with a capacitor having one side grounded. (The state of charge on the ungrounded plate of the capacitor corresponds to the state of vacuum in the chamber; the flow of charge from this plate to ground is like the flow of gas particles out of the chamber.)
>
> Series-connected chambers are represented by capacitors with resistors between them for the ports, as shown in Figure IV.2.

APPENDIX

- A pump is modeled with a resistor having one side grounded, with the conductance of this resistor representing the pumping speed.
- The ultimate pressure of a pump is accounted for by placing a current source in parallel with the resistor (both with one side grounded) representing the pump (see Fig. IV.3).
- A conductance-limiting element, such as a port, baffle, or valve, is represented with a resistor in the current path.
- A chamber leak is represented by a current source in parallel with, and feeding, the chamber's capacitor.
- Desorption within the chamber is represented by a time-varying current source also in parallel, typically an exponentially decaying source.

The motivation for developing an electrical network analog for vacuum systems is twofold. First, some folks, through prior experience, are able to more readily grasp the transient behavior of a circuit than that of a vacuum system. Then, on making the substitutions of Q for i, and so on, the behavior of the analogous vacuum system quickly becomes apparent. Second, sophisticated computer models of electrical networks are available. These have been employed to model the behavior of complex vacuum systems that do not lend themselves to simple analysis [Santeler, 1987; Wilson, 1987]. The real power of these computer models is to solve networks that are much more complicated than those considered as examples here. They can perform realistic simulations of real vacuum systems, by incorporating nonlinear elements having *variable* conductances.

A Survey of Past Definitions of Throughput

Attempts to define throughput in the literature have been surprisingly diverse. The following are examples of opaque, misleading, or incomplete definitions taken from widely known books (references are available from the author on request):

> It is that quantity of gas which flows through the intake cross section of the operating pump in unit time, and it is pressure-dependent. ... throughput is the energy per unit time crossing a plane.

> By definition the throughput is the mass flow rate of gas past any point in a vacuum system. ... Mass flow rate is measured in gs^{-1} or the number of molecules per second.

> Q is the mass flow rate (throughput) measured in torr-liters per sec.

> We define throughput by $Q = pdV/dt$ mbar ℓs^{-1}. ... it is convenient to define a gas flow rate Q as $Q = d(PV)/dt$.

> The throughput Q is closely related to the mass flow, while the volumetric flow S is the net volume of gas passing through a given area in unit time. ... the throughput is proportional to both the pressure and the volumetric flow rate and is usually defined as the product $Q = pS$.

> The conductance multiplied by the pressure is called the throughput Q.

IV.6 MATHEMATICAL SYMBOLS, CONSTANTS, AND THEIR UNITS

SI units are given first, followed by other units in widespread use.

a	Inner radius of tube (m)
c	Capacitance (F)
g	Electrical conductance (S; $\Omega^{-1} = S$)
i	Current (A)
j	Particle flux density (m$^{-2} \cdot$ s^{-1})
j_e	Current density (C $\cdot$ m$^{-2} \cdot$ s^{-1})
k	Boltzmann's constant (1.38×10^{-23} J/K; 8.62×10^{-5} eV/K)
n	Particle density (m^{-3})
q	Charge (C)
q_1, q_0	Coefficients in outgassing expression (torr $\cdot$ liter $\cdot$ s$^{-1} \cdot$ m^{-2} = 0.133 W $\cdot$ m^{-2})
t	Time (s)
v	Voltage (V)
v_{av}	Average speed (m/s)
z	Impingement rate (m$^{-2} \cdot$ s^{-1})
A	Area (m^2)
C	Conductance (m^3/s; liter/s = 10^{-3} m^3/s)
J	Particle flow (s^{-1})
L	Length of tube (m)
N	Number of particles
N_a	Total number of adsorbed particles
P	Pressure (Pa; torr $\approx$ 133 Pa; atm $\approx 1.01 \times 10^5$ Pa; μm = 10^{-3} torr)
Q	Throughput (W; torr $\cdot$ liter $\cdot$ s^{-1} = 0.133 W)
S	Pumping speed (m^3/s; liter/s = 10^{-3} m^3/s)
S_a	Pumping speed per unit area (m/s; liter $\cdot$ s$^{-1} \cdot$ m^{-2} = 10^{-3} m/s)
T	Temperature (K)
U	Internal energy (J; cal = 4.19 J)
V	Volume (m^3; liter = 10^{-3} m^3)
α_s	Sticking coefficient
η	Viscosity (kg $\cdot$ m$^{-1} \cdot$ s^{-1})
ρ	Electrical charge density (C/m^3)
τ	Mean residence time (s)
σ	Capture coefficient (dimensionless)

REFERENCES

AVS, 1967, "Graphic Symbols in Vacuum Technology," *J. Vac. Sci. Technol.* **4**, 139.

Bayard, R. T., and Alpert, D., 1950, "Extension of the Low Pressure Range of the Ionization Gauge," *Rev. Sci. Instr.* **21**, 571.

REFERENCES

Chambers, A., Fitch, R. K., and Halliday, B. S., 1989, *Basic Vacuum Technology*, Adam Hilger, Bristol.

Dennis, N. T. M., and Heppell, T. A., 1968, *Vacuum System Design*, Chapman and Hall, London.

Dushman, S., 1949, *Scientific Foundations of Vacuum Technique*, Wiley, New York.

Ehrlich, C. D., 1986, "A Note on Flow Rate and Leak Rate Units," *J. Vac. Sci. Technol.* **A4**(5), 2384.

Ehrlich, C. D., and Basford, J. A., 1992, "Critical Review: Recommended Practices for the Calibration and Use of Leaks," *J. Vac. Sci. Technol.* **A10**(11), 1.

Fasteneau, J., 1993, private communication.

Hashim, I., Raaijmakers, I. J., Park, S.-E., and Kim, K.-B., 1997, "Vacuum Requirements for Next Wafer Size Physical Vapor Deposition System," *J. Vac. Sci. Technol.* **A15**(3), 1305.

Holland, L., Steckelmacher, W., and Yarwood, J., 1974, *Vacuum Manual*, E. & F.N. Spon, London.

Kaminsky, M. S., and Lafferty, J. M., 1980, *Dictionary of Terms for Vacuum Science and Technology, Surface Science, Thin Film Technology, and Vacuum Metallurgy*, American Institute of Physics.

Li, M., and Dylla, H. F., 1993, "Model for the Outgassing of Water from Metal Surfaces," *J. Vac. Sci. Technol.* **A11**(4), 1702.

O'Hanlon, J. F., 1980, *A User's Guide to Vacuum Technology*, Wiley, New York.

Santeler, D. J., 1987, "Computer Design and Analysis of Vacuum Systems," *J. Vac. Sci. Technol.* **A5**(4), 2472.

Sasaki, Y. T., 1991, "A Survey of Vacuum Material Cleaning Procedures: A Subcommittee Report of the American Vacuum Society Recommended Practices Committee," *J. Vac. Sci. Technol.* **A9**(3), 2025.

Weissler, G. L., and Carlson, R. W., 1979, *Vacuum Physics and Technology*, Vol. 14 of Methods of Experimental Physics, Academic Press, Orlando, FL.

Wilson, S. R., 1987, "Numerical Modeling of Vacuum Systems Using Electronic Circuit Analysis Tools," *J. Vac. Sci. Technol.* **A5**(4), 2479.

V

EVAPORATION SOURCES

There are two basic kinds of evaporation source:-*quasiequilibrium* and *nonequilibrium* sources. With the former, the steady state prevails and the evaporant is virtually in equilibrium with its vapor, approaching the condition of equilibrium evaporation. With the latter, the vapor pressure above the source material is, by contrast, very much less than the thermal equilibrium vapor pressure; the mode of operation may also be a transient one.

The *effusion cell* (sometimes called a *Knudsen cell*) is perhaps the only quasiequilibrium source. As shown in Figure V.1, it is a container of an evaporant with a large surface area relative to that of the orifice through which the evaporant escapes. It is this extreme area difference that allows the vapor pressure within the cell to approach the thermal equilibrium value.

The electron-beam evaporator (sometimes called an *E-gun evaporator*) is a widely used example of a nonequilibrium source. As shown schematically in Figure V.2, an open crucible contains the evaporant, which is heated by an electron beam. While the device reaches a steady state during operation, the vapor pressure above the source is typically far below the thermal equilbrium value, and there is practically no return flux to the evaporant; a source operating in this mode has been called a *Langmuir source*. Other examples of steady-state but nonequilibrium sources include a molten droplet suspended from a hot filament, or the molten pool of evaporant in a resistively heated boat. Nonequilibrium sources may also operate in a transient mode (e.g., flash evaporators, laser ablation sources, and cathodic arc sources).

The vaporization rate is related both to the thermal equilibrium vapor pressure corresponding to the source temperature, and to the actual prevailing vapor pressure. It is expressed quantitatively in the historic Hertz–Knudsen–Langmuir equation for the net vaporization flux from a surface:

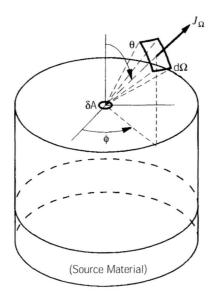

Figure V.1 Schematic diagram of an effusion cell.

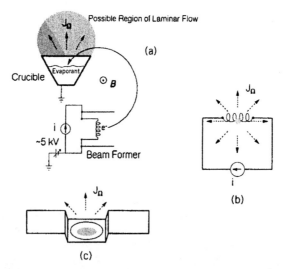

Figure V.2 Three types of free evaporation source: (*a*) a 270° bent-beam, work-accelerated, electron-beam evaporator is shown schematically in cross section, with a possible region of laminar flow that would create a virtual source above the crucible; (*b*) a simple hot filament evaporator; (*c*) a boat source.

$$j_{vap} = \alpha_v z_{eq} - \alpha_c z_i$$
$$= \frac{\alpha_c(P_{eq} - P_i)}{(2\pi m k T)^{1/2}}. \tag{V.1}$$

The second line of this equation is the central result of the collision theory model of evaporation as originated by Hertz and further developed by Knudsen and Langmuir. The presence of α_c, the condensation coefficient as defined in Chapter II, should be understandable; α_v (the vaporization coefficient) is needed in a general expression because vaporization fluxes are sometimes found to be less than z_{eq}. For most metals α_v may be assumed to be unity, but for other complex and anisotropic solids it can be very much less than unity [Rosenblatt, 1976]. Nevertheless, we have assumed α_v to be equal to α_c in the second expression. The other symbols should be familiar from Chapter I. The maximum possible net vaporization flux is equal to the thermal equilibrium impingement rate of the vapor at the temperature of the source (times α_v). The two cases of $z_i = 0$ and $z_i = z_{eq}$ correspond to free evaporation, and to the equilibrium evaporation already mentioned.

Both solids and liquids can be the source of vapor within an effusion cell, and the source material in an E-gun may or not be melted. Solids undergo sublimation, and liquids, evaporation. *Atomization* and *volatilization* are general terms that can apply to either vaporization process. *Atomization* refers to when a specifically *mon*atomic vapor particle is created, and the term has even been applied to sputtering.

The most comprehensive reference on evaporation sources has been the chapter by Glang [1970] in the well-known *Handbook of Thin Film Technology*; information of a more practical nature may be found in the booklet published by Temescal, *Physical Vapor Deposition* [Hill, 1986].

V.1 THE EFFUSION CELL AND NOZZLE-JET EVAPORATION SOURCES

The Ideal Effusion Cell

An effusion cell is shown schematically in Figure V.1. *Effusion* is the process by which particles escape through the orifice, in a collisionless fashion. The cell is said to produce a "Maxwellian effusive stream," or *molecular beam*. It is typically not a beam of molecules; the name refers to its property of being in the molecular flow regime, with interparticle collisions rather infrequent. "Molecular beam epitaxy" is so named because traditionally the evaporation sources produce such a molecular beam.

The following assumptions define the *ideal* effusion cell:

1. The liquid and vapor are in equilibrium within the cell.
2. The mean free path inside the cell is much greater than the orifice diameter.

3. The orifice is flat.
4. The orifice diameter is much less than the distance to the receiving surface.
5. The wall thickness is much less than the orifice diameter.

Assumption 1 implies that the vapor particles within the cell have a Maxwell–Boltzmann velocity distribution with a characteristic temperature equal to that of the condensed phase, and that the thermal equilibrium vapor pressure prevails within the cell. Assumption 2 means that the escaping particles do so in molecular flow. Assumption 3 suggests that the cell is a *flat* source, rather than a *point* source having spherical symmetry. Assumption 4 means that the source is *small*—the particles impinging at a particular point on the substrate come from a single point. Assumption 5 implies that no particles are reflected from, or emitted from, the orifice wall.

The flatness of the orifice is fundamental. It is not a point source, which is among several other geometries analyzed by Holland [1963], and which has a different, spherically symmetric, emission law. (Holland called the flat source the "directed surface source.") The cosine law of emission is characteristic of the small flat source.

The Cosine Law of Emission

From the kinetic theory of gases, the expression for the directional distribution of the flux of particles *onto* a container wall (the incident flux distribution) is

$$j_\Omega(\theta, \varphi) = \frac{z \cos \theta}{\pi}, \qquad (V.2)$$

where θ and φ are the polar and azimuthal angles, as shown in Figure V.1.

The angular distribution of emitted particles must be identical to that of impinging particles—they simply pass through the orifice untouched, being imaged, in fact, by this pinhole lens. The beam intensity is obtained from expression (V.2) by multiplying by the orifice area:

$$\begin{aligned} J_\Omega(\theta, \varphi) &= j_\Omega \, \delta A \\ &= \frac{z \, \delta A \cos \theta}{\pi}. \end{aligned} \qquad (V.3)$$

This is the well-known *cosine law of emission*. It gives the total number of particles per second per unit solid angle that are emitted in the direction specified by θ and φ. The cosine law is analogous to *Lambert's law* in optics, which gives the angular dependence of the radiation intensity emitted from the surface of an ideal blackbody. In both cases the cosine angular dependence originates in the expression for the projected area, in the direction of interest, of the small flat emitting surface.

There are other derivations of the cosine law in the literature, some of which the author finds mystifying although they end up with the correct result. The author greatly prefers the formulation just presented, which is based on a calculation of the angular distribution of incident flux from the kinetic theory of gases (see Chapter I). Knudsen [1915] first proposed the cosine law.

The Nonequilibrium Effusion Cell

Normally, real evaporation sources whose designs approximate the above ideal cell can provide only the very low fluxes that are suitable for a process such as semiconductor doping during epitaxial growth, or perhaps for very slow depositions measured in monolayers. An example of a cell that is capable of giving, by contrast, a practical film growth rate, is shown in Chapter I. There are several ways in which the assumptions of the ideal effusion cell are violated by practical cells. We examine these one at a time in this, and the following, sections.

As the orifice is enlarged, one finds that the beam intensity increases but does so with a sublinear dependence on δA. This occurs because the impingement rate z at the inside of the orifice drops below the thermal equilibrium value for the cell temperature. A model that predicts this effect is shown schematically in Figure V.3. It has three nonidealities: the thick orifice wall, the long tubular body between liquid surface and orifice, and the possibility of an evaporation coefficient less than 1. In the nonequilibrium cell model, these nonidealities reduce the actual beam intensity ($J_{\Omega\text{noneq}}$) to

$$J_{\Omega\text{noneq}} = F_N \cdot J_\Omega, \tag{V.4}$$

which defines the correction factor, F_N. Although there will be a reduction in the cell's beam intensity if the orifice wall is thick, it is the restriction due to the long cell *body* that causes that has been called "nonequilibrium" behavior. We

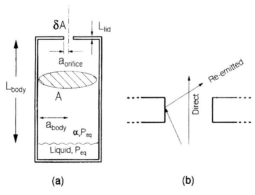

Figure V.3 Effusion cell nonidealities: (*a*) a nonequilibrium effusion cell, shown schematically in cross section; (*b*) with a thick orifice lid, diffuse and specular reflection off the sidewalls are possible.

present a model that contains both effects, representing them with Clausing factors.

The *Clausing factor* estimates the reduction of molecular flow through a circular pipe as compared to a circular orifice of the same area [Clausing, 1930]. For a *short* pipe, such as an orifice in a thick lid of length L and radius a, the Clausing factor may be approximated with $1/(1+L/2a)$. The conductance of the orifice then becomes $v_{av} A/[4(1 + L/2a)]$, the maximum theoretical conductance of an orifice times the Clausing factor. For a *long* pipe of circular cross section (with $L/a \geq 1.5$), the Clausing factor is given approximately by $(8a/3L)/(1+8a/3L)$. The conductance of the tube then becomes $[(8a/3L) \cdot v_{av}A]/[4(1 + 8a/3L)]$. This expression is often pertinent to the body of an effusion cell. These and other approximations are discussed at length by Dushman [1949]. Clausing factors do not predict any alteration in the *angular distribution* of the emitted flux.

We will now model the output of a nonequilibrium cell that has three nonidealities: (1) an evaporation coefficient less than unity, (2) a cell body conductance less than the theoretical maximum, and (3) an orifice conductance less than the theoretical maximum. The flow calculation refers to the diagram of Figure V.3, which is similar to that of a well-known analysis by Motzfeldt [1955].

We represent the effect of the evaporation coefficient by giving a conductance value to the liquid–vapor interface: $C_{evap} = \alpha_v v_{av} A/4$. The conductance of the cell body will be given as $C_b = W_b v_{av} A/4$, where W_b is the Clausing factor of the body. The conductance of the orifice will then be given as $C_{orif} = W_{orif} v_{av} fA/4$, where f is the fraction, $\delta A/A$. The effective conductance of the entire structure is calculated from

$$\frac{1}{C_{eff}} = \frac{1}{C_{evap}} + \frac{1}{C_{body}} + \frac{1}{C_{orif}}, \tag{V.5}$$

leading to

$$C_{eff} = \frac{1}{(1/\alpha_v) + (1/W_b) + (1/f\,W_{orif})} \cdot \frac{v_{av}A}{4}, \tag{V.6}$$

and so

$$F_N = \frac{1}{(1/\alpha_v) + (1/W_b) + (1/f\,W_{orif})}, \tag{V.7}$$

taking J_Ω here to be $zA\cos\theta/\pi$.

Example To what extent is the output reduced for a nonequilibrium cell of practical design? As an example, let's assume an L of 5 cm, an a of 1 cm, an orifice area equal to the body's cross-sectional area, and ignore any possible reduction in output due to vaporization coefficient. The only nonideality that then remains in the expression for F_N is W_b.

We show in Section V.3 that a vapor pressure of 10^{-2} torr within a source often leads to a practical deposition rate for many substances. This pressure corresponds to a diffusion length on the order of 0.5 cm, so flow within the cell is probably in the transition regime between molecular and viscous flow.

Nevertheless, for the purpose of this simple example, we assume the Clausing factor for the cell body that was given above for molecular flow:

$$W_b = \frac{8a/3L}{1 + 8a/3L} = 0.348$$

and so F_N in this case is 0.348. ∎

Alterations to the ideal cosine law beam intensity are often a more serious concern than simple reductions in output, because they usually lead to a decrease in film thickness uniformity. The modeling of real effusion cells then includes several additional cases. These include the near-ideal effusion cell, the open-tube cell, and the conical cell. The first case represents a cell with a small opening in a thin lid (Fig. V.4a), and negligible pressure gradients inside the cell. The second represents cells that resemble open test tubes with the melt near the bottom as in Figure V.4b. The third (Fig. V.4c) represents a currently popular effusion cell crucible design.

The Near-Ideal Effusion Cell

Inevitably, an orifice of nonnegligible thickness does modify the angular distribution of the beam. The effusion flux becomes more focused along the axis of the cell. In the near-ideal cell model, the beam intensity is modified in just this manner, but the total flux is not reduced. Unfortunately, experimental data directly showing a focused beam intensity are rather scarce; instead, the

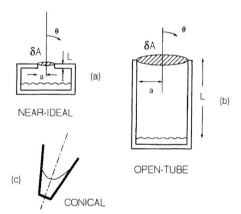

Figure V.4 Three focusing effusion cell models are shown schematically in cross section: (a) the near-ideal cell nearly reproduces the cosine law; (b) the open-tube cell is strongly focused into small emission angles; (c) models of the conical cell often show it tipped away from the vertical.

effects have been characterized with measurements of film thickness profiles across broad substrates, which indicate such a focusing effect.

Physically, this focusing occurs because the particles that pass the orifice along near-axial directions continue to do so freely while at the same time, the particles that encounter the sidewall of the orifice are thus hindered. The standard reference on the focusing effect is the paper of Dayton [1956], who interpreted and extended the original results of Clausing [1930]. Clausing and Dayton assumed that particles which encounter the sidewall condense there briefly, and are then reemitted according to a local cosine law of emission as in Figure V.3b. This analytic model resembles the calculation of the conductance of a tube in the molecular flow regime, which was discussed in Chapter II, but is even more involved since the angular distribution as well as the total flow is calculated. The geometry is shown in Figure V.4.

To express the focusing effect, Dayton adds a correction factor to the ideal beam intensity:

$$J_{\Omega NI} = T(\alpha, p) \cdot J_\Omega, \tag{V.8}$$

where

$$T(\alpha, p) = \begin{cases} 1 - \left[\frac{2}{\pi} \cdot (1 - \alpha) \cdot \left(\sin^{-1} p + p\sqrt{1 - p^2}\right)\right] \\ + \left[\left(\frac{4}{3\pi p}\right) \cdot (1 - 2\alpha) \cdot \{1 - (1 - p^2)^{3/2}\}\right] & \text{for } p < 1 \\ \alpha + \dfrac{4(1 - 2\alpha)}{3\pi p} & \text{for } p > 1. \end{cases} \tag{V.9}$$

The two intermediary parameters are functions of the cell geometry and emission angle:

$$\alpha\left(\frac{L}{a}\right) = \frac{\sqrt{(L/a)^2 + 4} - \dfrac{L}{a}}{2 + \dfrac{4}{\sqrt{(L/a)^2 + 4}}}, \tag{V.10}$$

and

$$p\left(\frac{L}{a}, \theta\right) = \frac{L}{2a} \tan \theta. \tag{V.11}$$

It is assumed that the substrate is sufficiently far away that the cell may be considered a small source—a single beam intensity value representing the cell is assigned to the center point of the outer end of the orifice.

In Figure V.5 we show the calculated beam intensities for L/a values ranging from 0 to 0.5; $L/a = 0$ yields the ideal cosine law. Otherwise, there is a modest beaming effect (a focusing of the output toward the $\theta = 0$ direction).

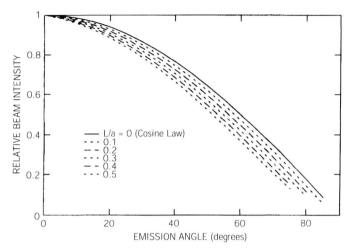

Figure V.5 The relative beam intensity of the near-ideal effusion cell is calculated with the theory of Dayton [1956]. Curves are shown for orifice length:radius ratios. Deviation from the ideal cosine law ($L/a = 0$) increases as the lid thickens.

Ruth and Hirth [1962] improved on the model of Dayton by adding a surface diffusion effect. In their calculations, adatoms were allowed to migrate for some distance on the sidewall of the orifice before their reevaporation, even diffusing onto the outer surface of the lid. Surface diffusion decreases the amount of focusing; Ruth and Hirth observed at the time that it was only their theory that agreed with the experimental data (of Günther [1957]). Ward et al. [1970] added specular reflection of escaping particles by the orifice wall to the above-mentioned mechanisms of direct transmission, diffuse reflection from the wall, and reemission after surface diffusion, and incorporated these in a Monte Carlo simulation. Still, Dayton's work remains a useful and significant theory that illustrates the basic principles and provides at least a starting point for modeling situations of specific interest.

The Open-Tube Effusion Cell

The Dayton theory also describes the focused output of the open-tube type of effusion cell. There is no lid, so a becomes the inner radius of the cell body, and L is the distance from the melt to the top of the tube (see Fig. V.4b). L will vary during evaporation.

Figure V.6 shows some calculate beam intensity curves for L/a values varying from 0 to 10. There can be a strong focusing of the output along the axis of the open tube cell. This has both advantages and disadvantages— focusing reduces the film thickness uniformity across a large substrate, but directs a greater fraction of the evaporant onto the substrate for a small one.

Adamson et al. [1989] published an interesting and detailed Monte Carlo simulation of the open-tube cell. They demonstrated an asymmetry in beam intensity versus θ when the axis of the cell (containing a liquid source) is tilted

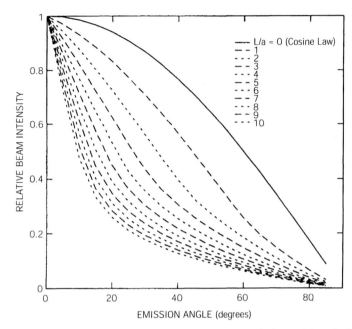

Figure V.6 The relative beam intensity of the open-tube effusion cell is calculated with the theory of Dayton [1956]. Curves are shown for tube length-to-tube radius ratios (L/a) from 0 to 10. Deviation from the ideal cosine law can be strong, with intense beaming of the emitted flux.

from the vertical. Some experimental beam intensity data, which agree well with Clausing's, and consequently Dayton's, theories, may be found in the well-known paper by Stickney et al. [1968].

The Conical Effusion Cell

The conical effusion cell crucible design has been popular, and is not accurately described by Dayton's model. An example was shown in Chapter I. The walls are not parallel to the cell axis, but tapered. The axis of the cell may be tipped from the vertical, such that the surface of a liquid source is not parallel to the cell opening. The evaporant often wets the sidewalls. These design features are shown schematically in Figure V.4.

The complexities described above are beyond the scope of Dayton's analytic model, but they have been treated successfully with Monte Carlo simulations of cell output. Furthermore, such simulations handle cases where the source is not "small," that is, where the distance to the substrate is not a lot greater than the size of the orifice. The studies of Curless [1985], Yamashita et al. [1987], and Wasilewski et al. [1991] are examples. As is typical, the authors chose not to publish beam intensity results, but rather the simulated thickness profile of the film across the substrate (from which the beam intensity *could* be deduced).

One drawback of computer modeling is that the results are not always as general as one might wish, because specific cases are analyzed, one at a time.

The Nozzle-Jet Source

The nozzle-jet source is an interesting and useful extension of the effusion cell. A small nozzle (diameter on the order of a few mm) is attached to a furnace, as shown schematically in Figure V.7. The temperature (and consequently the vapor pressure inside the furnace) is raised to a point where the flow within the nozzle becomes viscous (or at least transitional) in character.

The nozzle jet can be a source of relatively high deposition rate, which is useful for relatively thick film deposition. Because it is so small, it can be a virtually unidirectional source as seen by the substrate. This latter capability offers good "via- filling" in the metallization of integrated circuits — the filling of a hole in the substrate without coating the hole's sidewalls [Ramayarananan et al., 1987].

Here is a procedure for calculating the beam intensity of the nozzle-jet. It was developed by Jackson et al. [1985], who applied the nozzle jet theory of Giordmaine and Wang [1960] to the high rate evaporation of cadmium and zinc. Their beam intensity (adapted to the present notation and terminology) is

$$J_\Omega = J \cdot f(\theta), \quad (V.12)$$

where J is the total flow emitted from the nozzle (in particles per second). The normalized angular distribution is given by

$$f(\theta) = \frac{(8/\pi^{3/2}) \cdot \int_0^1 \sqrt{1-z^2} \cdot \left(\frac{1}{2}\sqrt{\pi}\mathrm{erf}(k(\theta)) \cdot z\right) dz \cdot \cos^{3/2}(\theta)}{\int_0^{2\pi} \int_0^{\pi/2} [(8/\pi^{3/2}) \cdot \int_0^1 \sqrt{1-z^2} \cdot \left(\frac{1}{2}\sqrt{\pi}\,\mathrm{erf}(k(\theta)) \cdot z\right) dz \cdot \cos^{3/2}(\theta)] \cdot \sin(\theta) d\theta\, d\varphi} \quad (V.13)$$

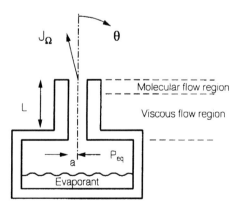

Figure V.7 The nozzle-jet source. There is a viscous flow region at the high-pressure end of the nozzle, and a molecular flow region at the low-pressure end.

The function $k(\theta)$ in Giordmaine and Wang's theory is

$$k(\theta) = \frac{A}{\tan(\theta) \cdot \sqrt{\cos(\theta)}}, \tag{V.14}$$

where A is a parameter depending on the nozzle radius (a), the particle diameter of the evaporant (d), and the flow conditions (r). Estimates for d are tabulated in Chapter II. The value of A is given by

$$A = 2a\sqrt{\frac{\pi d^2 r}{\sqrt{2}}}, \tag{V.15}$$

r is actually a density gradient within the nozzle that is estimated from the total flow:

$$r = \frac{3J}{2\pi a^3 v_{\text{mean}}}, \tag{V.16}$$

where v_{mean} is the mean speed of evaporant particles within a molecular flow region of the nozzle near its low-pressure end, as discussed next. Instead of obtaining A by calculating first v_{mean} and then r, Jackson et al. obtained A from an empirical curve that they developed, showing A as a function of Knudsen number.

It is difficult to calculate the output of the nozzle jet from first principles — the choice of evaporant, the furnace temperature, and the geometry of the nozzle. Typically one must assume values for J and v_{mean}, and then work backward to find the furnace temperature that produced J, without knowing in advance the flow conditions in the nozzle. A source pressure on the order of 1 torr is typical, for which the mean free path is ~ 0.1 mm. This is less than a typical nozzle diameter, and so the vapor entering the nozzle is probably in transitional or viscous flow. The pressure falls toward the outlet of the nozzle and near the low-pressure end of the tube the mean free path becomes much larger than the tube diameter. Thus, the nozzle may be divided into two regions: at the high-pressure end, where viscous flow prevails, and within the low pressure end, where there is a region of molecular flow. The density gradient enters the model because Giordmaine and Wang assumed that within the portion of the tube where molecular flow pertains, the density is given by rz, where z is the distance from the outlet of the tube. The preceding expression for r [Eq. (V.16)] is based on that of conductance in the molecular flow regime. The value of v_{mean} must be consistent with both the flow conditions and the temperature in the furnace; v_{mean} *will* be consistent if both J and v_{mean} are experimentally obtained, but if the initial value of v_{mean} is a guess in a modeling effort, the value may have to be refined in order to be consistent.

Example Suppose that a nozzle-jet source is operating at an internal aluminum vapor pressure of 1.4 torr, yielding a mass flow rate of 10^{-7} kg/s out of the nozzle at a velocity of 1200 m/s. The nozzle has an inner diameter of 3 mm and a length of 30 mm. Calculate the beam intensity of the source. The furnace temperature corresponding to this vapor pressure is 1900 K.

The following material parameters for aluminum are needed:

$$d = 3.5 \text{ Å}, \quad m = 26.98 \text{ amu}$$

The total particle flow rate is

$$J = \frac{10^{-7} \text{ kg/s}}{26.98 \times 1.66 \times 10^{-27} \text{ kg}} = 2.2 \times 10^{18} \text{ particles/s}.$$

The density gradient may now be calculated:

$$r = \frac{3 \times 2.2 \times 10^{18} \text{ particles/s}}{2\pi \times (1.5 \text{ mm})^3 \times 1.2 \times 10^3 \text{ m/s}} = 2.6 \times 10^{23} \text{ m}^{-4}.$$

Next, the parameter A is

$$A = 2 \times 1.5 \text{ mm} \times \left[\frac{\pi \times (3.5 \times 10^{-10} \text{ m})^2 \times 2.6 \times 10^{23} \text{ m}^{-4}}{2^{1/2}} \right]^{1/2} = 0.80.$$

The value of $k(\theta)$ may now be calculated for any emission angle that one wishes and thence, $f(\theta)$. ■

We show in Figure V.8 the calculated beam intensity of a nozzle jet operating as a high-rate evaporation source for aluminum. A focusing of the beam along the axis of the nozzle is seen, as compared to a Knudsen cell. The curve was calculated as described in the previous example.

Giordmaine and Wang [1960] found, in fact, that the theoretical beam intensity profile is sharper than experimental profiles and they identified certain aspects of their assumptions that might cause this. On purely empirical grounds, Jackson et al. [1985] improved the fit by substituting the following expression for $k(\theta)$: $A/\{\tan(\theta)\sin^{1/2}(\theta)\cos^{1/2}(\theta)\}$. The former expression for $k(\theta)$, however, has a basis in theory. Jackson et al. published a variety of experimental beam intensity distributions obtained under differing conditions.

The focusing of the nozzle jet is generally less than that of an open-tube effusion cell for the same L/a ratio. This is because the molecular flow portion of the nozzle is only a fraction of L.

V.2 FREE EVAPORATION SOURCES

Free Evaporation

Consider now the actual surface of the source material within an effusion cell. It is in equilibrium with its vapor. An infinitesimal surface element of that

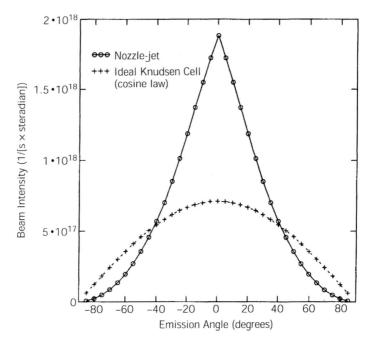

Figure V.8 The calculated beam intensity of a nozzle jet evaporation source emitting aluminum vapor at a mass flow rate of 10^{-7} kg/s. The other experimental conditions are described in the accompanying example. For comparison, the beam intensity of an ideal Knudsen cell having the same total output is also shown.

condensed phase emits a flux that also follows exactly the incident flux distribution, j_Ω, which was derived in Chapter II. This is because "by the second law of thermodynamics, the molecules leaving a surface at equilibrium must reflect those incident upon it in all particulars on the average" [Rosenblatt, 1976].

Free evaporation from a surface occurs if the impinging flux is eliminated, leaving only the emission flux. In the *ideal Langmuir source model*, it is assumed that both the total emission flux value and its angular distribution remain unchanged. This assumption means that an infinitesimal area of a surface in free evaporation obeys the cosine law of emission. This model may be used to calculate the deposition flux of large area deposition sources (see Chapter VIII).

Furthermore, the model implies that an electron-beam-heated evaporation source can provide a beam intensity according to the cosine law. If its molten pool is small compared to the distance to the substrate, then it is represented with δA, just as the orifice of the effusion cell. The flat molten pool of evaporant in a heated boat may also be represented with the cosine law.

However, E-guns may deviate from the ideal cosine law in several ways. One obvious way is that they may not be "small area." Another is that the molten pool may not be really flat. In yet another way, a special application of electron beam-heated evaporators is in "high rate" deposition in a range far beyond the capacity of the typical effusion cell (deposition rates of many micrometers per second). Under such conditions, the vapor density near the source is so great that there is a spatial region in which laminar flow prevails. The vapor density naturally decreases with distance from the source and at some point there is a transition to molecular flow. Smith [1969] modeled such a device with a "virtual source" in the space above the actual source, from which a true molecular beam emanates according to the cosine law.

There are other types of free evaporation sources, as well. A molten droplet suspended from a hot filament, as shown schematically in Figure V.2b, may be approximated as a spherical evaporation source. Levitation heating of a molten droplet [Van Audenhove, 1965] is another technique that gives an approximately spherical source. These surfaces are spherical, not flat like the orifice of an effusion cell, and they do not obey the cosine law. They may be modeled with the ideal point source.

The Ideal Point Source Model

The ideal point source is spherically symmetric. The beam intensity is simple to derive from this fact. If we assign to the point source a surface area δA, then the beam intensity (per unit solid angle) is uniform, given by

$$J_\Omega(\theta, \varphi) = \frac{\alpha_v z \delta A}{4\pi}. \tag{V.17}$$

This isotropic beam intensity is suggested in Figure V.2b.

How E-Gun Evaporators Work

An electron-beam evaporation source is shown schematically in Figure V.2a. It is an example of a 270° bent-beam, work-accelerated gun. The basic parts are the thermionic emitter, the high-voltage DC power supply, a permanent magnet or electromagnetic coil (not shown) producing the transverse magnetic field, and the crucible containing the evaporant.

The principles of operation are as follows:

1. A beam of energetic electrons bombards the evaporant, where their kinetic energy is transformed into heat on impact. Their kinetic energy corresponds to the potential of the DC supply, typically several kilovolts.
2. This electron beam is steered from the emitter assembly (filament plus beam former structures) to the crucible by the transverse magnetic field. The steering is accomplished through the Lorentz force

$$\mathbf{F} = -q\mathbf{v} \times \mathbf{B}, \tag{V.18}$$

where q is the magnitude of the electronic charge, $\mathbf{v}$ is the electron velocity, and $\mathbf{B}$ is the magnetic flux density.

3. The electrons depart from different parts of the filament and in different directions, and are focused by beam former structures, so that virtually all of them are traveling to the right when they emerge from the emitter assembly (see Fig. V.2a). This focusing is accomplished by the electrostatic forces of repulsion between the electrons and the negatively biased emitter assembly.

4. The electron current density j_e leaving the hot filament is due to thermionic emission, as expressed by Richardson's equation:

$$j_e = AT^2 \exp\frac{-q\Phi}{kT}, \qquad (V.19)$$

where A is Richardson's constant, q is the magnitude of the electronic charge, and Φ is the work function.

Expression (V.19) gives the maximum current that can be drawn at the selected filament temperature. In many E-guns, the emitter is operated at this saturation level, such that the beam current is independent of the potential difference between filament and work; the beam current is thus controlled by selecting the filament current, which determines the filament temperature. Below this saturation level, the current density depends on the accelerating voltage according to the Langmuir–Child law. The initial kinetic energy of the thermionically emitted electrons is $3kT_{\text{filament}}/2$, and is negligible compared to the ultimate kinetic energy that they acquire after being accelerated out of the emitter assembly by the large negative bias.

Beam steering and focusing in the 270° bent-beam E-gun are quantitatively explored in Figure V.9. Figure V.9a shows an idealized two-dimensional model, which consists of a (1) "square" design, in that the lateral and vertical displacements of the filament from the crucible are equal, (2) a uniform transverse magnetic field of 205 G, (3) a filament represented by a point charge and emitting electrons at a temperature of 3000 K, (4) a beam former represented by three additional point charges equal to that of the filament, (5) four starting points for the electron trajectories, and (6) three possible initial directions of travel for the emitted electrons. The values of the point charges correspond to an emitter voltage of ~ -4.5 kV.

The electron trajectories were calculated by numerical integration of the equations of motion, and some results are shown in Figure V.9b, for the initial position *below* the filament. Regardless of the initial direction of travel, the electron is focused and steered into the crucible quite accurately. It was found that the imaging of the electrons onto the crucible is even more precise for the initial position to the right of the filament, and slightly less precise for the initial position above the filament. For the initial position to the left of the filament, the electrons followed unstable trajectories and are lost. An efficient beam former structure would presumably capture those electrons as well.

FREE EVAPORATION SOURCES

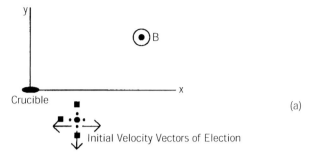

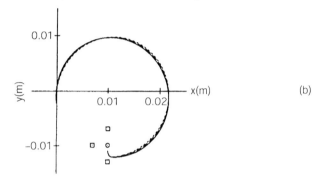

Figure V.9 (*a*) An idealized model of the 270° bent-beam E-gun, in cross section at the midplane. The crucible is at the origin of the coordinate system. The constant magnetic field is perpendicular to the page. The emitting filament is represented by a point charge at the circle, and beam former structures are represented by three point charges at the squares. Four starting points surrounding the filament (small dots) were used and for each of them, three initial directions of travel were investigated. (*b*) Results are shown for the starting point below the filament. The three traces are for electrons with initial velocities to the right, to the left, and downward. There is excellent steering and focusing of the electrons into the crucible.

There are numerous other adaptations of the principle of electron-beam heating for evaporation sources, including the pendant drop and self-accelerated guns, and plasma electron-beam guns. One advantage of the E-gun is that it may be used to deposit refractory materials that require higher temperatures than are practical with an effusion cell. Another is that the evaporant may often serve as its own crucible within the actual crucible, maintaining purity of the source material.

Beam Intensity of the E-Gun Evaporator

The electron-beam evaporation source can serve as a relatively high intensity source, because it is not necessary to raise a furnace or other enclosure to the temperature of the evaporant. Typically, the evaporant is the hottest object in the source (except for the filament emitter).

The evaporant flux can be so dense near the evaporant surface that it is in laminar flow, as suggested schematically in Figure V.2. This has two effects that cause the beam intensity to become focused along the vertical axis of the source in a manner similar to the nozzle jet or the open-tube effusion cell. First, there is an upward drift component of particle velocity due to the directed laminar flow, which is superposed on the Maxwellian velocity distribution. Second, a *virtual source* is created in the space *above* the actual evaporant surface, from which the effusive flux effectively emanates. To give some perspective, for high-rate aluminum deposition Bhatia [1994] found the virtual source location to be 1.2 cm above the actual source.

Example An E-gun evaporator is depositing aluminum at a rate of 100 Å/s onto a horizontal substrate that is located 30 cm directly above the crucible. The flat evaporant surface has a radius of 0.5 cm. It is known independently that the source temperature is 1840 K. Is there a region of laminar flow? How large is it?

The following material parameters for aluminum are needed:

Particle density of film $n_f = 6.02 \times 10^{28}$ m^{-3}; $m = 26.98$ amu; $d = 2.85$ Å

A variety of results from Chapter VIII will also be used. This deposition rate corresponds to a condensation flux $(v_n \cdot n_f)$ at the substrate of

$$j_c = 100 \text{ Å/s} \times 6.02 \times 10^{28} \text{ m}^{-3} = 6.02 \times 10^{20} \text{ m}^{-2}\text{s}^{-1}.$$

Assuming a unity condensation coefficient, this is also the value of the incident flux at the substrate. Treating the incident flux as the impingement rate from the kinetic theory of gases, the effective vapor pressure at the substrate $(j_i [2\pi m k T]^{1/2})$ is

$$P_{Al} = 6.02 \times 10^{20} \text{ m}^{-2}\text{s}^{-1} [2\pi\, 26.98 \text{ amu} \times 1.38 \times 10^{-23} \text{ J/K} \times 1840 \text{ K}]^{1/2}$$
$$= 0.38 \text{ mtorr}.$$

Now the incident flux $(J_\Omega \cos \beta / R^2)$ varies as $1/R^2$, where R is the distance from the source (neglecting the possible occurrence of the virtual source). Thus the vapor pressure as a general function of height directly above the source is given by

$$P_{Al}(R) = 0.38 \text{ mtorr} \times \left(\frac{0.3\,m}{R}\right)^2.$$

Using the radius of the evaporant surface as the defining length, laminar flow will occur in the region close to the source where the mean free path is less than 0.5 cm (as a rough estimate), or equivalently where the vapor pressure is greater than 106 mtorr (using the expression for mean free path from kinetic theory). Inverting the immediately preceding expression for $P(R)$ shows that the critical vapor pressure is reached at a height of 1.8 cm above the source. There is indeed a region of laminar flow below this height. Its exact shape, however, is not well known. ∎

The phenomenon of the virtual source was first observed in the inadvertent coating of chamber viewports. These viewports had no direct view of the E-gun source material; yet, they became coated "as though the port were looking at a much larger vapor source than the actual evaporating surface" [Smith, 1969].

The beam intensity of E-gun evaporators has been described with a $\cos^n(\theta)$ law [Graper, 1973]:

$$J_\Omega \sim (1 - C)\cos^n(\theta) + C, \quad (V.20)$$

with the exponent n varying from 2 to at least 6, as J increases. To our knowledge, no one has come forth with a theoretical model predicting this θ dependence; it should be viewed as entirely empirical.

V.3 PULSED LASER DEPOSITION

Laser-Induced Vaporization

Pulsed laser deposition (PLD) is a "flash evaporation" method. A condensible vapor is produced when a powerful laser beam strikes a target, is absorbed, and vaporizes a thin surface region. The vaporized region of the target is typically several hundred to 1000 angstroms thick [Hubler, 1992], as an order-of-magnitude (OM) estimate. A conical plume of evaporant is created, extending along the direction normal to the target surface. A characteristic speed of the evaporant particles (which can be both neutrals and ions) is $\sim 3 \times 10^5$ cm/s, corresponding to a kinetic energy of ~ 3 eV (as an OM estimate and depending on the particle mass) [Cheung and Sankur, 1988]. The film growth rates can approach 0.5 μm/min.

The typical hardware arrangement is shown schematically in Figure V.10a, and a photograph of a plume may be found in Chapter I. The laser waveform is shown schematically in Figure V.10b. One of the most frequently used lasers is the KrF excimer laser, operating at 248 nm [Brannon, 1993]. Some representative laser parameters are as follows:

- A pulse on the order of 25 ns in duration (the pulse duration, δt)
- At a power density j of 2.4×10^8 W/cm² at the target
- While illuminating an area of the target (δA) of typically 0.1 cm²
- At a repetition rate (f) of 50 Hz.

The *fluence* of this typical pulse ($j\,\delta t$) is thus 6 J/cm². The incident energy per pulse is 0.6 J. The instantaneous power is 2.4×10^7 W, and the average power is 30 W.

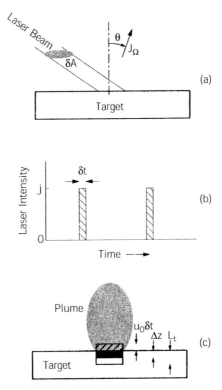

Figure V.10 (a) The pulsed laser deposition arrangement; (b) the ideal laser waveform; (c) the heated volume L_t, the vaporized layer Δz, the initial vapor volume $\Delta z + u_0 \delta t$, and the plume.

The elementary physical picture of laser-induced vaporization is as follows. The laser pulse strikes the target — a fraction of the energy is reflected and the rest is absorbed. During the pulse, there is heat conduction downward from the surface into the target, to a depth given by the thermal diffusion length (L_t). Except for weakly absorbing targets (such as electrical insulators), the absorption depth of the laser light is typically much less than this diffusion length. Thus, the *heated volume* at the surface of the target, where the laser energy is effectively deposited, is an elliptical cylinder whose cross-sectional area is the projected area (δA) of the laser beam and whose thickness is L_t (see Fig. V.10c). The solid material within the volume is raised to the melting point and then melts. If there is any energy remaining, evaporation will commence as a fraction of the atoms within the heated volume receive the heat of evaporation. As the result of a large pressure gradient, this Knudsen layer then expands out of the shallow crater, primarily in the direction normal to the target surface. The initial temperature of the Knudsen layer is the melting point of the target; the vapor cools enormously during this adiabatic expansion while at the same time acquiring a large flow velocity in the surface normal direction.

The laser pulse may photoemit electrons from the target surface and also photoionize atoms in the expanding plume before the pulse terminates, forming a plasma [Singh and Narayan, 1990]. When the density of the vapor plume falls sufficiently, the adiabatic expansion ceases at a terminal flow velocity; the vapor particles then continue on to the substrate in free molecular flow.

The target can take many forms, such as a powder, a single crystal, sintered pellets, or even a liquid. The *yield* (atoms deposited on the substrate per laser photon) is sensitive to the surface morphology of the target, which can change as the target ages. A rough surface can have a significantly lower reflectivity than a very smooth surface, and thus absorb substantially more laser power. Brannon [1993] suggests that the "easiest and best" target materials for PLD are good thermal insulators (a thermal diffusivity $< 10^{-3}\,\mathrm{cm^2/s}$) with strong light absorption (an absorption coefficient $> 10^5\,\mathrm{cm^{-1}}$). A low fluence requirement is almost always considered advantageous.

Removal of material from the target exhibits a threshold as a function of laser fluence, and then increases approximately linearly as a function of fluence above threshold (finally saturating at a maximum value). Some representative data are shown in Figure V.11 for a wide variety of materials [Stafast and von Przychowski, 1989; Dam et al., 1994].

Two of the most successful PLD applications have been the preparation of thin films of YBCO high-temperature superconductors and of calcium hydroxyapatite biocompatible coatings [Hubler, 1992]. The technique seems unusually effective in recreating in the thin film the stoichiometric composition of complex, multicomponent target materials; the vaporization is so fast that segregation is nearly impossible. Sometimes preserving stoichiometry is assisted by performing PLD with a high partial pressure (in the millitorr range) of reactive gas, such as oxygen. PLD allows this by virtue of the absence of hot filaments or other hot components. Reactive PLD has been successfully applied to the production of TiN [Mihailescu et al., 1993] and BN [Doll et al., 1991] thin films.

A Simple Heating Model

We develop here a simple, idealized description of vaporization by the laser pulse and characterize the deposition method with a time-averaged beam intensity. The purpose of this modeling is to convey a basic physical understanding of the process, so we avoid clouding the picture by attempting to include secondary phenomena associated with the technique (however, we mention some of these at the end of this section). Many of the quantities that we will use to describe the processes are order of magnitude estimates only, and should not be taken too seriously. However, they serve to establish the basic physical phenomena of PLD. A variety of methods have been developed to estimate the surface temperature rise due to a laser pulse [Sobol, 1995; Ready, 1971]; ours is among the simplest.

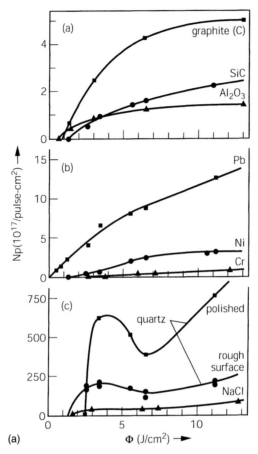

Figure V.11 Experimental data showing material removal from a target as a function of laser fluence. (*a*) Some behavior representing a wide variety of materials. (Reprinted by permission from H. Stafast and M. von Przychowski, "Evaporation of Solids by Pulsed Laser Irradiation," *Appl. Surf. Sci.* **36**, 150, copyright 1989 by Elsevier Science Publishers); (*b*) Thin-film deposition rate for a $YBa_2Cu_3O_7$ target. The threshold for stoichiometric ablation of $YBa_2Cu_3O_7$ is $1 \, J/cm^2$. The step is due to "the particular character of non-stoichiometric ablation below threshold" and is not seen with elemental targets. [Reprinted by permission from B. Dam et al., "Laser Ablation Threshold of $YBa_2Cu_3O_{6+x}$," *Appl. Phys. Lett.* **65**(12), 1581, copyright 1994 by the American Institute of Physics.]

The optical absorption length is the inverse of the optical absorption constant of the sample:

$$L_0 = \frac{1}{\alpha}. \tag{V.21}$$

where L_0 is a measure of the depth of penetration of the laser light. For metal targets, as well as strongly absorbing semiconductors, L_0 is on the order of

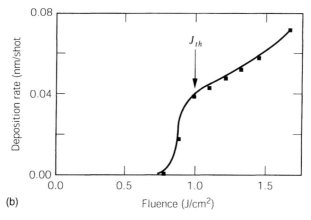

(b)

Figure V.11 *(Continued)*

100 Å if an ultraviolet laser is used. The thermal diffusion length, on the other hand, is given by

$$L_t = \sqrt{\frac{2\delta t \cdot \kappa}{cn_{\text{mol}}}}, \qquad (V.22)$$

where κ is the thermal conductivity, c is the molar heat capacity, and n_{mol} is the molar density of the target. The term κ/cn_{mol} is the *thermal diffusivity*. It is interesting to note that the depth of this heated volume is fixed by the pulse duration of the laser pulse, not by the power.

These two lengths establish two regimes of laser heating of solids — weakly absorbing, for which $L_0 \gg L_t$, and strongly absorbing (the opposite case). For a weakly absorbing target the depth profile of the deposited energy is the same as that of the light intensity, an exponential decay with depth ($\propto e^{-\alpha z}$). For a strongly absorbing target, however, the deposited energy profile is smeared out by fast heat conduction that carries the energy much deeper into the target than the light itself penetrates. This heat conduction occurs on the time scale of the pulse itself. The behavior of insulators and wide-bandgap semiconductor targets usually is in the former category, while that of metals and narrow-bandgap semiconductors is usually in the latter. We model here only the strongly absorbing regime, assuming that the heated volume is isothermal.

Example Estimate the thermal diffusion lengths for copper and silicon targets for a laser pulse duration of 25 ns. The necessary material parameters are summarized below:

	κ (W · m^{-1} · K^{-1})	c (J · mol^{-1} · K^{-1})	n_{mol} (mol/m^3)
Copper	401	24.5	1.41×10^5
Silicon	148	20.1	8.29×10^4

The thermal diffusion lengths are

$$L_{t,Cu} = \left[\frac{(2 \times 25 \times 10^{-9}\,\text{s} \times 401\,\text{W}\cdot\text{m}^{-1}\cdot\text{K}^{-1})}{(24.5\,\text{J}\cdot\text{mol}^{-1}\cdot\text{K}^{-1} \times 1.41 \times 10^5\,\text{mol/m}^3)} \right]^{1/2} = 2.41\,\mu\text{m}$$

and

$$L_{t,Si} = \left[\frac{(2 \times 25 \times 10^{-9}\,\text{s} \times 148\,\text{W}\cdot\text{m}^{-1}\cdot\text{K}^{-1})}{(20.1\,\text{J}\cdot\text{mol}^{-1}\cdot\text{K}^{-1} \times 8.29 \times 10^4\,\text{mol/m}^3)} \right]^{1/2} = 2.11\,\mu\text{m}$$

These values are much greater than typical absorption depths of UV radiation in these materials. ∎

To continue with our model, vaporization of the target material occurs only if the laser fluence raises the solid material within the heated volume to the melting point, melts it, and then imparts the heat of sublimation to some fraction of the melt. The threshold laser fluence is found from

$$(1 - R)(j\delta t)_{\text{th}} = \Delta_{\text{fus}} H \cdot n_{\text{mol}} L_{\text{t}} + (T_m - 298\,\text{K}) \cdot c \cdot n_{\text{mol}} L_{\text{t}}, \quad (\text{V}.23)$$

where R is the reflectivity of the target and T_m is the melting point. We are assuming here that the target is initially at room temperature (298 K).

Energy above and beyond the threshold fluence goes toward producing the vapor. Assuming that the fluence is above threshold, the number of vapor particles created by each laser pulse is

$$\Delta N = \frac{(1 - R)(j\delta t - (j\delta t)_{\text{th}}) \cdot \delta A \cdot 6.02 \times 10^{23}/\text{mol}}{\Delta_{\text{evap}} H} \quad (\text{V}.24)$$

The corresponding thickness of material which is removed from the target is

$$\Delta z = \frac{\Delta N / (6.02 \times 10^{23}/\text{mol})}{\delta A \cdot n_{\text{mol}}} \quad (\text{V}.25)$$

Example Suppose that the laser pulse fluence for a copper target is 6 J/cm². Estimate the number of target atoms vaporized by each pulse and the corresponding thickness of the target that is removed. Assume a pulse duration of 25 ns and a projected area for the laser beam of 0.1 cm². Take the enthalpy of melting for copper to be 13.1 kJ/mol, the enthalpy of evaporation to be 325 kJ/mol, and the reflectivity to be 0.37. The melting point is 1357 K.

First we will calculate the threshold fluence in order to determine whether vaporization occurs. (Some data and results from the previous example are used.)

$$(j\delta t)_{\text{th}} = [13.1\,\text{kJ/mol} + (1357\,\text{K} - 298\,\text{K}) \times (24.5\,\text{J}\cdot\text{mol}^{-1}\cdot\text{K}^{-1})]$$

$$\times \frac{(1.41 \times 10^5\,\text{mol/m}^3) \times 2.41\,\mu\text{m}}{(1 - 0.37)} = 2.11\,\text{J/cm}^2.$$

Thus, the laser fluence exceeds threshold. The number of vaporized copper atoms per pulse is

$$\Delta N = (1 - 0.37) \times [(6 - 2.11)\,\text{J/cm}^2] \times 0.1\,\text{cm}^2 \times \frac{6.02 \times 10^{23}\,\text{mol}^{-1}}{325\,\text{kJ/mol}} = 4.55 \times 10^{17}.$$

The thickness of the vaporized region of the target is

$$\Delta z = 4.55 \times 10^{17} / [0.1\,\text{cm}^2 \times (1.41 \times 10^5\,\text{mol/m}^3) \times (6.02 \times 10^{23}\,\text{mol}^{-1})] = 0.535\,\mu\text{m}.$$

The amount of material removed is plotted as a function of laser fluence in Figure V.12. ∎

Our simple model probably overestimates the amount of material removed, because lateral heat conduction and radiation loss are neglected, and the volume is assumed to be isothermal. Furthermore, one might expect that not all the vaporized particles will end up in the vapor plume, because some may recondense on the crater wall. Another factor is the omission of effects of the vapor plume. The plume develops rapidly and may shadow the target during the latter part of the laser pulse, thus effectively reducing the pulse duration. Consequently, the simple model probably overestimates the threshold laser fluence, which is proportional to L_t. We should note, though, that a threshold

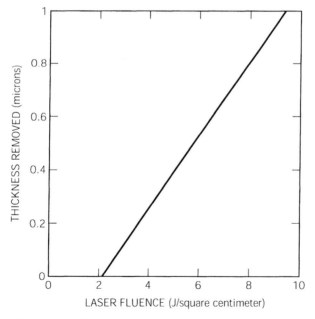

Figure V.12 Theoretical thickness of copper target removed per laser pulse, as a function of laser fluence. Values were calculated with the parameters given in the examples of this section. A threshold fluence of 2.11 J/cm² is predicted.

fluence for copper of several joules per square centimeter has been reported [Viswanathan and Husla, 1986]; the experimental threshold for silicon is near 1 J/cm^2 for 20-ns pulses [Shinn et al., 1986].

During and immediately after the laser pulse, this vapor is very dense and has the shape of a flat disk of area δA and thickness L_t plus some amount due to expansion before the laser pulse terminates (see Fig. V.10c). We will assume that the vapor front moves during the pulse at an average speed (from kinetic theory) corresponding to the melting temperature of the target, $u_0 = (8kT_m/\pi m)^{1/2}$. (For copper, this speed is 6.72×10^4 cm/s.) The initial thickness of the plume (its thickness at the end of the pulse) is then $L_t + u_0 \delta t$, and the initial density is therefore

$$n_0 = \frac{\Delta N}{\delta A (L_t + u_0 \delta t)}. \tag{V.26}$$

The actual value of n_0 for the preceding example of copper PLD is 2.62×10^{21} cm^{-3}. This value probably overestimates typical experimental values, for reasons stated above.

Now, the vapor does not merely effuse away from the target according to the cosine law, but because of its high pressure and short mean free path, undergoes one-dimensional adiabatic expansion. The flat shape causes the adiabatic expansion to be directed normal to the target surface [Singh and Narayan, 1990]. The probability distribution function for velocity of the vapor has been modeled as that of the ideal gas with a flow velocity (u) superposed:

$$F(v_x, v_y, v_z) \propto e^{m[v_x^2 + v_y^2 + (v_z - u)^2]/2kT}, \tag{V.27}$$

where v_z is the velocity component perpendicular to the target surface.

Eventually, the density within the plume decreases to the point where free molecular flow prevails and the velocity distribution becomes fixed at its terminal flow velocity and temperature. It is difficult to predict this terminal flow velocity because equations for continuous flow do not apply very well to the vapor pulse [Saenger, 1991]. Experimentally, it is observed that the kinetic energy of the particles in the plume increases with fluence; this may be presumed to be due to an initial density increasing with fluence. (The terminal velocity of the expansion increases with initial density.)

It is possible to characterize PLD with a beam intensity based on the expression for the effusion cell; ΔN target particles are liberated at the pulse frequency f, and the expression must contain a parameter that describes the beaming of the velocity distribution due to the adiabatic expansion. A time-averaged (and properly normalized) beam intensity can be written as

$$J_\Omega = \frac{\Delta N \cdot f \cdot \cos^p(\theta)}{2\pi/(p+1)}, \tag{V.28}$$

Saenger [1991] obtained the following empirical expression for p in numerical simulations of film thickness profiles:

$$p = 4 + 2.13 \frac{\gamma^{1/2} M_t}{\gamma^{1/2} M_t + 1} + \gamma M_t^2, \qquad (V.29)$$

where γ is the ratio of heat capacities (c_p/c_v) of the vapor, $\frac{5}{3}$ for the ideal monatomic gas, and M_t is the terminal Mach number of the expansion (u_t divided by the local velocity of sound). As one might expect, the flow velocity causes the beam intensity to be peaked in the forward direction as compared to the ideal cosine law. Experimentally, beaming parameter values fall in the range 8–12 [Hubler, 1992] for deposition of high-temperature superconductors.

Example Calculate the theoretical beam intensity for the copper PLD source of the previous examples. Assume a terminal Mach number of 2.

The p parameter according to Equation (V.29), with $\gamma = \frac{5}{3}$, is ~ 12. The expression for the time-averaged beam intensity of the PLD source is then

$$J_\Omega(\theta) = \frac{4.55 \times 10^{17} \times 50\,\text{Hz} \times \cos^{12}\theta}{2\pi/13} = 4.71 \times 10^{19} \cos^{12}\theta \cdot \text{steradian}^{-1} \cdot \text{s}^{-1}.$$

This expression is plotted in Figure V.13, along with that of the ideal Knudsen cell ($p = 1$) for comparison. ∎

Other Phenomena

Now that we have presented these simple descriptions, we would like to put them in the context of the real complexities of *laser ablation* (general laser-induced material removal), which encompasses various nonidealities and a complex set of interacting phenomena. The laser beam is probably not uniform. If melting occurs prior to vaporization, the reflectivity of the target probably drops. Cumulative heating occurs (as a result of repetitive pulses). Strong nonlinear absorption may develop in materials that are normally weakly absorbing. The plume can absorb the laser pulse, shadowing the target and at the same time becoming a plasma through photoionization. Real plumes contain atoms (both excited and ground- state), molecules, ions, electrons, and atom clusters. The adiabatic expansion may not terminate before reaching the substrate. Experimentally, a variety of kinetic energy distributions (other than the one given above) and temperatures have been observed [Saenger, 1993]. A background gas in the deposition chamber can *thermalize* the plume (see Chapter VIII). Tilting of the plume both toward and away from the laser beam, which seems to depend on surface morphology, has been observed. With compound targets, the different elements in the plume can have different angular distributions [Saenger, 1991].

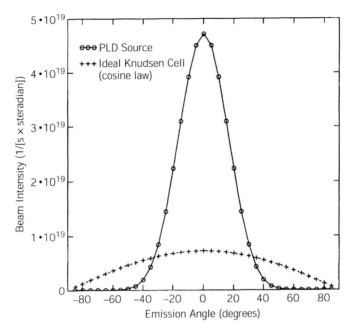

Figure V.13 The theoretical time-averaged beam intensity of a PLD source emitting 4.55×10^{17} copper atoms per pulse at a repetition rate of 50 Hz. The other experimental conditions are described in the accompanying examples. A beaming parameter of 12 was assumed, corresponding to a terminal Mach number of 2. For comparison, the beam intensity of an ideal Knudsen cell ($p = 1$) having the same total output is also shown.

The mechanisms of laser ablation encompass a wide range of *nonthermal* phenomena. Expulsion of micrometer-sized solid particles and liquid droplets from the target is perhaps the principal difficulty in applying PLD to thin-film deposition. It has been suggested that the solid particles are emitted when a target is porous or weakened from previous irradiation, so that poorly connected particles are ejected without being vaporized. *Splashing* is one proposed mechanism for droplet ejection [Ready, 1963]. In splashing, a subsurface region of the target becomes superheated and then blows off a molten surface layer as in boiling. A *piston effect* has also been proposed as a cause of droplet ejection. Here, the escaping vapor exerts a recoil force on the liquid below it, forcing the liquid to the side and then out of the crater. Other nonthermal processes of material removal include photostimulated desorption and fractoemission [Donaldson et al., 1988].

Of the general modeling effort directed toward laser–solid interaction, Cheung and Horwitz [1992] wrote that "none of the models accounts for all the physics, nor explains all the observations. In fact, such an attempt may be unrealistic." In spite of these complexities, relatively simple models based on energy balance calculations, such as the one presented in the previous section, are useful for predicting and understanding the critical performance parameters of pulsed laser deposition.

V.4 MATERIALS ASPECTS OF EVAPORATION SOURCES

Evaporation Temperatures of the Elements

A source vapor pressure of 10^{-2} torr often leads to a useful film deposition rate for a reasonably practical deposition geometry (see the example in Chapter I). Others have suggested the generalization that a minimum practical evaporation temperature is the one for which the vapor pressure is $\sim 10^{-2}$ torr [Maisel and Glang, 1970]. This temperature is typically near the melting point. Figure V.14, which is a correlation of practical evaporation temperature (as just defined) versus melting temperature for many common elements, confirms this general behavior and also reveals several exceptions. Chromium, arsenic, and manganese notably exhibit a relatively high vapor pressure, such that films of these metals can be produced by sublimation rather than evaporation. Carbon exhibits the greatest deviation from the rule.

The vapors of many common metals are mostly monatomic; we saw this to be true for silicon, in a previous chapter. Carbon, phosphorous, arsenic,

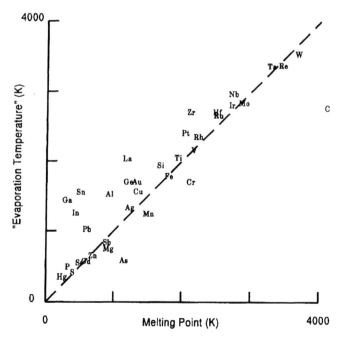

Figure V.14 A practical evaporation temperature for many metallic elements is the temperature for which the vapor pressure is 10^{-2} torr. A rule of thumb is that this temperature is equal to the melting point.

antimony, bismuth, and numerous elements to the right of these in the Periodic Table have a predominantly polyatomic vapor phase. The best sources of information on vapor pressures are critically evaluated, tabulated data (e.g., the JANAF tables [Chase et al., 1986]). The composition of the vapor phase is also obtainable from such data.

The Problem of Composition Change in the Evaporation of Alloys

If an effusion cell contains a binary alloy as the source material, the effusive stream will contain both components. Each component's flux will follow the cosine emission law, but the total emission of each component will be proportional to its own impingement rate.

For an ideal solution, as we saw in Chapter III, the vapor pressure of component A is reduced by a factor equal to its numerical fraction:

$$P_A(T) = X_A P_{Aeq}(T), \qquad (V.30)$$

where P_{Aeq} is the value it would have if pure A were present. The impingement rate of A at the orifice of the effusion cell is therefore $X_A P_{Aeq}/(2\pi m A k T)^{1/2}$. The ratio of the impingement rates of the two components on the orifice is

$$\frac{z_A}{z_B} = \frac{X_A P_{Aeq}}{X_B P_{Beq}} \left(\frac{m_B}{m_A}\right)^{1/2} \equiv \frac{K X_A}{X_B}, \qquad (V.31)$$

where K is a *segregation parameter* containing the vapor pressures and the mass ratio, whose definition should be clear from this equation.

There will be a gradual alteration of the composition of the melt unless K is exactly unity. The relative rate of removal of A particles and B particles from the melt is given by

$$\frac{dN_A}{dN_B} = \frac{z_A}{z_B}, \qquad (V.32)$$

which leads to

$$\frac{dN_A}{dN_B} = \frac{K X_A}{X_B} = \frac{K N_A}{N_B}, \qquad (V.33)$$

where N_A and N_B are the total number of particles of each component contained within the melt.

The coordinated variation of N_A and N_B may be calculated from $dN_A/N_A = K\, dN_B/N_B$:

$$N_A = C(N_B)^K, \qquad (V.34)$$

where C is a constant of integration that depends on the initial composition of the melt.

We have numerically evaluated the preceding equation V.34, and calculated X_B and z_A/z_B as functions of the fraction of the total number of particles of the melt remaining. It was assumed for this example that $K = 0.5$ and initially $X_B = 0.5$. The melt gradually becomes enriched in component A, with the process accelerating as the melt is consumed. The vapor stream initially contains twice as many B as A particles, but as time progresses A comes to dominate its composition. The results of the calculation are depicted in Figure V.15.

The process leads to a compositionally stratified film because of the continually changing composition of the emitted flux. This is usually a drawback and is the reason why individual sources are typically used to produce evaporated alloy films. One can see from the plot that there is only one instant during the evaporation when the composition of the emitted flux is equal to the initial alloy composition ($z_a/z_b = 1$).

One should realize that *this model pertains to all evaporation sources because all source materials contain impurities to some degree*. Indeed, if one of the components is a low-level impurity within a selected source material, and is more volatile, the process may lead to a quick purification of the melt (while giving the initial films a relatively high impurity content). Oxygen may be removed from silicon in this manner. On the other hand, if an impurity is substantially less volatile than the host material, the initial vapor stream, especially, is of higher purity than the source. The distillation of water (and stronger beverages) works likes this, with the evaporant leaving the impurities in the source.

We have assumed complete homogenization of the melt during the evaporation, but if the source were a solid, a thin *altered layer* would form on the surface, which would be depleted in the more volatile component. The composition of the altered layer would adjust automatically until the

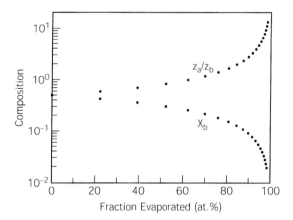

Figure V.15 The results of a numerical simulation of alloy evaporation; components A and B were initially present in equal amounts, but the segregation parameter was 0.5. Consequently, as time passes, the source and the efflux become enriched in A.

steady-state composition of the vapor stream were equal to that of the bulk. A similar "target conditioning" process occurs in the sputtering of multi-component targets, which, of course, remain solid.

Crucible Interactions

For most evaporation sources, a crucible is needed for containing the evaporant, and it must do so in a well-behaved fashion. The chief property desired for a crucible is stability—mechanical and chemical. The typically high temperatures render the task difficult.

The crucible itself must not vaporize, decompose, or outgas to a significant degree. As suggested by Figure V.14, refractory materials (those with a high melting temperature) are used in order to minimize the vapor pressure of the crucible material. Examples include graphite, tantalum, molybdenum, tungsten, SiO_2, Al_2O_3, BeO, and ZrO_2. Boron nitride, another popular crucible material, does actually thermally decompose at a significant rate above ~ 1600 K, preferentially losing N_2 [Chatillon and Massies, 1990]. Outgassing can be minimized by making sure that the crucible is nonporous, and by outgassing it deliberately before deposition, when under vacuum.

The crucible must not react with the evaporant. Unfortunately, this is in conflict with the first "law" of high-temperature chemistry, "at high temperatures everything reacts with everything else" and with the second, "the higher the temperature, the more seriously everything reacts with everything else." The third "law" expresses the danger, "the products might be anything" [Spear, 1976].

In choosing a crucible, the following questions should be considered:

- Will the evaporant melt when in use? If so, this would greatly accelerate any reaction with the crucible because of the rapid mass transport that is possible in liquids as compared to solid-state diffusion. (Chromium is one example of a metal that provides a high deposition rate by sublimation, rather than evaporation.)
- Is the crucible material soluble to a significant degree in the evaporant? If so, then one should expect dissolution of the crucible, and at a rapid rate if the evaporant is molten. (A graphite crucible is soluble in silicon to about 1.0 at% just above the melting point of silicon.)
- Does the molten evaporant wet the crucible? If not, then this would minimize their reactivity.
- Is there a chemical compound involving the evaporant and a component of the crucible? Is there a negative free energy change for this reaction? If so, then it may be expected to occur to some degree.

Example Consider the evaporation of iron oxide, Fe_2O_3, from a nickel crucible. Could the crucible react with the iron oxide evaporant?

We determine the thermodynamic driving force for the following reaction:

$$3Ni + Fe_2O_3 \rightarrow 3NiO + 2Fe$$

If the standard free energy of reaction is negative, the reaction is favored; if it is positive, the reaction should not occur.

The free energies of formation of the compounds involved are

$$\Delta_f G°_{Fe_2O_3} = -177 \text{ kcal/mol} \quad \Delta_f G°_{NiO} = -52 \text{ kcal/mol}.$$

The standard free energies of formation of the pure elements are, of course, zero. Now the standard free energy of reaction is given by $\Delta_r G° = \sum i \Delta_f G°_I$, where i is the stoichiometric coefficient of species I. The standard free energy of the preceding reaction, then, is $3 \times (-52 \text{ kcal/mol}) - 1 \times (-177 \text{ kcal/mol}) = +21 \text{ kcal/mol}$. Thus, the reaction is unfavorable. However, a conclusion of crucible stability based on this reaction alone would be incorrect. Another reaction should be considered:

$$Ni + 3Fe_2O_3 = NiO + 2Fe_3O_4.$$

To analyze this second reaction, we need the standard free energy of formation of the additional iron oxide, which is

$$\Delta_f G°_{Fe_3O_4} = -242 \text{ kcal/mol}.$$

Here the standard free energy of reaction is $1 \times (-52 \text{ kcal/mol}) + 2 \times (-242 \text{ kcal/mol}) - 3 \times (-177 \text{ kcal/mol}) = -5 \text{ kcal/mol}$. The second reaction, by contrast, is favorable. At elevated temperatures it might occur, even in the solid state. ∎

The example shows that all possible products must be considered before stating that a reaction with the crucible will not occur. One may, of course, consult the handbooks for recommendations of compatible evaporant–crucible combinations [Maissel and Glang, 1970].

V.5 SUMMARY OF PRINCIPAL EQUATIONS

Hertz–Knudsen–Langmuir equation	$j_{vap} = \alpha_v z_{eq} - \alpha_c z_i$ $= \alpha_v (P_{eq} - P_i)/(2\pi m k T)^{1/2}$
Incident flux distribution	$j_\Omega(\theta, \phi) = z \cos\theta / \pi$
Cosine law of emission	$J_\Omega(\theta, \varphi) = z\, \delta A \cos\theta / \pi$
Correction factor for nonequilibrium effusion cell	$F_N = [1/(1/\alpha_v) + (1/W_b) + (1/f\, W_{orif})]$
Beam intensity of ideal point source	$J_\Omega(\theta, \varphi) = z\, \delta A / 4\pi$

Beam intensity of E-gun evaporator	$J_\Omega(\theta,\varphi) \propto J \cdot \cos^n(\theta)$
Optical absorption length	$L_0 = 1/\alpha$
Thermal diffusion length	$L_t = \sqrt{(2\delta t \cdot \kappa)/c\, n_{mol}}$
Number of particles	$\Delta N = [(1-R)(j\,\delta t - (j\,\delta t)_{th}) \cdot \delta A$ $\cdot 6.02 \times 10^{23}/\text{mol}]/\Delta_{evap}H$
Thickness of target evaporated	$\Delta z = [\Delta N/(6.02 \times 10^{23}/\text{mol})]/$ $(\delta A \cdot n_{mol})$
Initial density of plume	$n_0 = \Delta N/[\delta A(L_t + u_0 \delta t)]$
Beam intensity of PLD source	$J_\Omega = [\Delta N \cdot f \cdot \cos^p(\theta)]/[2\pi/(p+1)]$
Beaming parameter for PLD source	$p = 4 + 2.13[\gamma^{1/2}M_t/(\gamma^{1/2}M_t + 1)]$ $+\gamma M_t^2$
Standard free energy of reaction	$\Delta_r G^\circ = \sum i \Delta_f G_I^\circ$

V.6 MATHEMATICAL SYMBOLS, CONSTANTS, AND THEIR UNITS

SI units are given first, followed by other units in widespread use.

a	Radius (m)
c	Heat capacity ($J \cdot \text{mol}^{-1} \cdot K^{-1}$)
d	Distance from cathode to anode (m)
f	Area fraction; pulse frequency (Hz)
$f(\theta)$	Normalized angular distribution (steradian^{-1})
j	Deposition flux ($m^{-2} \cdot s^{-1}$); power flux ($W \cdot m^{-2} \cdot s^{-1}$)
j_e	Current density (A/m^2)
j_Ω	Incident flux distribution ($m^{-2} \cdot s^{-1} \cdot$ steradian^{-1})
k	Boltzmann's constant (1.38×10^{-23} J/K; 8.62×10^{-5} eV/K)
$k(\theta)$	Parameter in Giordamine–Wang theory (dimensionless)
m	Particle mass (kg; amu $= 1.66 \times 10^{-27}$ kg)
n	Particle density (m^{-3}); exponent (dimensionless)
p	A function in Dayton theory (dimensionless); beaming parameter for PLD source (dimensionless)
q	Magnitude of the electronic charge (1.6×10^{-19} C)
r	Density gradient (m^{-4})
t	Time (s)
u	Speed (m/s)
v	Speed (m/s)
v_a	Applied potential (V)
v_n	Deposition rate
x	Position (m)
y	Position (m)
z	Position normal to surface (m); impingement rate ($m^{-2} \cdot s^{-1}$)
A	Chemical species; area (m^2); Richardson's constant (1.20×10^6 A/m^2); parameter in Giordmaine–Wang theory (dimensionless)

B	Chemical species; magnetic flux density (tesla; gauss = 10^{-4} tesla)
C	Integration constant (dimensionless); Clausing factor (dimensionless)
F	Force (newton)
F_N	Correction factor (dimensionless)
G_i	Gibbs free energy of component i (J/mol)
H	Enthalpy (J/mol)
J	Total particle flow (s^{-1})
J_Ω	Beam intensity per unit solid angle (steradian$^{-1} \cdot$s^{-1})
K	Segregation parameter (dimensionless)
L	Length; wall thickness (m)
M_t	Terminal Mach number (dimensionless)
N_i	Number of particles i
P	Pressure (Pa; torr = 133 Pa; atm = 1.01×10^5 Pa)
R	Molar gas constant (8.31 J$\cdot$mol$^{-1} \cdot$K^{-1}); distance from evaporant surface to substrate (m); reflectivity (dimensionless)
T	Temperature (K); Dayton's correction factor (dimensionless)
X_i	Numerical fraction of component i
α	Vaporization or condensation coefficient (dimensionless); a function in Dayton's theory (dimensionless); optical absorption coefficient (m^{-1})
γ	Ratio of heat capacities (dimensionless)
δA	Area (m^2)
δt	Pulse duration (s)
ΔN	Number of particles
ε_0	Permittivity of free space (8.85×10^{-12} F$\cdot$m^{-1})
θ	Polar angle (rad)
κ	Thermal conductivity (W$\cdot$m$^{-1} \cdot$K^{-1})
φ	Azimuthal angle (rad)
Φ	Work function (J)
Ω	Solid angle (steradian)

REFERENCES

Adamson, S., O'Carroll, C., and McGilp, J. F., 1989, "Monte Carlo Calculations of the Beam Flux Distribution from Molecular-Beam Epitaxy Sources," *J. Vac. Sci. Technol.* **B7**(3), 487.

Bhatia, M. S., 1994, "Novel In Situ Method for Locating Virtual Source in High-rate Electron-beam Evaporation," *Appl. Phys. Lett.* **65**(2), 251.

Brannon, J., 1993, *Excimer Laser Ablation and Etching*, AVS Monograph Series, The American Vacuum Society, New York.

Chase, M. W., Davies, C. A., Downey, J. R., Frurip, D. J., McDonald, R. A., and Syverud, A. N., 1986, *JANAF Thermochemical Tables*, 3rd ed., American Institute of Physics, New York.

Chatillon, Ch., and Massies, J., 1990, "Practical Aspects of Molecular Beam Epitaxy," in A. Chamberod and J. Hillairet, eds., *Materials Science Forum*, Metallic Multilayers, Vols. 59 and 60, Trans Tech Publications, Zürich, p. 229.

Cheung, J. T., and Sankur, H., 1988, "Growth of Thin Films by Laser-Induced Evaporation," *CRC Crit. Rev. Solid State Mat. Sci.* **15**(1), 63.

Cheung, J., and Horwitz, J., 1992, "Pulsed Laser Deposition History and Laser-Target Interactions," in G. K. Hubler, ed., *MRS Bull.* **XVII**(2).

Clausing, P., 1930, "Über die Strahlformung bei der Molekularströmung," *Z. Physik* **66**, 471.

Curless, J. A., 1985, "Molecular Beam Epitaxy Beam Flux Modeling," *J. Vac. Sci. Technol.* **B3**(2), 531.

Dam, B., Rector, J., Chang, M. F., Kars, S., deGroot, D. G., and Griessen, R., 1994, "Laser Ablation Threshold of $YBa_2Cu_3O_{6+x}$," *Appl. Phys. Lett.* **65**(12), 1581.

Dayton, B. B., 1956, "Gas Flow Patterns at Entrance and Exit of Cylindrical Tubes," in E. S. Perry and J. H. Durant, eds., *1956 National Symposium on Vacuum Technology Transactions*, Pergamon Press, New York.

Detorre, J. F., Knorr, T. G., and Hall, E. H., 1966, "Evaporation Processes," in C. F. Powell, J. H. Oxley, and J. M. Blocher, eds., *Vapor Deposition*, Wiley, New York.

Doll, G. L., Sell, J. A., Taylor, C. A., and Clarke, R., 1991, "Growth and Characterization of Epitaxial Cubic Boron Nitride Films on Silicon," *Phys. Rev. B* **43**, 6816.

Donaldson, E. E., Dickinson, J. T., and Bhattacharya, S. K., 1988, "Production and Properties of Ejecta Released by Fracture of Materials," *J. Adhesion* **25**, 281.

Dushman, S., 1949, *Scientific Foundation of Vacuum Technique*, Wiley, New York.

Glang, R., 1970, "Vacuum Evaporation," in L. I. Maissel and R. Glang, eds., *Handbook of Thin Film Technology*, McGraw-Hill, New York.

Giordmaine, J. A., and Wang, T. C., 1960, "Molecular Beam Formation by Long Parallel Tubes," *J. Appl. Phys.* **31**(3), 463.

Graper, E. B., 1973, "Distribution and Apparent Source Geometry of Electron-beam-heated Evaporation Sources," *J. Vac. Sci. Technol.* **10**(1), 100.

Günther, K. G., 1957, "Über die Intensitätsverteilung von Dampf- und Molekularstrahlen," *Z. angew. Phys.* **9**, 550.

Hill, R. J., ed., 1986, *Physical Vapor Deposition*, The BOC Group, Inc., Berkeley, CA.

Holland, L., 1963, *Vacuum Deposition of Thin Films*, Chapman and Hall, London.

Hubler, G. K., 1992, guest ed., "Pulsed Laser Deposition," *MRS Bull.* **XVII**(2).

Jackson, S. C., Baron, B. N., Rocheleau, R. E., and Russell, T. W. F., 1985, "Molecular Beam Distributions from High Rate Sources," *J. Vac. Sci. Technol.* **A3**(5), 1916.

Knudsen, M., 1915, "Das Cosinusgesetz in der Kinetischen Gastheorie," *Annal. Physik IV* **48**, 1113.

Maissel, L. I., and Glang, R., 1970, *Handbook of Thin Film Technology*, McGraw-Hill, New York.

Mihailescu, I. N., Chitica, N., Teodorescu, V. S., Luisa De Giorgi, M., Leggieri, G., Luches, A., Martino, M., Perrone, A., and Dubreuil, B., 1993, "Excimer Laser Reactive Ablation: An Efficient Approach for the Deposition of High Quality TiN Films," *J. Vac. Sci. Technol.* **A11**(5), 2577.

Motzfeldt, K., 1955, "The Thermal Decomposition of Sodium Bicarbonate by the Effusion Method," *J. Phys. Chem.* **59**, 139.

Ramayaranan, R., Polasko, K., Skelly, D., Wong, J., Mei, S.-N., and Lu, T.-M., 1987, "Unidirectional Deposition of Aluminum Using Nozzle Jet Beam Technique," *J. Vac. Sci. Technol.* **B5**(1), 359.

Ready, J. F., 1963, "Development of Plume of Material Vaporized by Giant-pulse Laser," *Appl. Phys. Lett.* **3**, 11.

Ready, J. F., 1971, *Effects of High Power Laser Radiation*, Academic Press, New York.

Rosenblatt, G. M., 1976, "Evaporation from Solids," in N. B. Hannay, ed., *Treatise on Solid State Chemistry*, Vol. 6A, Surfaces I, Plenum Press, New York.

Ruth, V., and Hirth, J. P., 1962, "The Angular Distribution of Vapor Flowing from a Knudsen Cell," in Rutner, E., Goldfinger, P. and Hirth, J. P., eds., *Condensation and Evaporation of Solids*, Proc. Internat. Symp. Condensation and Evaporation of Solids, Dayton, OH, Sept. 12–14, 1962, Gordon & Breach, New York.

Saenger, K. L., 1991, "On the Origin of Spatial Nonuniformities in the Composition of Pulsed-Laser-Deposited Films," *J. Appl. Phys.* **70**(10), 5629.

Saenger, K. L., 1993, "Pulsed Laser Deposition: Part II, A Review of Process Mechanisms," *Proc. Adv. Mat.* **3**, 63.

Shinn, G. B., Steigerwald, F., Stiegler, H., Sauerbrey, R., Tittel, F. K., and Wilson, W. L., 1986, "Excimer Laser Photoablation of Silicon," *J. Vac. Sci. Technol.* **4**, 1273.

Singh, R. K., and Narayan, J., 1990, "Pulsed-laser Evaporation Techniques for Deposition of Thin Films: Physics and Theoretical Model," *Phys. Rev. B* **41**(13), 8843.

Smith, H. R., 1969, "Deposition Distribution and Rates from Electron-Beam-Heated Vapor Sources," *Society of Vacuum Coaters Annual Technical Conf. Proc.*, Washington, DC (5–6 March), 50.

Sobol, E. N., 1995, *Phase Transformations and Ablation in Laser-Treated Solids*, Wiley, New York.

Spear, K. E., 1976, "High-Temperature Reactivity," in N. B. Hannay, ed., *Treatise on Solid State Chemistry*, Vol. 4, Reactivity of Solids, Plenum Press, New York.

Stafast, H., and von Przychowski, M., 1989, "Evaporation of Solids by Pulsed Laser Irradiation," *Appl. Surface Sci.* **36**, 150.

Stickney, R. E., Keating, R. F., Yamamoto, S., and Hastings, W. J., 1968, "Angular Distribution of Flow from Orifices and Tubes at High Knudsen Numbers," *J. Vac. Sci. Technol.* **4**(1), 10.

Van Audenhove, J., 1965, "Vacuum Evaporation of Metals by High Frequency Levitation Heating," *Rev. Sci. Instr.* **36**(3), 383.

Viswanathan, R., and Hussla, I., 1986, "Ablation of Metal Surfaces by Pulsed Ultraviolet Lasers under Ultrahigh Vacuum," *J. Opt. Soc. Am. B* **3**, 796.

Ward, J. W., Bivins, R. L., and Fraser, M. V., 1970, "Monte Carlo Simulation of Specular and Surface Diffusional Perturbations to Flow from Knudsen Cells," *J. Vac. Sci. Technol.* **7**(1), 206.

Wasilewski, Z. R., Aers, G. C., Spring Thorpe, A. J., and Miner, C. J., 1991, "Studies and Modeling of Growth Uniformity in Molecular Beam Epitaxy," *J. Vac. Sci. Techno.* **B9**(1), 120.

Yamashita, T., Tomita, T., and Sakurai, T., 1987, "Calculations of Molecular Beam Flux from Liquid Source," *Jpn. J. Appl. Phys.* **26**(7), 1192.

Colorplate I.4 In a pulsed laser deposition system, an intense laser beam creates a condensible vapor by vaporizing the surface of the target (on the right side of the photo). The visible plume appears when the emitted vapor is ionized by the laser, forming a plasma. The substrate is at left-center in the photo. (Reprinted with permission from D.B. Chrisey and M.A. Savell, *MRS Bulletin*, Vol. XVII, No. 2. Copyright 1992, Materials Research Society.)

(a)

(b)

Colorplate I.5 In a sputter deposition source, ions from a glow-discharge bombard a target made of the source material, which is the cathode of the discharge. Atoms ejected from the target condense on a substrate, which is often the anode of the discharge. (a) Shown here are a pair of magnetron sources above a substrate (courtesy of Denton Vacuum, LLC). (b) Two ion-beam sources on the far left (mostly obscured by the substrate) are bombarding two sputtering targets near the center of the photo. Material accumulates on the substrate on the left. The growing film is being simultaneously bombarded during deposition, with an additional ion source on the far right (out of the photo). (Courtesy of Ion Tech, Inc.)

Colorplate VI.18 A DC glow discharge in argon. The anode (grounded) is on the right and the cathode (supported by a white teflon insulator) on the left. The orange wire is a Langmuir probe whose bare tip enters the positive column of the discharge. The positive column is the largest luminous region, which extends about 75% of the way from the anode toward the cathode. The pressure was 100 mtorr and the voltage applied between cathode and anode was 1 kV. The discharge current density was 0.22 mA/cm^2.

Colorplate VI.19 An RF glow discharge in air at a pressure of 3 torr [Knipp and Wang, 1991]. The anode (grounded) is on the left and the floating cathode (mounted on a white Teflon insulator) is on the right. It was driven with a 13.56-MHz generator delivering 50 W of power to the discharge.

VI

PRINCIPLES OF SPUTTERING DISCHARGES

> Compared to some other technical fields, results in plasma physics can be less precise.
> — Kaufman and Robinson [1987]

> There is a fourth phase of matter — the plasma phase — which consists of separated nuclei and electrons . . . . The plasma phase is by far and away the most common — stars are plasmas–but it is also by far and away the most difficult to discuss quantitatively. For this reason we exclude it from further consideration.
> — Walton [1989]

Sputtering in its many forms has become perhaps the physical vapor deposition process of most widespread use. In diode sputtering, energetic ions (usually argon ions) from the plasma of a gaseous discharge bombard a target that is the cathode of the discharge. Target atoms are ejected and impinge on a substrate (the anode), forming a coating. Material removal from the cathode itself, or a sample placed on the cathode, also has many practical applications in sputter etching.

The purpose of this chapter is to provide a physical and semiquantitative understanding of sputtering discharges. By employing some simplification and generalization, we hope to convey a basic and true understanding of how sputtering discharges work. First, we will present three fundamental diode sputtering arrangements, which utilize a DC discharge, a capacitive radio frequency (RF) discharge, and either a DC or a capacitive RF discharge plus a magnet in the form of a planar magnetron. In Section VI.2, a "practical plasma" (an argon plasma that is representative of those used in thin-film sputter deposition systems) is presented, with attention to the current densities and the various potentials that occur within the deposition chamber. In Section

VI.3, the operation of a DC sputtering discharge is discussed, followed by an RF discharge model. We will model planar discharges, and perform only one-dimensional analyses. The current–voltage characteristics of DC and RF discharges are presented in the end-of-chapter Appendix, together with a description of their luminous glow regions.

A *plasma* is a partially ionized gas. There are at least three distinct particle populations: singly charged positive ions, electrons, and neutral gas particles. Figure VI.1 gives a perspective on the range of electron densities and temperatures of plasmas. Plasmas were named by Irving Langmuir, who observed that the positive column of his DC discharges (a certain spatial region) conformed to the shape of the containing vessel [Langmuir, 1929]; the root of "plasma" is a Greek word meaning "to mold." A *discharge* is a current flowing through a low-pressure gas. Some plasmas are also *glow discharges*, in that they glow with light emission from excited atoms. Our practical plasma is such a glow discharge. Within plasmas relevant to sputter deposition, the ion and electron densities are essentially equal except in volume space charge regions called *sheaths* that surround any physical object that is in contact with the plasma.

The reader should be ready to encounter in practice, many deviations from the simplified models of this chapter. Plasmas are especially hard to understand, and to model. At the end of Section VI.5, subsection "exceptions to the above," we mention some possible differences between real sputtering discharges and the models of this chapter, but it is far from inclusive.

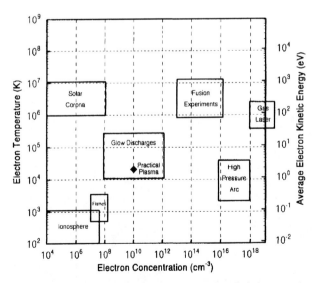

Figure VI.1 Plasma regimes. Some approximate ranges of electron concentration and temperature for various types of plasma.

The author has benefitted from the books by Chapman [1980] (written with a sense of humor), Brown [1966] (a broad, easy-to-read introduction to gaseous discharges), and Cobine [1941] (older, but good reading). A review of plasma etching presents basic sputtering plasma concepts, as well [Manos and Flamm, 1989]. Lieberman and Lichtenberg [1994] give a broad and fairly detailed treatment of plasma discharges as applied to microelectronic device fabrication. The classic review of the Langmuir probe is perhaps that of Chen [1965]. There are also some excellent probe reviews concerned with the "how to" of probe measurements [Clements, 1978; Ruzic, 1994].

VI.1 SPUTTERING ARRANGEMENTS

The purposes of this section are to describe the fundamental hardware configurations of practical deposition systems in widespread use, and to suggest representative operating parameters.

DC Sputtering

Figure VI.2 shows in schematic fashion the essential arrangements for DC sputter deposition with a parallel-plate discharge. The power supply is simply a high-voltage DC source (supplying several kilovolts). The sputtering target is the cathode of the discharge, tens to several hundred square centimeters in area; the anode may be the substrate and/or the vacuum chamber walls. The cathode–anode separation is typically a few centimeters. Argon is the most common sputtering gas, at a pressure on the order of 1 torr.

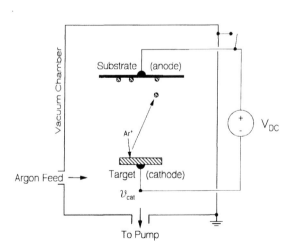

Figure VI.2 DC sputtering utilizes a DC gaseous discharge. Ions strike the target (the cathode of the discharge), which is the deposition source. The substrate and/or the vacuum chamber walls may be the anode.

The plasma is created and sustained by the DC source via mechanisms that pertain to the abnormal glow regime — secondary electron emission at the cathode and impact ionization of neutral gas atoms–as discussed in the Appendix. An ion current density to the cathode of $\sim 1\,\text{mA/cm}^2$ is typical unless it is a magnetron discharge as discussed below. Nonmagnetron DC discharges are not commonly used for film deposition.

RF Sputtering

Figure VI.3 shows schematically the essential arrangements for RF sputtering with a capacitive, parallel plate discharge. The power supply is a high voltage RF source (offering perhaps 0.5–1 kV amplitude, $\geq 0.1\,\text{MHz}$ frequency). 13.56 MHz is often used. The mean ion current density to the target is on the order of 1 mA/cm^2, while the amplitude of the total RF current is substantially (an order of magnitude or more) higher. A blocking capacitor (C) is placed in the circuit to develop the all-important DC self-bias, and a matching network is utilized to optimize power transfer from the RF source to the plasma. The dimensions are nominally the same as in DC sputtering.

RF sputtering offers advantages over DC; For instance, lower voltages and lower sputtering gas pressures may be used, with higher deposition rates obtained. Sputtering of an electrically insulating target becomes possible.

The plasma is created and maintained by the RF source, by the same atomistic processes which occur in the DC discharge. Magnetron RF discharges are used much more widely than nonmagnetron arrangements.

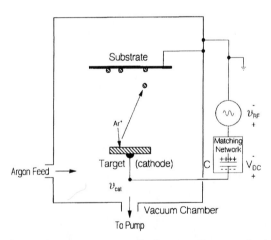

Figure VI.3 In RF sputtering, there are typically a small area cathode (the target) and a large area anode, in series with a blocking capacitor (C). The capacitor is actually part of an impedance-matching network that improves the power transfer from the RF source to the plasma discharge.

The Magnetron

The magnetron, developed in the 1970s, is the design of high-deposition-rate sputtering sources. The magnetron is a magnetically assisted discharge, as portrayed in Figure VI.4. As in the sputtering arrangements already discussed, there is a perpendicular (to the target surface) electric field—but a permanent magnet (or electromagnet) is added, to create lines of magnetic flux that are perpendicular to this electric field and thus parallel to the surface of the target. The magnetic field concentrates and intensifies the plasma in the space immediately above the target, as a result of *trapping* of electrons near the target surface. The magnetron effect results in enhanced ion bombardment and sputtering rates for both DC and RF discharges. In general, a magnetron discharge is much more efficient, with either DC or RF excitation, than one that does not utilize magnetic trapping.

How does this trapping occur?–There is a drift of the electrons in the $-\bar{E} \times \bar{B}$ direction, a "Hall effect," as shown in Figure VI.4, but superposed on this drift is a cycloidal motion as depicted in Figure VI.4c. The gyro radius of their orbits is given by

$$\rho = \frac{mv_\perp}{qB}, \qquad (\text{VI}.1)$$

where $v_\perp$ is the component of the electron velocity that is perpendicular to the flux lines. In sputtering systems, this radius is typically on the order of a few millimeters or less; thus, the confinement near the target surface can be quite effective. An electron encircles the lines of flux until it is scattered by another

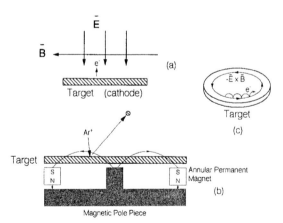

Figure VI.4 The magnetron: (*a*) in the planar magnetron sputtering arrangement, whether DC or RF, a static magnetic field is created parallel to the surface of the sputtering target to retain secondary electrons in that region; (*b*) an annular design, such as the one shown here schematically in exploded cross section, is often employed; (*c*) the electrons drift in the $-\bar{E} \times \bar{B}$ direction, actually executing a cycloidal path.

particle. Practically speaking, a magnetron exists when the $-\bar{E} \times \bar{B}$ path is closed and the electrons remain trapped for several trips around the loop.

Example A DC planar magnetron has a cathode fall of 600 V. Estimate the value of the magnetic flux density required to trap secondary electrons within 1 cm of the cathode surface.

The cathode fall ($V_p - v_{cat}$) is a concept explained in Section VI.3. Its value corresponds to the maximum kinetic energy that the ions have when they strike the surface of the cathode, and to the kinetic energy that the secondary electrons have (assuming no collisions) when they have crossed the cathode sheath (a space-charge region at the cathode surface) and enter the main body of the plasma. Thus, we may estimate $v_\perp$ for the electrons from $m^- v_\perp^2/2 = q(V_p - v_{cat})$, assuming a collisonless sheath. This relationship yields

$$v_\perp = \left(\frac{2 \times 1.6 \times 10^{-19} \text{ C} \times 600 \text{ V}}{9.11 \times 10^{-31} \text{ kg}}\right)^{1/2} = 1.45 \times 10^7 \text{ m/s}.$$

In most practical instances, the cathode sheath thickness is less than 1 cm, so the question implies that the gyro radius must be on the order of 0.5 cm. We estimate the magnetic flux density from

$$B = \frac{m^- v_\perp}{q\rho} = \frac{9.11 \times 10^{-31} \text{ kg} \times 1.45 \times 10^7 \text{ m/s}}{1.6 \times 10^{-19} \text{ C} \times 0.5 \text{ cm}} = 165 \text{ G}$$

Typically a flux density of 200–500 G is used to achieve strong, practical confinement. ∎

Figure VI.3 portrays the *planar* magnetron arrangement; many other configurations have been successfully utilized, particularly cylindrical or post cathodes. DC planar magnetrons are typically operated with argon at pressures of 0.5–30 mtorr and DC voltages of 300–700 V [Waits, 1978], as compared to discharge voltages 10 times larger when magnetic confinement is not employed. The resulting ion current densities to the cathode are in the range 4–60 mA/cm^2.

RF planar magnetrons may be operated at much lower voltages (often under 500 V amplitude) than RF sputtering discharges that do not employ magnetic trapping. RF magnetrons are about half as efficient as DC magnetrons, but the RF excitation is necessary for sputtering insulators.

Other Sputtering Arrangements

There have been many successful modifications to the basic sputtering arrangements described above. These include the usage of substrate bias to induce bombardment of the substrate and/or growing film, and thermionically assisted discharges (in which additional electrons are supplied to the plasma by a hot filament).

Ion-beam sputtering is an interesting and useful alternative. A low voltage discharge is established within an ion source. The entire source is biased to a high DC potential, again on the order of 1 kV. An ion beam is extracted from this source and directed onto a sputtering target, possessing a maximum kinetic energy equal to the DC bias of the source.

VI.2 A PRACTICAL SPUTTERING PLASMA AND ITS CURRENT DENSITIES AND POTENTIALS

In this section we define a plasma that is representative of those used in sputter deposition of thin films. Then, we consider the behavior of the Langmuir probe, a conducting electrode immersed in the plasma. Finally, we describe the sheath — a volume space-charge region — that surrounds the probe and is also found between the neutral body of the plasma and both cathode and anode. We are not concerned here with the processes that create and sustain the plasma — they are discussed briefly in the Appendix with the voltage–current characteristics.

A Practical Sputtering Plasma

A pressure of 1 torr of argon will be assumed for our practical plasma; this corresponds to an argon atom density of $3 \times 10^{16} \mathrm{cm}^{-3}$ at room temperature. The densities of electrons and ions will be assumed to be $10^{10} \mathrm{cm}^{-3}$. This gives an ionization fraction of $\sim 3 \times 10^{-7}$ and the label "weakly ionized plasma." The three particle populations each have their own characteristic temperatures. Often the neutrals are close to room temperature (300 K), and they have a classical Maxwell–Boltzmann energy distribution. For the charged particles, the characteristic temperatures are above ambient, reflecting higher average kinetic energies for electrons and ions as compared to neutrals. For the practical plasma, representative values are $T^+ = 500$ K (the ion temperature) and $T^- = 23{,}000$ K (the electron temperature).

The cause for T^- being much greater than T^+ is that the electrons are much more mobile than the Ar^+ ions. This is because they are lighter by a factor of about 73,000. Because they are more mobile, the electrons pick up much more kinetic energy from an applied electric field than do the ions, in the brief times they have between collisions. Thus, their higher average kinetic energy gives them the higher characteristic temperature.

In Table VI.1 we summarize the basic parameters and properties of our practical plasma, some of which are yet to be introduced.

Example Suppose that a free argon ion is within a radio frequency electric field of 5 V/cm amplitude, at 13.56 MHz.

1. Calculate its acceleration, velocity, and displacement due to the field, assuming that there are no collisions with other particles. The

TABLE VI.1 A Practical Plasma

Assumed Plasma Parameters

The gas is argon (atomic weight 39.948 amu)
The charged particles are singly charged argon ions (Ar^+) and electrons (e^-; 5.49×10^{-4} amu)
The densities are
$$n^+ = n^- = 10^{10} \, cm^{-3};$$
$$n = 3 \times 10^{16} \, cm^{-3}$$

The pressure is therefore 0.93 (≈ 1) torr
Using an argon atom diameter of 3.83 Å, the mean free path is 0.051 mm
The characteristic temperatures are $T = 300$ K, $T^+ = 500$ K, $T^- = 23{,}000$ K

Assumed Properties and Features

The Boltzmann distribution prevails for each particle type
Consequently, the maximum possible random impingement rates (according to kinetic theory) of the charged particles are $z^+ = 1.28 \times 10^{14}$ cm$^{-2}\cdot$s^{-1}, $z^- = 2.35 \times 10^{17}$ cm$^{-2}\cdot$s^{-1}, and the corresponding current densities are $+qz^+ = 2.05 \times 10^{-5}$ A$\cdot$cm^{-2}, $-qz^- = -3.76 \times 10^{-2}$ A$\cdot$cm^{-2}
The plasma consists of a plasma body (within which the net charge density is zero) separated from all physical objects by sheaths (within which there is a volume space charge)
The plasma body is a "good" electrical conductor (i.e., it is equipotential)
The plasma body is large compared to the Debye length, which is 15.4 μm for the preceding parameters.

---

acceleration, velocity, and displacement of the ion will all be sinusoids of this same frequency. We will calculate their amplitudes. Using Newton's second law, acceleration = force/mass, the amplitude of the ion's acceleration is given by

$$a_0^+ = \frac{qE_0}{m^+} = \frac{1.6 \times 10^{-19} \, C \times 5 \, V/cm}{40 \times 1.66 \times 10^{-27} \, kg} = 1.2 \times 10^9 \, m/s^2.$$

Integrating once to get the amplitude of the ion's velocity, we obtain

$$v_0^+ = \frac{a_0^+}{\omega} = \frac{1.2 \times 10^9 \, m/s^2}{2\pi \times 13.56 \, MHz} = 14 \, m/s.$$

Integrating a second time to get the ion's displacement, we have

$$x_0^+ = \frac{v_0^+}{\omega} = \frac{14 \, m/s}{2\pi \times 13.56 \, MHz} = 1.7 \times 10^{-7} \, m.$$

2. The pressure is 10^{-2} torr—is the assumption of no collisions a good one? From Chapter II, the mean free path at this pressure is on the order of 0.8 cm. This is much larger than the amplitude of the ion's displacements, so the assumption is valid.
3. Will the temperature of the ion population be perturbed by this electric field? Temperature is a measure of the average kinetic energy of the particles. Now, the rms speed of the argon ions is

$$\frac{v_0^+}{\sqrt{2}} = 10 \text{ m/s}.$$

By contrast, the average speed of argon neutrals at room temperature $[(8kT/\pi m)^{1/2}$ from Chapter II] is 400 m/s. Thus the electric field will have a very modest effect on the temperature of the ion population.

4. Estimate the rms speed of an electron as it moves in response to the given field, again assuming no collisions, and estimate the effective temperature of the electron gas. The amplitude of the electron velocity is greater than that of the ions by a factor of $m^+/m^- = 40 \times 1.66 \times 10^{-27}$ kg/9.11×10^{-31} kg $= 73{,}000$. Thus $v_0^- = 73{,}000 \times 14$ m/s $= 1.0 \times 10^6$ m/s and the rms speed of the electron is $v_0^-/\sqrt{2} = 7.3 \times 10^5$ m/s. To estimate the temperature of the electrons, we will take their average energy to be

$$E_{rms} = \frac{m^-(v_{rms}^-)^2}{2} = 2.4 \times 10^{-19} \text{ J}.$$

Using the expression from the kinetic theory of gases, the average energy is given by $3kT^-/2$. Thus, the electron temperature may be estimated with

$$T^- = \frac{2E_{rms}}{3k} = \frac{2 \times 1.2 \times 10^{-19} \text{ J}}{3 \times 1.38 \times 10^{-23} \text{ J/K}} = 11600 \text{ K}$$

(ignoring the slight difference between v_{av} and v_{rms} in the kinetic theory). ∎

The Ideal Langmuir Probe

The temperatures and densities of the ions and the electrons can be measured, in principle, with a *Langmuir probe*, which is portrayed schematically in Figure VI.5. This probe is a metal object, inserted into the DC plasma and biased as the experimenter may desire, relative to ground potential. (It is common to connect both the anode of the discharge, and the chamber walls, to ground.) Our ideal probe is flat (because its effective area is constant), whereas most real probes are cylindrical (for which the effective area varies with the applied voltage). Understanding the current–voltage characteristic of the probe is crucial, because in addition to providing a means to determine plasma

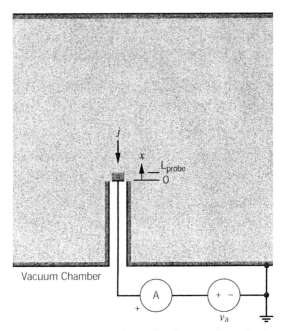

Figure VI.5 Schematic diagram showing a flat Langmuir probe immersed in a DC plasma. The vacuum chamber wall serves as a large-area, low-resistance contact. There are sheaths at the probe-plasma interface and between the plasma and the wall.

parameters, it is a key to understanding the electrical interaction of all physical objects with the plasma, including the cathode and the anode.

A positive current flows to the probe when ions strike it; conversely, a negative current flows to the probe when electrons strike. When an ion actually strikes, it is instantly neutralized by an electron from the metal, and desorbs as a neutral. The overall effect is a positive current to the probe. On the other hand, when an electron strikes the probe, it simply enters the metal and continues to travel on through the external circuit. The overall effect of the impinging electron is thus a negative current to the probe.

A theoretical current density vs. applied voltage characteristic is shown in Figure VI.6. The potential quantities are v_a (that applied to the probe) and V_p (the plasma potential). It will be our convention to designate applied potentials with a lowercase v, and fundamental plasma potentials with the uppercase V.

With the probe held at the plasma potential, there is a net negative current flowing through the probe to ground. This is a tremendously important effect and is basic to understanding much of the behavior of a plasma. It is due to a large difference between the impingement rates for electrons ($n^- v_{av}^-/4$) and ions ($n^+ v_{av}^+/4$) with $v_{av} = (8kT/\pi m)^{1/2}$ in each case, as estimated from kinetic theory. Even if the electrons and ions had the same characteristic temperature, because of the relatively small mass of the electrons it would be true that $v_{av}^- \gg v_{av}^+$. The difference is even greater in reality because of the experimental

A PRACTICAL SPUTTERING PLASMA

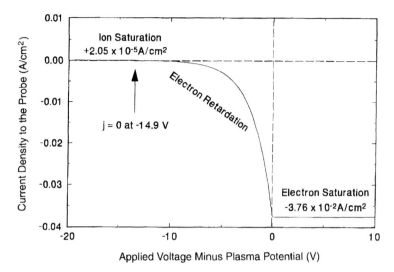

Figure VI.6 The theoretical current–voltage characteristic of the ideal Langmuir probe. The arrow indicates the floating potential, which is 14.9 V below the plasma potential for our practical plasma.

fact that the electron temperature greatly exceeds the ion temperature. For our practical argon plasma, the electron impingement rate (z^-) is $2.35 \times 10^{17}\,\text{cm}^{-2} \cdot \text{s}^{-1}$ and the ion impingement rate (z^+) is $1.28 \times 10^{14}\,\text{cm}^{-2} \cdot \text{s}^{-1}$.

It is a very important fact that the probe–plasma interface is electrically rectifying. The rectifying behavior occurs because if $v_a \neq V_p$, one of the two types of charged particle will be repelled from the probe. If $v_a > V_p$, positive ions will be repelled. The ion flux to the probe is reduced by a Boltzmann factor, $\exp[-q(v_a - V_p)/kT^+]$. The electron flux, on the other hand, does not increase, and remains at the random impingement rate, z^-. The converse is true for $v_a < V_p$. The two ideal saturation current densities, at large positive and negative applied voltages, are simply equal to the electron and ion fluxes individually (multiplied by $\pm q$).

If the probe is made to be floating (meaning the current must be exactly zero), it acquires a floating potential which is less than V_P. The mechanism is as follows. Because of the excess electron impingement, the floating probe becomes negatively charged, lowering its electrostatic potential. This retards further electron impingement and culminates in the condition of zero net current to the floating probe in steady state.

When $V_f < v_a < V_p$, the probe is in the electron retardation regime. Assuming a Maxwellian energy distribution, the electron current density to the probe decreases according to $\exp[-q(V_p - v_a)/kT^-]$. Including the ions, the current density for applied voltages less than V_p actually is

$$j_{\text{ideal}<} = qz^+ - qz^- \exp\frac{-q(V_p - v_a)}{kT^-}. \tag{VI.2}$$

Data in this applied voltage range may be used to determine the electron temperature (see the next section). Experimental adherence to this predicted behavior has confirmed that the electron energy distribution of many practical plasmas is reasonably close to Maxwellian. It is not uncommon, however, to have two distinct electron populations, "hot" and "cold," with their own characteristic temperatures. It is, in fact, possible to obtain the energy distribution of the electrons by differentiating $j_{\text{ideal}<}(v_a)$ with respect to v_a. The ideal current density in the *ion saturation* regime is qz^+ (2.05×10^{-5} A/cm^2).

The current density for applied voltages greater than V_p is

$$j_{\text{ideal}>} = -qz^- + qz^+ \exp\frac{-q(v_a - V_p)}{kT+}. \tag{VI.3}$$

The ideal current density in the *electron saturation regime* is $-qz^-$ (-3.76×10^{-2} A/cm^2). Equations (VI.2) and (VI.3) were used to plot the theoretical $j-v_a$ characteristic in Figure VI.6. Equation (VI.3) was scarcely needed — the current density is virtually constant in the range $v_a > V_p$.

Some assumptions behind the calculated curve of Figure VI.6 should be mentioned:

1. Any current drawn by the probe is negligible compared to the anode–cathode current. Thus, as we apply a probe potential and draw a probe current, the particle densities and temperatures are unperturbed.
2. There is a low-resistance electrical contact to the plasma to complete the circuit of the Langmuir probe, such as the relatively large-area interface with the vacuum chamber walls.
3. The surface area of the sheath of the probe is fixed. This can be a reasonable approximation for a flat probe, but the surface area of a cylindrical probe fundamentally varies with the potential drop across the sheath.
4. We have neglected the enhancement of the ion current density related to the Bohm sheath criterion, which is discussed below.

In summary

- The Langmuir probe is strongly rectifying.
- The current density flowing to the probe is determined by the difference between v_a and V_P.
- The theoretical current–voltage characteristic has three fundamental regimes: ion saturation at large negative potentials with respect to V_p, electron saturation at virtually any positive potential with respect to V_p, and the electron retardation regime separating ion saturation from $V_p = 0$.

An ion retardation regime exists in principle for $v_a > V_p$ but is typically too small to be seen in either theoretical or experimental plots.

We would like to emphasize the importance of understanding the Langmuir probe's current–voltage characteristic by reminding the reader that these same current transport processes often apply at other interfaces to the plasma, such as at the sputtering target, at the substrate, and at the vacuum chamber walls — except that there is one more concept to introduce, the *enhanced ion current density*. In Section VI.3, we use our knowledge of Langmuir probe behavior to understand the potential distribution of a DC discharge, and to predict the DC self-bias of a capacitive RF discharge.

An Experimental Langmuir Probe Characteristic

For the experimental DC glow discharge discussed in the Appendix, a Langmuir probe current–voltage characteristic was obtained (Fig. VI.7). The probe was biased with respect to the grounded anode. The results generally resemble the theoretical curve; however, both saturation currents, in contrast to the simple theory, vary slightly with probe potential. The electron saturation current is about $-5\,\text{mA}$; for the probe area of $\sim 0.18\,\text{cm}^2$, this corresponds to an electron current density of $-28\,\text{mA/cm}^2$, which is rather close to the $-38\,\text{mA/cm}^2$ of the practical plasma. A representative value for the ion saturation current would be $\sim 50\,\mu\text{A}$, corresponding to an ion current density of $\sim 0.28\,\text{mA/cm}^2$. This is 14 times larger than the assumed ion current density of the ideal probe. This large increase in the ion saturation current is typical, is very important, and will be discussed next.

The Enhanced Ion Current Density

In a sputtering discharge the most important practical aspect of the plasma behavior is the bombardment of the cathode by ions. It is an experimental fact, and a fortunate one for physical vapor deposition, that in a sputtering discharge the ion current density to the cathode is on the order of $\sim 1\,\text{mA/cm}^2$, not the $20.5\,\mu\text{A/cm}^2$ (qz^+), which is the theoretical value of the current density in the ion saturation regime of the Langmuir probe immersed in the practical plasma. This is true even when the actual sputtering plasma closely resembles the practical plasma. The experimental DC discharge whose behavior is presented in the Appendix had a current density of $0.22\,\text{mA/cm}^2$, which could easily have been increased. This enhanced current density also appears in experimental Langmuir probe data as the saturation ion current density.

Ions are actually pulled toward an electrode that is held below the plasma potential. The attractive force on the ions causes the collected ion current density to greatly exceed that which would be due merely to random impingement, qz^+. An accepted expression for the enhanced ion current density is

$$j_{\text{ion}} \approx qn^- \sqrt{\frac{k(T^+ + T^-)}{m^+}} \tag{VI.4}$$

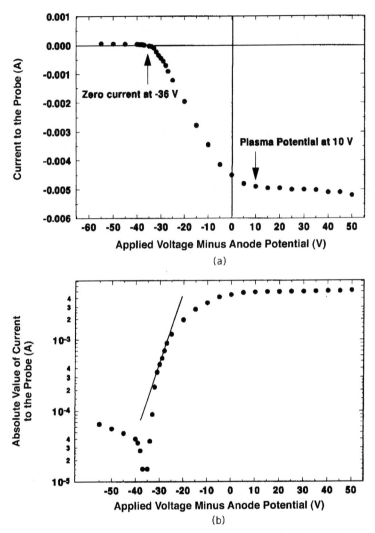

Figure VI.7 Experimental Langmuir probe current–voltage data. The probe was inserted into a DC glow discharge in air at a pressure of 250 mtorr. The discharge current was 6 mA with anode grounded and cathode biased to -600 V; the electrode areas were 45.5 cm^2. The cylindrical probe was AWG (American Wire Gauge) 22-gauge copper wire, length 0.7 cm, giving a probe area of 0.18 cm^2. The DC discharge was in the normal glow regime. The drawn straight line was used to estimate the electron temperature.

(see Manos and Dylla in Manos and Flamm [1989]). The occurrence of this enhanced current density is intimately tied to the *Bohm criterion*, a requirement that the ion velocity on approaching the sheath at an electrode be greater than a certain minimum value.

A PRACTICAL SPUTTERING PLASMA

Example Calculate the enhanced ion current density for the practical plasma.

$$j_{ion} = 1.6 \times 10^{-19} \text{ C} \times 10^{10} \text{ cm}^{-3} \left[\frac{1.38 \times 10^{-23} \text{ J/K}(500 \text{ K} + 23{,}000 \text{ K})}{40 \times 1.66 \times 10^{-27} \text{ kg}} \right]^{1/2}$$

$$= 0.35 \text{ mA/cm}^2. \qquad \blacksquare$$

Thus, for a real plasma the current density to a probe or other electrode has qz^+ replaced by j_{ion}:

$$j_< = j_{ion} - qz^- \exp \frac{-q(V_p - v_a)}{kT_-} \qquad \text{(VI.5)}$$

and

$$j_> = -qz^- + j_{ion} \exp \frac{-q(v_a - V_p)}{kT_+}. \qquad \text{(VI.6)}$$

The *floating potential* is that for which the net current to the probe is zero. Its value may be estimated by equating the electron current, reduced by a Boltzmann factor, to the enhanced ion current density. This leads to the following expression:

$$V_p - V_f = \left(\frac{kT_-}{q} \right) \ln \left[\frac{qz^-}{j_{ion}} \right]. \qquad \text{(VI.7)}$$

Example Calculate the floating potential for the practical plasma, assuming the enhanced ion current density:

$$V_p - V_f = \frac{(1.38 \times 10^{-23} \text{ J/K}) \times 23{,}000 \text{ K}}{1.6 \times 10^{-19} \text{ C}} \ln \left(\frac{37.6}{0.35} \right) = 9.3 \text{ V}.$$

Thus, V_f is 9.3 V below V_p. $\qquad \blacksquare$

We now determine the *plasma potential*. In the theoretical current–voltage characteristic of the Langmuir probe, the sharp "knee" of the curve actually marks where $v_a = V_p$. The plasma potential itself cannot be calculated from the parameters of the electron and ion populations. It is set by the currents and potentials required to sustain the plasma. In Section VI.3 we show how this happens for a DC discharge.

Example For the experimental probe data in Figure VI.7, what is the plasma potential? What is the floating potential? What is the electron temperature? What is the electron density? What is the ion temperature?

We infer from Figure VI.7a a plasma potential about 10 V above the anode potential. The "knee" is not so sharp as in the theoretical curve, but +10 V is about where the experimental curve begins to rise above the electron saturation current.

The floating potential is at $-36\,\text{V}$ below the anode potential, because the measured current is zero there. The Langmuir probe theory suggests that the electron temperature may be determined from the slope of a plot of $\ln[j_< - j_{\text{ion}}]$ versus $(V_p - V_a)$, which should be linear in the electron retardation regime. We estimate the electron temperature with the semilog plot of the same data, shown in Figure VI.7b. We utilize values of current much greater than that corresponding to ion saturation ($\sim 50\,\mu\text{A}$) so that it is not necessary to subtract out the ion saturation current. A straight line was drawn by eye through selected data points that seem to line up. From the slope of this line, an electron temperature of 47,840 K is deduced:

$$T^- = \frac{q}{k[d\,\ln(j)/dv]} = \frac{1.16 \times 10^4\,\text{K/V}}{0.242\,\text{V}^{-1}} = 47,840\,\text{K}.$$

The electron density may be estimated from T^- and the electron saturation current density $-qz^- = -qn^-(kT^-/2\pi m^-)^{1/2}$, which yields

$$n^- = \frac{-28\,\text{mA/cm}^2}{-1.6 \times 10^{-19}\,\text{C} \times [(1.38 \times 10^{-23}\,\text{J/K}) \times 47,840\,\text{K}/(2\pi \times 9.11 \times 10^{-31}\,\text{kg})]^{1/2}}$$
$$= 5.1 \times 10^9\,\text{cm}^{-3}.$$

The direct experimental determination of T^+ from the Langmuir probe current–voltage characteristic is typically impossible. However, by taking n^+ equal to n^-, it is possible (in principle at least) to estimate the ion temperature from j_{ion}:

$$T^+ = \left(\frac{m^+}{k}\right) \times \left[\frac{j_{\text{ion}}}{-qn^-}\right]^2 - T^- = \frac{40 \times 1.66 \times 10^{-27}\,\text{kg}}{1.38 \times 10^{-23}\,\text{J/K}}$$
$$\times \left[\frac{0.28\,\text{mA/cm}^2}{-1.6 \times 10^{-19}\,\text{C} \times 5.1 \times 10^9\,\text{cm}^{-3}}\right]^2 - 47,840\,\text{K}$$
$$= 8813\,\text{K}.$$

This experimental estimate of a T^+ is perhaps not very accurate because it is a small difference between large numbers, which are themselves not very accurately known. ∎

Langmuir probe measurements can be difficult, hard to interpret, and ambiguous; expert reviews of the subject should be consulted. Ruzic [1994] offers suggestions for practical probe construction and advice on analyzing the data.

The Probe Sheath

A *sheath* is a volume space charge region (a region where the electrical charge density, ρ, is not equal to zero) that surrounds physical objects within the

A PRACTICAL SPUTTERING PLASMA

plasma and separates the plasma from the vacuum chamber walls, as illustrated in Figure VI.5.

Solutions to Poisson's equation express how the electrostatic potential varies spatially within the sheath, under the influence of the local charge density:

$$\nabla^2 V = \frac{-\rho}{\varepsilon_0}, \quad \text{(VI.8)}$$

where ε_0 is the permittivity of free space and ∇^2 is the Laplacian operator.

If the probe potential is very close to V_p, the Debye length gives the characteristic thickness of the sheath. It is obtained from an approximate solution of Poisson's equation, where it is assumed that the potential difference across the sheath is small compared to kT^+ (and consequently very much less than kT^-). When the local potential is less than V_p (one could just as well do this analysis for $v_a > V_p$), the ion density is enhanced and the electron density is depleted, each in amounts given by Boltzmann factors:

$$n^+(x) = n^+ \exp \frac{q(V_p - V(x))}{kT^+} \quad \text{(VI.9)}$$

and

$$n^-(x) = n^- \exp \frac{-q(V_p - V(x))}{kT^-} \quad \text{(VI.10)}$$

where x is the distance from the probe, as shown in Figure VI.5.

The charge density is given simply by $q(n^+(x) - n^-(x))$:

$$\rho(x) = q \left[\frac{n^+ \exp q(V_p - V)}{kT^+} - n^- \exp \frac{-q(V_p - V)}{kT^-} \right]. \quad \text{(VI.11)}$$

The Debye length is obtained from an approximate solution of Poisson's equation. The exponentials are approximated using $\exp(x) \approx 1 + x$:

$$\rho(x) = qn^+ \left(\frac{q(V_p - V)}{kT^+} + \frac{q(V_p - V)}{kT^-} \right) = \frac{q^2 n^+ (V_p - V)}{kT'}. \quad \text{(VI.12)}$$

where T' is the reduced temperature, $T^+T^-/(T^+ + T^-)$. (We have used the fact that, in the neutral body of the plasma, the parameter n^+ gives both the electron and ion densities.) Poisson's equation in one dimension becomes

$$\frac{d^2 V(x)}{dx^2} = \frac{-q^2 n^+ [V_p - V(x)]}{\varepsilon_0 kT'}$$

$$= \frac{[V_p - V]}{L_d^2}, \quad \text{(VI.13)}$$

where the Debye length is given by

$$L_d = \left(\frac{\varepsilon_0 kT'}{q^2 n^+}\right)^{1/2}. \quad (\text{VI.14})$$

For the parameters of our practical plasma, L_d is 15.4 μm.

Assuming the following boundary conditions, namely, that (1) $V = V_p$ at $x = \infty$ (L_d is much less than the assumed plasma dimensions, a few centimeters) and that (2) $V = v_a$ at $x = 0$, the potential as a function of distance (x) from the probe is given by

$$V(x) = v_a + (V_p - v_a)\left(1 - \exp\left(\frac{-x}{L_d}\right)\right). \quad (\text{VI.15})$$

The Debye shielding regime is one of very small potential differences, much smaller than the range of the Langmuir probe characteristic. Another approximate solution to Poisson's equation has been obtained, which gives estimates of the sheath thickness at a probe (L_{probe}) for potentials on the order of $V_p - V_f$ [Ruzic, 1994]. Its result is

$$L_{\text{probe}} = L_d \cdot 1.02 \cdot \left(\sqrt{\frac{V_p - v_a}{kT^-/q}} - \frac{1}{\sqrt{2}}\right) \cdot \left(\sqrt{\frac{V_p - v_a}{kT^-/q}} + \sqrt{2}\right). \quad (\text{VI.16})$$

For the practical plasma, $kT^-/q = 1.98$ V, so L_{probe} is typically a few L_d.

VI.3 GASEOUS DISCHARGES FOR SPUTTERING

There are two main purposes of this section: (1) to explain the electrostatic potentials of the DC sputtering discharge (as portrayed in Fig. VI.8) and (2) to account for the development of the DC self-bias at the blocking electrode of a capacitively coupled, RF sputtering discharge. In addition, we estimate the thicknesses of the sheaths of these discharges, and show that the sputtering projectiles that bombard the cathode are actually a mix of energetic neutral gas atoms and ions.

For both DC and RF discharges, the total current flowing is usually considered one of the independent variables, along with pressure and geometry of the discharge. The applied voltage required to maintain this current is a dependent variable. This current is created by secondary electron emission from the cathode and impact ionization of neutrals by these electrons, as elaborated in the Appendix.

A DC Discharge Model

The Langmuir probe–plasma interface is electrically rectifying—it is a diode. Likewise, a pair of electrodes are back-to-back diodes as portrayed in the

GASEOUS DISCHARGES FOR SPUTTERING

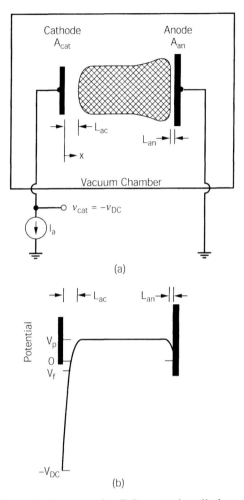

Figure VI.8 (*a*) Schematic diagram of a DC sputtering discharge; (*b*) the potential profile (not to scale, as $V_p - V_f$ is typically much less than V_{DC}).

electrical circuit schematic of Figure VI.9a. The *cathode* is that electrode which attracts *cat*ions (the positive ions) from the plasma, while the *anode* attracts *an*ions (the electrons). We will see that the anode receives a negative current while having a small sheath voltage and narrow sheath; at the same time, the cathode receives a positive current while having a large sheath voltage and a wide sheath.

Consider the pair of electrodes shown in Figure VI.8a. For a practical discharge, the current density to the cathode will be on the order of j_{ion}, perhaps enhanced by secondary electron emission, and for current continuity, the current density to the anode must be the negative of this value (assuming equal areas):

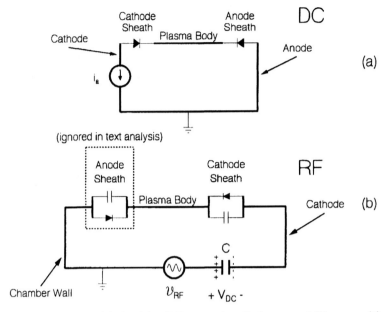

Figure VI.9 Circuit models for (*a*) a DC sputtering discharge and (*b*) a capacitive RF sputtering discharge.

$$j_{an} = -j_{cat}. \tag{VI.17}$$

To achieve these currents, we must have

$$v_{cat} = -V_{DC} \ll V_f \tag{VI.18}$$

but

$$V_f < v_{an} < V_p. \tag{VI.19}$$

Thus, we are led to the frequent observation that *the plasma is the "most positive" body in the main current path of the discharge.*

Applying to the anode the current density expression for $v_a < V_p$, $j_{an} = j_{ion} - qz^- \exp[-q(V_p - v_{an})/kT^-]$, we obtain the following expression for the plasma potential relative to the anode:

$$V_p - V_{an} = \frac{-kT^-}{q} \ln\left[\frac{j_{an} - j_{ion}}{-qz^-}\right]. \tag{VI.20}$$

When the anode is grounded as in Figure VI.8, $v_{an} = 0$.

Example Consider a DC sputtering discharge whose argon ion current density to the cathode is j_{ion}. The anode's area is the same. Assuming the parameters of the practical plasma, calculate the plasma potential with respect to the grounded anode.

$$V_p - v_{an} = (-1.38 \times 10^{-23} \text{ J/K}) \times \frac{23,000 \text{ K}}{1.6 \times 10^{-19} \text{ C}} \times$$
$$\ln \frac{-0.35 \text{ mA/cm}^2 - 0.35 \text{ mA/cm}^2}{-37.6 \text{ mA/cm}^2} = 7.9 \text{ V}. \qquad \blacksquare$$

It is customary practice to design the anode to be of much larger surface area than the cathode; in fact, the vacuum chamber walls, if metal, serve quite well. This increases V_p as compared to the case of equal area electrodes.

The body of the plasma is assumed to be equipotential in this simple model. Then Kirchhoff's voltage law applied to the entire DC discharge circuit requires that the potential drop across the cathode sheath be

$$\text{Cathode fall} = V_p + V_{DC}. \qquad (\text{VI.21})$$

This is the maximum possible kinetic energy with which the ions strike the cathode in a sputtering discharge, a sum greater than the applied voltage V_{DC}! The sputtering power density to the target is given by

$$p_{ion} = j_{ion} \cdot (V_p + V_{DC}). \qquad (\text{VI.22})$$

Interestingly, we will see below that we do *not* have j_{ion} ions per unit area bombarding the target, each with a kinetic energy of $q(V_p + V_{DC})$. We have more particles than this, and of a lower kinetic energy. However, the power density in (VI.22) remains nominally correct.

In summary, most of the space between the electrodes is electric-field-free. Two relatively high-field regions, the sheaths, separate the plasma body from the anode and the cathode. The resulting electrostatic potential as a function of position between the electrodes is shown schematically in Figure VI.8b. It is apparent that ions bombard the anode, as well as the cathode. Their very small energy (qV_p), however, causes virtually no sputtering of this electrode. (A more accurate model of the potential distribution of the DC discharge, which is beyond the scope of the present discussion, has a small positive slope in the potential within the body of the plasma, culminating in a sharp, narrow, small peak just before the anode sheath [Lieberman and Lichtenberg, 1994]. These details are not necessary to our qualitative understanding of the DC discharge.)

We have said little about how the plasma of this discharge is created or maintained — how is this accomplished? The origin of free electrons is located on the surface of the cathode, where electrons are liberated by secondary electron emission when ions strike this surface. These secondary electrons are then accelerated across the cathode sheath and into the body of the plasma. Here they strike neutral gas atoms, creating an ion–electron pair; this is how the plasma is sustained. These processes are discussed more fully in the Appendix.

The Cathode and Anode Sheaths

The experimental fact is that the sheath thickness at the cathode is much greater than L_d when a DC or RF current density of "practical" magnitude (on the order of j_{ion} for the practical plasma) is flowing. In this practical current regime, the charge density within the sheath is actually created by the ions that are carrying the current, the *injected* carriers. This behavior is called *space-charge-limited current flow*, but it is perhaps more accurate to consider the current as controlling the space charge, rather than what the name seems to suggest.

For the sheath at the cathode of a DC discharge, as shown in Figure VI.8a, we solve Poisson's equation and calculate the sheath thickness for space-charge-limited current under the following assumptions:

1. There is continuity of current across the sheath.
2. The charge density is due entirely to the relatively slow-moving ions, because the electrons, which are much more mobile, are effectively removed from the sheath by the high field there.
3. The ions moving through the sheath do not experience collisions (i.e., it is a *collisionless* sheath).

The ion current density flowing toward the cathode is given by

$$j_{ion} = -\rho(x) v_x(x), \tag{VI.23}$$

where v_x is the component of velocity that is perpendicular to the cathode surface and x is the distance from the cathode. The ballistic velocity of the ions is a function of position within the sheath according to conservation of energy:

$$\frac{m^+[v_x(x)]^2}{2} = q[V_p - V(x)]. \tag{VI.24}$$

Thus, the charge density is given by

$$\rho(x) = -j_{ion} \left\{ \frac{m^+}{2q[V_p - V(x)]} \right\}^{1/2} \tag{VI.25}$$

and Poisson's equation in one dimension becomes

$$\frac{d^2[V_p - V(x)]}{dx^2} = \frac{-j_{ion}\{m^+/2q[V_p - V(x)]\}^{1/2}}{\varepsilon_0}. \tag{VI.26}$$

For convenience, we have referenced the potential to V_p and multiplied through by -1 to arrive at Equation (VI.26). The boundary conditions are that $V = V_p$ at $x = L_{sc}$ (the sheath thickness), and that $V = -V_{DC}$ at $x = 0$ for the given value of j_{ion}. The solution for the potential as a function of position in the sheath is hard to guess, simply from looking at the preceding differential equation VI.26:

$$\frac{4}{3}[V_p - V(x)]^{3/4} = \left(\frac{4j_{ion}}{\varepsilon_0}\right)^{1/2} \left(\frac{m^+}{2q}\right)^{1/4} [L_{sc} - x]. \quad (VI.27)$$

The sheath thickness is obtained by substituting $V(x) = -V_{DC}$ at $x = 0$:

$$L_{sc}(\text{collisionless}) = \left(\frac{4\varepsilon_0}{9j_{ion}}\right)^{1/2} \left(\frac{2q}{m^+}\right)^{1/4} [V_p + V_{DC}]^{3/4}. \quad (VI.28)$$

There seem to be three independent variables (j_{ion}, V_{DC}, and L_{sc}) in this one equation. How are they fixed, for example, at the cathode of an actual DC discharge? That is, how does the plasma decide on a unique combination of $[V_p + V_{DC}]$ and L_{sc} that gives the right current? The starting point is that the plasmatologist chooses a j which then becomes the total current density of the discharge; j_{ion} is then $j/[1 + \gamma]$ (correcting for secondary electron emission at the cathode — see Appendix). The $[V_p + V_{DC}]$ that is required to sustain this value of j is supplied by the DC current supply, but determined by the processes that create and sustain the plasma as elaborated in the Appendix. The required potential difference for a sputtering discharge is typically a "fairly high" value, one that puts the discharge into the abnormal glow regime (see Appendix). Then, L_{sc} is fixed by the principles embodied in the above relation. In summary, the plasmatologist chooses j, which determines j_{ion} and the required $(V_p + V_{DC})$. Finally, j_{ion} and $(V_p + V_{DC})$ together set L_{sc}.

There is also a *mobility-limited* version of space-charge-limited ion current. It is needed whenever the ions suffer collisions in passing through the cathode sheath. Such collisions will occur unless the mean free path is substantially greater than the sheath thickness. The velocity of the ions is determined by a drift velocity expression instead of by conservation of energy as in the derivation above (VI.24). An analysis similar to the preceeding one [Lieberman and Lichtenberg, 1994] shows that in this instance L_{sc} is given by

$$L_{sc}(\text{collisional}) = \left(\frac{2\varepsilon_0}{3j_{ion}}\right)^{2/5} \left(\frac{5}{3}\right)^{3/5} \left(\frac{2q\lambda_i}{\pi m^+}\right)^{1/5} [V_p + V_{DC}]^{3/5}. \quad (VI.29)$$

Here, λ_i is the mean free path of the ions. As a rough estimate, it is equal to the mean free path of neutrals. (For argon ions in argon gas, the cross section for elastic scattering is 2.5×10^{-15} cm^2 as an order of magnitude estimate [Lieberman and Lichtenberg, 1994]).

Example What is a typical value for the sheath thickness at the cathode of a sputtering discharge?

Suppose $V_p - V_{DC} = 1000$ V and the ion current density flowing across the cathode sheath to the cathode is 2 mA/cm^2. Let's assume the practical plasma is present, for which the mean free path is 0.051 mm. We will calculate first the collisional sheath thickness, not knowing at the outset whether the collisional or collisionless expression should be used. If it is not substantially less than 0.051 mm, then we do indeed have a collisional sheath. Now

$$L_{sc}(\text{collisional}) = \left(\frac{2 \times 8.85 \times 10^{-12}\,\text{F/m}}{3 \times 2\,\text{mA/cm}^2}\right)^{2/5} \times \left(\frac{5}{3}\right)^{3/5} \times$$

$$\left(\frac{2 \times 1.6 \times 10^{-19}\,\text{C} \times 0.051\,\text{mm}}{\pi \times 40 \times 1.66 \times 10^{-27}\,\text{kg}}\right)^{1/5} \times (1000\,\text{V})^{3/5} = 2.0\,\text{mm}.$$

This value is much larger than the mean free path, so we do indeed have a collisional sheath. (The theoretical collisionless sheath thickness is 3.7 mm.) ∎

The *anode* sheath thickness may be estimated from the expression of the previous section for the sheath thickness at a Langmuir probe.

As a final note, the plasma is not quite electric-field-free outside a sheath. The cathode sheath as such is surrounded by a *presheath*, a region of relatively weak but nonnegligible electric field. One can appreciate this fact by recognizing that the cathode fall ($\sim 1\,\text{kV}$) is enormous compared to the ion temperature ($kT^+ = 0.04$ eV) and that the physical boundary of the sheath will not be as sharp as in the theoretical model. It is in the spatial volume of the presheath where the enhanced ion current is collected. The collection mechanism is a presheath acceleration of ions toward the sheath, which has been called *Bohm presheath diffusion* (although motion under the influence of an electric field is usually called *drift*, not *diffusion*).

The Sputtering Projectiles that Bombard the Cathode

As we saw in the example above, the cathode sheath of a sputtering discharge is very likely a collisional one, since at 1 torr the mean free path is only 0.051 mm. There is a process that occurs readily in the cathode sheath that does not alter the total current density or the potential drop across the sheath, but drastically changes the nature, number, and kinetic energy of the bombarding particles.

This process is symmetric charge exchange, as illustrated in Figure VI.10. An energetic argon ion, as it passes a neutral argon atom, takes an electron from the neutral. With this event, however, there is very little energy or momentum transfer. The result is that an energetic ion en route to the cathode, plus a cold neutral, are converted into an energetic neutral plus a cold ion:

$$\text{Ar}^+(\text{hot}) + \text{Ar}(\text{cold}) \rightarrow \text{Ar}(\text{hot}) + \text{Ar}^+(\text{cold}).$$

For argon ions in argon gas, the cross section for symmetric charge exchange is approximately the same as that for elastic scattering (as an order of magnitude

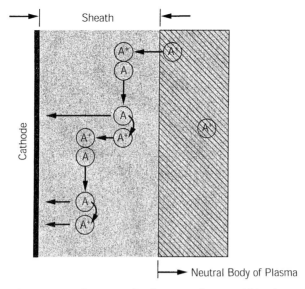

Figure VI.10 The process of symmetric charge exchange within the cathode sheath. When two charge exchange events occur, as shown here, a single ion entering the sheath is converted into two neutrals plus an ion, all of which strike the cathode (but with kinetic energies corresponding to only a fraction of the cathode fall).

estimate, the cross section is 2.5×10^{-15} cm^2 [Lieberman and Lichtenberg, 1994]).

If this charge exchange occurs part way through the cathode sheath, the original energetic argon ion has a kinetic energy corresponding to some fraction of the cathode fall. After it has been neutralized, it no longer can be accelerated by the electric field within the sheath, but travels on to the cathode with whatever kinetic energy that it had at the moment of encounter with the cold neutral. The newly created cold ion, however, *can* be accelerated by the field, and acquires a kinetic energy that is, at a maximum, the remaining fraction of the cathode fall. If the mean free path for symmetric charge exchange is small compared to the cathode sheath thickness, the new ion probably will experience its own charge exchange event before reaching the cathode. Immediately after the second charge exchange event, there will be two *hot* neutrals plus *one* cold ion. Depending on the remaining distance to the cathode, this new cold ion may itself become "hot," or may even become neutralized by a third charge exchange event. The ion current density value remains unchanged throughout this process.

The overall result is that the enhanced ion current density is converted into a flux of ions plus neutrals, which can be predominantly neutrals if the mean free path is much less than the cathode sheath thickness. The average kinetic energy of the projectiles, on striking the cathode, becomes less than that of the cathode fall:

$$j_{ion} \times q[V_p + V_{DC}] \to [j_{ion} + z_{neutral}] \times KE_{av}$$

with

$$KE_{av} < q[V_p + V_{DC}].$$

Fortunately for sputter deposition, the sputter yield of ions and neutrals of the same species is the same. However, the overall sputtering rate of the cathode may be enhanced or diminished by charge exchange, depending on the projectile energy dependence of the sputter yield.

An RF Discharge Model

Initial interest in RF excitation (kHz to Mhz) was for the purpose of depositing electrically insulating material [Anderson et al., 1962]. To place an insulator onto the cathode of a *DC* discharge would make the discharge inoperable because no current could flow through the insulator. The situation is quite different with RF—on average, no net current *does* flow. However, through the rectifying property of the sheath, a large DC *self-bias* of such an electrically blocking electrode develops—which we will attempt to explain below. This self-bias causes virtually continual ion bombardment of the blocking electrode; thus, it is a *cathode*. In fact, if the sputtering target is a conductor it is deliberately configured to be a floating electrode as the surface of an insulating target is. This is portrayed in Figure VI.3, where an RF voltage is applied through an external *blocking capacitor*. The purpose of our RF discharge model is to explain the mechanism of the DC self-bias, which is the key to efficient RF sputtering. The model is simplistic, but it does convey an understanding of what actually happens:

Our very simple circuit model for the complete RF discharge is presented in Figure VI.9*b*. The plasma body is represented by a conductor, and we will again assume that it is equipotential. Each sheath is represented with a diode, because of the rectifying property, and with a capacitor. Each sheath has an intrinsic electrical capacitance because it consists, after all, of a vacuum region separated by two conducting bodies. The capacitance was omitted from the DC discharge model because it is irrelevant to DC behavior.

For our analysis of the DC bias, we will take the position that since the surface area of the target is much smaller than that of the vacuum chamber (which serves as the anode of the discharge), the potential drop across the anode sheath is negligible. Furthermore, we will ignore the displacement current (due to the cathode sheath capacitance) so that all the current that flows in the circuit is due to ions and electrons moving across the cathode sheath—conduction current. (In fact, most of the discharge current *is* displacement current, but considering the displacement current obscures the analysis of the basic effect we want to explain here, the DC self-bias. The conduction and displacement currents are superposed in the real discharge and it is in this instance valid to calculate them, and consider their effects, separately. We will need the displacement current when we calculate the RF sheath thickness in a later section.)

The input waveform is portrayed in Figure VI.11a:

$$v_{RF}(t) = V_{RF} \sin \omega t. \qquad (VI.30)$$

We will assume that the amplitude of v_{RF} is on the order of 500 V, much greater than $V_p - V_f$ for the practical plasma. We show in Figure VI.11b the

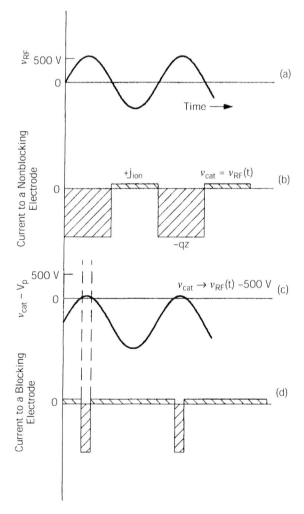

Figure VI.11 DC self-bias. Potentials and currents for RF excitation of $(a-b)$ nonblocking and $(c-d)$ blocking electrodes.

current that would flow to the target *if the blocking capacitor were not there*. Because the plasma potential is quite close to ground, the current density to the electrode alternates between $\sim j_{ion}$ and $-qz^-$ as v_{RF} changes sign. Clearly, the net current averaged over one period is *not* equal to zero.

However, with the blocking capacitor in place, the situation is quite different. In this instance it must be true that *in sinusoidal steady-state, the net current to each electrode, averaged over one RF cycle, must be zero*. Although $j_{ion} \gg qz^+$, there is still a large difference between the magnitudes of the positive and negative currents. It is accumulation of negative charge and the requirement of zero net current over an RF cycle that leads to the DC self-bias of the blocking capacitor. The self-bias achieves an operating point on the current–voltage characteristic that ensures zero net current over each RF cycle. We now consider the current flow in detail, and estimate the magnitude of the self-bias.

In Figure VI.11c we show schematically the effect of the DC self-bias on the key potential difference, that of the target (cathode) minus that of the body of the plasma. Referring to the circuit diagram in Figure VI.9b, the cathode potential is

$$v_{cat}(t) = v_{RF}(t) - V_{DC}. \tag{VI.31}$$

Most of the time the potential of the target is *below* that of the plasma. In order to achieve zero net current over each RF cycle, it is necessary that during most of the cycle the electrode's potential be below V_f. The self-bias of the blocking capacitor thus develops because negative charge accumulates on the right-hand plate:

$$V_{DC} \to V_{RF}, \tag{VI.32}$$

which is the amplitude of the applied voltage. All this may be visualized with the help of Figure VI.11d, which shows the current flowing to the target *with the self-bias*. Averaging the current to zero means that the area under the curve in Figure VI.9d must be zero for one complete period in steady state. The curve is drawn in such a manner as to suggest that condition. A curve similar to Figure VI.11c was first published, to our knowledge, by Butler [1961].

We simulated the development of the self-bias with a simple electrical circuit consisting of a diode and capacitor in series, with a sinusoidal source. The circuit diagram was that of Figure VI.9b, for the RF discharge, with the elements representing the anode sheath replaced with a conductor. The transient response of this circuit is portrayed in Figure VI.12. In the simulation, it takes about five periods for the DC bias to fully develop.

How is the value of the blocking capacitor chosen? What restrictions are there on its value? There are at least three:

1. Its charging time during electron bombardment must be not much greater than the RF period. This is to ensure that the self-bias quickly develops.

GASEOUS DISCHARGES FOR SPUTTERING

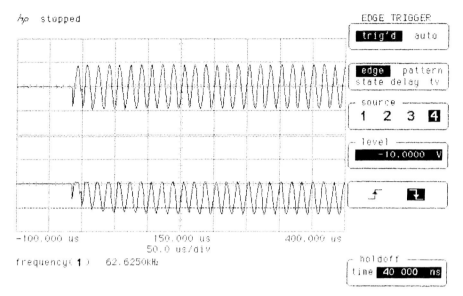

Figure VI.12 Results of a circuit simulation of an RF discharge, showing the development of the self-bias. The upper trace is the input signal; the lower trace is the diode voltage. The circuit diagram is that of Figure VI.9*b*, with the anode sheath replaced with a short and the cathode sheath capacitance omitted.

2. Its *dis*charging time during ion bombardment must be substantially greater than the RF period. This is to prevent the neutralization of significant amounts of negative charge on the cathode.
3. Most of the applied RF potential must be dropped across the cathode sheath rather than the blocking capacitor; otherwise, the value of the cathode fall could never become very great.

These conditions result in typical values of 10–1000 pF.

We have neglected the potential difference between the plasma and the other electrode, the vacuum chamber wall, because of its relatively large area. The exact area-dependence is a matter of some debate, as neither theory nor experiment have been definitive to date; however, it is generally found that the smaller of the two develops the larger self-bias. A well-known analysis by Koenig and Maissel [1970] gave the following area-ratio effect:

$$\frac{V_{DC1}}{V_{DC2}} = \left(\frac{A_2}{A_1}\right)^4. \tag{VI.33}$$

The analysis of Kaufman and Rossnagel [1988] gave an exponent of 2, rather than 4. In any case, the vacuum chamber walls are frequently employed as the large-area electrode; with a relatively small target, sputtering of the chamber walls is practically avoided. Koenig and Maissel assumed that the entire applied RF is capacitively divided between the cathode and anode sheaths; it seems that we still do not possess a precise analysis of this, or of exactly how the applied RF voltage divides between the two electrodes and the external capacitor. Certainly, however, it is found that the smaller the electrode, the greater its impedance, and the greater its resulting self-bias.

Also, it is observed in practice that the distance between the electrodes has a significant effect on the plasma impedance. This effect is omitted by our modeling, because we assume the plasma to be a perfect conductor and therefore equipotential.

How is the RF discharge itself maintained? The concensus seems to be that it is primarily by electron impact ionization of neutrals. The primary source of these electrons is secondary electron emission by ions which strike the self-biased electrode.

The RF Sheaths

The sheaths of a capacitive RF discharge are formed in a manner entirely different from those of the DC discharge. The governing principle is not space-charge-limited current flow through the sheaths, but oscillatory free electron motion which creates a displacement current within the sheaths.

As suggested in Figure VI.13, the electrons in the body of the plasma oscillate en masse between the two electrodes, while the ions are virtually stationary. The width of the ion cloud is essentially d, the electrode separation. The width of the electron cloud, on the other hand, is less than d, because any electron that comes in contact with an electrode is collected. Since the local ion and electron densities are equal within the bounds of the electron cloud, while outside the electron cloud there are, of course, no electrons, the net result is a positive charge for the plasma body as a whole, once again making it the most positive body in the discharge.

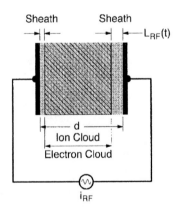

Figure VI.13 Oscillation of the free electron cloud in an RF discharge creates the sheaths at the two electrodes. To effect a symmetric discharge, the blocking capacitor was omitted and the electrode areas made the same.

Let's assume a total current density for the discharge of $j_{RF} \sin \omega t$. In a symmetric, homogenous model of the RF discharge [Lieberman and Lichtenberg, 1994], each of the sheath thicknesses is of the form

$$L_{RF}(t) = L_{RF} - L_{RF} \cos \omega t. \tag{VI.34}$$

(Of course, there must be a 180° phase difference between the right- and left-hand sheaths.) Each sheath thickness oscillates with the period of the RF current, and goes to zero once during each cycle. (The sheath must collapse to zero briefly in order to transfer electrons to the electrode, which neutralize the positive charge that accumulates by continuous ion bombardment.) The width of the electron cloud is therefore $d - L_{RF}$.

The amplitude of the sheath thickness is given by

$$L_{RF} = \frac{j_{RF}}{qn^+ \omega}. \tag{VI.35}$$

(This is based on the total current of the discharge being a displacement current within the sheath.)

Example Estimate the sheath thickness of a practical RF discharge.

We will assume a total current density of 15 mA/cm^2 (representative of some RF data presented in the Appendix) and the ion density of the practical plasma:

$$L_{RF} = \frac{15 \text{ mA/cm}^2}{1.6 \times 10^{-19} \text{ C} \times 10^{10} \text{ cm}^{-3} \times 2\pi \times 13.56 \text{ MHz}} = 0.11 \text{ cm}. \quad \blacksquare$$

The RF sheath is much more complicated than the simple homogeneous model presented here, but the general physical picture is true and the estimate of sheath thickness as a function of time is qualitatively correct.

VI.4 SUMMARY OF PRINCIPAL EQUATIONS

Gyro (or Larmor) radius	$\rho = m v_\perp / qB$
Ideal Langmuir probe current density for applied voltages $< V_p$	$j_{ideal<} = qz^+ - qz^- \exp[-q(V_p - v_a)/kT^-]$
Ideal Langmuir probe current density for applied voltages $> V_p$	$j_{ideal>} = -qz^- + qz^+ \exp[-q(v_a - V_p)/kT^+]$
Enhanced ion current density	$j_{ion} \approx qn^- \sqrt{k(T^+ + T^-)/m^+}$
Electrode current density for applied voltages $< V_p$	$j_< = j_{ion} - qz^- \exp[-q(V_p - v_a)/kT^-]$
Electrode current density for applied voltages $> V_p$	$j_> = -qz^- + j_{ion} \exp[-q(v_a - V_p)/kT^+]$
Floating potential relative to plasma potential	$V_p - V_f = (kT^-/q) \ln [qz^-/j_{ion}]$

Poisson's equation	$\nabla^2 V = -\rho/\varepsilon_0$
Ion density in a sheath	$n^+(x) = n^+ \exp[q(V_p - V(x))/kT^-]$
Electron density in a sheath	$n^-(x) = n^- \exp[-q(V_p - V(x))/kT^+]$
Charge density in a sheath	$\rho(x) = q\{n^+ \exp[q(V_p - V)/kT^+] - n^- \times \exp[-q(V_p - V)/kT^-]\}$
Debye length	$L_d = [\varepsilon_0 kT'/q^2 n^+]^{1/2}$
Sheath potential for Debye shielding	$V(x) = v_a + (V_p - v_a)[1 - \exp(-x/L_d)]$
Sheath thickness at a probe	$L_{probe} = L_d \cdot 1.02$ $\times [(\sqrt{(V_p - v_a)/(kT^-/q)}) - (1/\sqrt{2})]$ $\times [(\sqrt{(V_p - v_a)/(kT^-/q)}) + \sqrt{2})]$
Plasma potential	$V_p - v_{an} = \frac{-kT^-}{q} \ln[(j_{an} - j_{ion})/-qz^-]$
Cathode fall	Cathode fall $= V_p + V_{DC}$
Sputtering power density to target	$p_{ion} = j_{ion} \cdot (V_{DC} + V_p)$
Sheath potential for space-charge-limited ion current	$\frac{4}{3}[V_p - V(x)]^{3/4}$ $= (4j_{ion}/\varepsilon_0)^{1/2}(m^+/2q)^{1/4}[L_{SC} - x]$
Collisionless sheath thickness for space-charge-limited ion current	$L_{sc}(\text{collisionless}) = (4\varepsilon_0/9j_{ion})^{1/2}$ $\times (2q/m^+)^{1/4}[V_p + V_{DC}]^{3/4}$
Collisional sheath thickness for space-charge-limited ion current	$L_{sc}(\text{collisional}) = (2\varepsilon_0/3j_{ion})^{2/5}$ $(\frac{5}{3})^{3/5}(2q\lambda/\pi m^+)^{1/5}[V_p + V_{DC}]^{3/5}$
RF cathode potential with DC self-bias	$v_{cat}(t) = v_{RF}(t) - V_{DC}$
DC self-bias	$V_{DC} \rightarrow V_{RF}$
Area-ratio effect for self-bias	$V_{DC1}/V_{DC2} = (A_2/A_1)^4$
RF sheath thickness	$L_{RF}(t) = L_{RF} - L_{RF} \cos \omega t$
Amplitude of RF sheath thickness	$L_{RF} = j_{RF}/(qn^+ \omega)$
Townsend discharge total current (impact ionization only)	$I_{td,total} = I_{sat} e^{S_i \cdot n \cdot d}$
Secondary electron emission coefficient	$\gamma \approx 0.016(E_i - 2W)$
Total discharge current before breakdown (but including secondary electron emission)	$I_{b,total} = I_{be}(d) = \{I_{sat} e^{S_i \cdot n \cdot d}/ [1 - \gamma(e^{S_i \cdot n \cdot d} - 1)]\}$

VI.5 APPENDIX

The Voltage–Current Characteristic of a DC Discharge

With gas discharges, it is customary to consider current the independent variable and voltage, the dependent variable. Figure VI.14 shows a schematic representation of the voltage–current characteristic of a DC discharge for a

APPENDIX 185

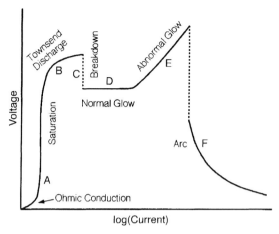

Figure VI.14 Schematic representation of applied voltage as a function of current for a DC discharge.

pressure of ~ 0.1–1.0 torr in argon. It is an amalgam of several classic diagrams from the historical literature [Brown, 1966; Holland, 1956; Cobine, 1941; Llewellyn-Jones; 1966; von Engel, 1965].

Current flow depends on the creation of populations of free electrons and ions. These charged particles may be generated by photoemission of electrons from the cathode, by ionization of neutrals by cosmic rays, by electron impact ionization of neutrals, or by secondary electron emission when ions strike the cathode (only the latter two are relevant to sputtering discharges). The basic features of this curve and their underlying mechanisms are described in the following paragraphs.

Ohmic Conduction and Saturation In the very low field (and current density) regime, there is a small free electron density due to some external agent, such as photoemission at the cathode by stray radiation. Current values below the nA/cm^2 range are typical, even if a strong ionizing radiation source such as a UV lamp is used as the external agent. Sometimes this free electron density fluctuates; hence this regime has also been called *random bursts*. Under these conditions the discharge is quite far from a sputtering plasma. At higher fields, but while still in a very low current range, the above-mentioned current saturates as a function of applied voltage:

$$I_{sat} = \text{const.} \qquad (VI.36)$$

In saturation, the electric field sweeps all the available electrons out of the volume between the electrodes as fast as they are created, pulling them to the anode where they are collected. The discharge is not yet *self-sustaining*, as it depends on an external agent for ionization.

Townsend Discharge If the applied voltage is increased still further, the current will eventually rise beyond the previous saturation value. This is due to the onset of a new process for the generation of additional free electrons and for the first time some ions: impact ionization of neutrals by electrons. As electrons are accelerated toward the anode, some of them acquire enough kinetic energy to ionize neutral gas atoms, en route. It was named after J. S. Townsend, an early investigator.

The impact ionization process is described with an ionization cross section (S_i). There is a minimum kinetic energy required for this process, given by the ionization potential of the neutral atom (e.g., 15.8 eV for argon). The cross section is zero for kinetic energies below the ionization potential, and exhibits a maximum at higher energies. The maximum value for argon is $\sim 3 \times 10^{-16}\,\text{cm}^2$ at $\sim 100\,\text{eV}$ [Chapman, 1980].

We assume here that I_{sat} is a photoemitted electron current coming off the cathode. Then impact ionization causes the electron current to increase with distance from the cathode (x) according to

$$\frac{dI_{tde}}{dx} = S_i \cdot n \cdot I_{tde}, \tag{VI.37}$$

leading to

$$I_{tde}(x) = I_{sat} e^{S_i \cdot n \cdot x}, \tag{VI.38}$$

The total current flowing through the discharge is constant, independent of x. Thus, it may be evaluated as the electron current at d, where d is the distance from the cathode to the anode:

$$I_{td,total} = I_{sat} e^{S_i \cdot n \cdot d}. \tag{VI.39}$$

This expression actually contains an implicit voltage–current relationship for the Townsend discharge. The voltage sensitivity is within S_i — as the discharge voltage increases, the average kinetic energy of the electrons increases, and so S_i increases and then $I_{td,total}$ increases as in region B of Figure VI.14.

Since the total current flowing at any point between the electrodes is constant, the ion current is given by $I_{td,total} - I_{tde}(x)$:

$$I_{tdi}(x) = I_{sat} e^{S_i \cdot n \cdot d}[1 - e^{S_i \cdot n \cdot (x-d)}]. \tag{VI.40}$$

The maximum ion current occurs at $x = 0$: $I_{tdi}(0) = I_{sat}[e^{S_i \cdot n \cdot d} - 1]$. At the anode the ion current is zero. The Townsend discharge is not yet self-sustaining because if the ionizing radiation were removed, the current would cease. *We still have not achieved the charged particle densities and characteristic temperatures of our practical plasma, and there are no sheaths.*

Breakdown Next, we have breakdown. But first, "Why don't the ions participate in the impact ionization process we have just described?" The answer is that they are too heavy. They do not acquire kinetic energy from the electrostatic field nearly as readily as do the electrons. In the Townsend discharge, they are unable to ionize neutrals, for the most part. However, in breakdown, ions begin to have a major impact (in fact, lots of them). This happens through secondary electron emission at the cathode. When breakdown occurs, it is because the ions finally have enough kinetic energy to create a charged particle — by knocking a free electron out of the cathode material. The bulk of the ions are now created by these secondary electrons as they perform electron impact ionization in transit to the anode.

This new process is described with a secondary electron emission coefficient, or yield (γ). This yield is simply the number of secondary electrons emitted per ion incident upon the cathode. Raizer [1991] has proposed an approximate, energy-independent, empirical expression:

$$\gamma \approx 0.016(E_i - 2W), \tag{VI.41}$$

where E_i is the ionization potential of the cathode atom (usually somewhere around 15 eV) and W is the work function of the cathode (typically near 5 eV). In general, γ is also dependent on the ion identity and energy, and the surface condition of the cathode. γ has an energy threshold on the order of 10 eV, and increases slowly with ion energy. For argon ions striking copper, γ is about 0.1 at an ion energy of 500 eV [Chapman, 1980].

Breakdown is the sudden drop in voltage required to sustain the discharge at a fixed current (region C in Fig. VI.14). It is caused by the onset of this secondary electron emission at the cathode and is accompanied by a drastic spatial rearrangement of the discharge. With the current held constant, the required discharge voltage decreases because the charged particle densities increase, the neutral body of the plasma becomes equipotential because it is a good conductor, sheaths form at the electrodes, and the electric field becomes concentrated in the cathode sheath. The potential difference there, the cathode fall, is actually greater than the applied voltage (see Section VI.3). One might not expect a drastic drop in overall discharge voltage to occur at the onset of secondary electron emission, with a yield only on the order of 0.1. However, secondary electron emission, impact ionization, and sheath formation cooperate to achieve this very thing. After breakdown, the potential distribution becomes as illustrated in Figure VI.8.

We now show mathematically how breakdown occurs. We formulate an expression for the current flowing in a Townsend discharge, with secondary electron emission current at the cathode added to the electron current already there (which is produced by photoemission). There is an electron current striking the anode [$I_{be}(d)$], and an ion current striking the cathode [$I_{bi}(0)$]. The electron current at the anode is given by the original saturation current enhanced by electron impact ionization (as derived above) plus the secondary electron current, also enhanced by electron impact ionization in passage from cathode to anode:

$$I_{be}(d) = I_{sat}e^{S_i \cdot n \cdot d} + \gamma I_{bi}(0)e^{S_i \cdot n \cdot d}. \quad (VI.42)$$

At this stage in the analysis, $I_{bi}(0)$ is not yet known. All the ions of the discharge are created by electron impact ionization. Thus, the ion current $I_{bi}(0)$ arriving at the cathode is equal to the *increase* in the original electron saturation current enhanced by impact ionization $\{I_{sat}[e^{S_i \cdot n \cdot d} - 1]\}$ plus the *increase* in electron current due to secondary electron emission at the cathode, as enhanced by impact ionization $\{\gamma I_{bi}(0)[e^{S_i \cdot n \cdot d} - 1]\}$. Equating $I_{bi}(0)$ to the sum of these and solving for $I_{bi}(0)$, we have

$$I_{bi}(0) = \frac{I_{sat}[e^{S_i \cdot n \cdot d} - 1]}{1 - \gamma[e^{S_i \cdot n \cdot d} - 1]}. \quad (VI.43)$$

The total current of the discharge is then evaluated as $I_{be}(d)$:

$$I_{b,total} = I_{be}(d) = \frac{I_{sat}e^{S_i \cdot n \cdot d}}{1 - \gamma[e^{S_i \cdot n \cdot d} - 1]}. \quad (VI.44)$$

Before breakdown, γ is zero, or at least small, and $I_{b,total} \to I_{sat}e^{S_i \cdot n \cdot d}$, its value for the Townsend discharge without secondary electron emission.

When breakdown occurs the denominator goes to zero and $I_{b,total}$ becomes large and indeterminate, and independent of I_{sat}. *Breakdown is a transition to a self-sustaining discharge*, in that the discharge no longer depends on any external ionizing agent. The breakdown voltage is actually set by the condition $\gamma[e^{S_i \cdot n \cdot d} - 1] = 1$. The voltage sensitivity is within γ and S_i—as the applied voltage increases, the average kinetic energy of the electrons increases, and so both γ and S_i increase.

Normal Glow On breakdown, the DC discharge enters the normal-glow regime. The current can assume a range of values (see region D in Fig. VI.14), but the discharge voltage is constant (more accurate models have it decreasing slightly with increasing current). How can the current increase at constant voltage? In the normal-glow regime, secondary electron emission is localized at the edges and any other irregularities of the cathode because the electric field is locally enhanced there. Thus, the current density at the cathode is not uniform, but concentrated at asperities. The current increases as the ion bombardment spreads over the whole cathode surface, but still at almost constant applied voltage.

Essentially all the action that creates and sustains the plasma transpires in or near the cathode sheath. The ions are not generated uniformly throughout the plasma volume now, but by impact ionization principally in this *cathode-fall* region, which is the spatial region where electrons can be accelerated to ionizing energies. *At this current level, the discharge does have the general characteristics of the practical plasma.*

Abnormal Glow When finally the entire cathode area is used, the current can increase further only if γ and/or S_i increase. Thus, for further increases in current, the DC applied voltage must increase. Most DC sputtering processes are actually conducted in this abnormal-glow regime. It is the practical DC sputtering regime because it affords the highest possible sputtering rate under practical conditions. A DC magnetron typically obeys the current–voltage relation

$$I \propto (V_{DC})^n, \tag{VI.45}$$

in the abnormal glow, with n in the range of 5–15 [Rossnagel and Kaufman, 1988].

The Arc The final regime in Figure VI.14 is the arc. The arc occurs as a result of extreme heating of the cathode by ion bombardment, to the point of thermionic emission of electrons.

As Experimental Voltage–Current Characteristic We show in Figure VI.15 an experimental I–V curve for an argon discharge at 1 torr. It is possible to identify the Townsend discharge, break-down, and the normal glow regime, and at least the start of the abnormal-glow regime. The discharge had aluminum electrodes of 45.5 cm² area. The actual apparatus may be seen in Figure VI.18.

The Voltage–Current Characteristic of an RF Discharge

We show in Figure VI.16 some experimental data for a RF discharge from a fine fundamental study by Godyak et al. [1991]. The data are for argon at 1

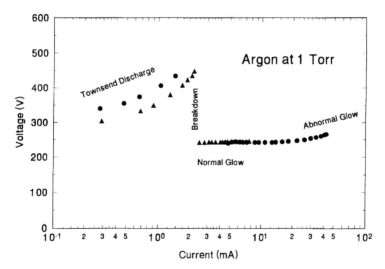

Figure VI.15 Some experimental data showing the voltage–current characteristic of a DC discharge [Hayes, 1996]. The discharge was in argon at a pressure of 1 torr. Two data sets are shown, which did not quite overlap as expected.

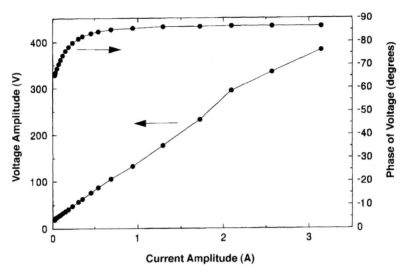

Figure VI.16 Experimental data for an RF discharge in argon at 1 torr [Godyak et al., 1991]. Both the voltage amplitude and its phase angle with respect to the current are shown as functions of the current amplitude.

torr. The driving current was a sine wave of 13.56 MHz frequency. Their cathode and anode were of equal 160 cm^2 areas.

The measured discharge voltage was also a sine wave, phase-shifted with respect to the current. Both the amplitude and phase shift are shown in Figure VI.16 as functions of the current amplitude. The sign of the phase shift indicates that the discharge has a predominantly capacitive impedance, with the phase of the voltage close to $-90°$ except at low current levels. (Consequently, the actual power delivered to the discharge is much less than the simple product of the current and voltage amplitudes.) The voltage amplitude is approximately proportional to that of the current. This behavior is qualitatively similar to that at other pressures in the range 0.003–3 torr. Much more data on the behavior of RF discharges can be found in the original reference by Godyak et al. cited above.

The DC Glow

While "discharge" means an electrical current flowing through a gas, a "glow discharge" is one that glows by light emission from excited gas atoms. Early researchers concerned themselves with the visible aspects of the glow discharge, resulting in the well-known classifications of the various luminous and dark regions reproduced in Figure VI.17. Some or all of these may be seen in a DC discharge in the normal- or abnormal-glow regimes.

Excited neutrals are created in "Townsend processes" similar to impact ionization. Light emission occurs when the excited neutrals return to their

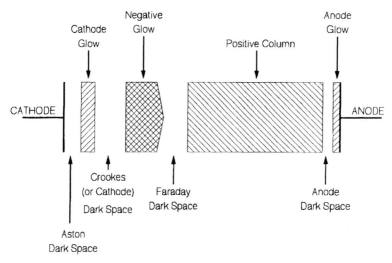

Figure VI.17 The classical luminous and dark regions of a DC glow discharge, based on classifications by early researchers [Holland, 1956; Maissel and Glang, 1970; Brown, 1966; Nasser, 1971, Llewellyn-Jones, 1966; Cobine, 1941; von Engel, 1965].

ground state via radiative transitions. Distinctive colors are produced that are characteristic of the gas specie, for example, the brilliant red of the familiar neon sign. Of course, light emission is irrelevant to the ion bombardment function of a plasma in sputtering.

In the *Aston dark space*, the electrons emitted from the cathode have very low energy, and probably cannot either excite or ionize any gas particles. Ions in this region are moving very rapidly toward the cathode. It is very thin and typically hard to distinguish.

The *cathode glow* "appears to cling to the cathode surface," according to Cobine [1941]. It is a "velvety coating [resulting from] the loss of excitation energy of the positive ions on neutralization" [Brown, 1966].

According to Maissel and Glang [1970], the *Crookes dark space* or *cathode dark space* is a "region of primary interest ... its thickness is approximately the mean distance traveled by an electron from the cathode before it makes an ionizing collision." Two kinds of electron are present: those with very high velocity, having been accelerated across the cathode fall; and those with fairly small velocity, having been liberated by impact ionization of neutral atoms. Some say that this region emits little light because neither can effectively excite neutrals. Others say that it is dark because few electrons are present. Commonly referred to as the "cathode sheath" of processing plasmas, it is typically several times thicker than the actual sheath.

In the *negative-glow* region, secondary electrons from the cathode have been slowed by a collision in the cathode dark space, and now possess a much larger cross section for excitation of neutrals, resulting in enhanced visible emission. Other explanations include: secondary electrons liberated in the Crookes dark space have moved farther away from the cathode and begin to excite gas particles, resulting in visible emission.

In the *Faraday dark space*, the electrons that lost their energy in the region of negative glow are not now able to excite or ionize, and the field here is so low that they do not rapidly gain more kinetic energy.

The *positive column* essentially fills the remainder of the luminous column — this is the primary luminous region in neon signs. It is said that electrons pick up enough kinetic energy from the small electric field there to once again become able to excite neutrals (only a few electronvolts would be required). The positive column begins when the electrons, moving toward the anode, once again attain the excitation energy. Perhaps the main plasma body between the sheaths is not strictly equipotential as we have assumed in earlier discussions in this chapter.

An *anode glow* sometimes appears near the anode. A few ions do bombard the anode, with kinetic energy qV_p. This kinetic energy is sufficient to emit secondary electrons from the anode, which then excite neutrals, producing the anode glow.

We present in Figure VI.18 a photograph of a DC discharge in air between two aluminum electrodes at a pressure of 100 mtorr [Barth, 1991]. Several of

Figure VI.18 A DC glow discharge in argon. The anode (grounded) is on the right and the cathode (supported by a white teflon insulator) on the left. The gray wire is a Langmuir probe whose bare tip enters the positive column of the discharge. The positive column is the largest luminous region, which extends about 75% of the way from the anode toward the cathode. The pressure was 100 mtorr and the voltage applied between cathode and anode was 1 kV. The discharge current density was 0.22 mA/cm^2. See color insert.

the classical regions may be seen in the photograph. No Aston dark space is apparent, but the very thin and bright cathode glow is visible on the cathode (wrapping around its edges), and then the cathode dark space to its right. There is a distinct boundary between the cathode dark space and the negative-glow region. A broad Faraday dark space (about 10% of the electrode spacing) exists between the negative glow and the positive column. The positive column occupies 75% of the space between the cathode and the anode. (Also seen extending into the positive column is a crude Langmuir probe, consisting of an insulated wire with the end bared.) Very close to the anode on the far right is a subtle anode dark space and anode glow.

Changing the pressure or electrode separation also affects the luminous features of the glow discharge. As the pressure is decreased, the negative glow and the Faraday dark space expand at the expense of the positive column. The positive column can also be consumed by moving the electrodes sufficiently close to each other.

Gu and Lieberman [1988] measured the axial distribution of light intensity for a DC planar magnetron discharge. They found a peak in the luminosity near the cathode, identifying its position with the outer edge of the sheath, as determined by space-charge-limited ion current. Increasing magnetic field increases the peak light intensity and moves it closer to the cathode.

The RF Glow

Studies of the structure of the RF glow, even empirical ones, seem to be sparse. Koenig and Maissel [1970] describe the luminous aspects of an RF sputtering discharge rather simply — there are "dark spaces" at the two electrodes that separate them from the "glow space."

We show in Figure VI.19 a photograph of a RF glow discharge in air at a pressure of 3 torr [Knipp and Wang, 1991]. The structures of the glow regions surrounding each electrode differ slightly — the grounded anode has two luminous regions that are separated by a dark space; the cathode has only one.

Exceptions to the Above

There are myriad ways in which real sputtering discharges differ in detail from the simplified models of this chapter. Examples are as follows:

1. The plasma density is not spatially uniform. In sputtering discharges it is likely to be significantly more intense near the cathode, and in fact the glow is often observed to be brighter in that region. With magnetrons, the plasma density is by design the most intense in the electron trapping region created by the magnetic field.
2. The charged particle densities and their temperatures are most certainly *not fixed*. They increase as the discharge voltages are raised, or as the power delivered to the plasma is increased. And they are not

Figure VI.19 An RF glow discharge in air at a pressure of 3 torr [Knipp and Wang, 1991]. The anode (grounded) is on the left and the floating cathode (mounted on a white Teflon insulator) is on the right. It was driven with a 13.56-MHz generator delivering 50 W of power to the discharge. See color insert.

spatially uniform within the body of the plasma; for example, both charged particle densities are reduced at the sheath edge.
3. Non-Maxwellian electron energy distributions are common. For an RF discharge in argon, Turner et al. [1993] reported two findings: At a pressure of 0.2 torr, the electron distribution was fairly close to Maxwellian with a temperature of 34,800 K, while at 0.02 torr the distribution was "bi-Maxwellian." This latter distribution consisted of two populations of electrons having temperatures of 11,600 and 34,800 K. Chapman [1980] said that the distribution is found to be "more Maxwellian than it should be," based on detailed analyses of pertinent plasma processes.
4. If a Langmuir probe draws a large electron current, it may significantly perturb the plasma.
5. Cathode heating and/or the flux of sputtered particles can reduce the local density of the sputtering gas quite significantly at the cathode surface [Rossnagel, 1989]. Hoffman [1985] has described a "sputtering wind" of gas particles in a circulation pattern within the deposition chamber.

Thus, the reader should be prepared to encounter, in practice, many deviations from the simplified models of this chapter. It is hoped, however, that the broad physical picture of sputtering discharges which we have presented is generally applicable.

VI.6 MATHEMATICAL SYMBOLS, CONSTANTS, AND THEIR UNITS

SI units are given first, followed by other units in widespread use.

a	Acceleration (m/s^2)
d	Electrode separation (m)
j	Current density (A/m^2)
k	Boltzmann's constant (1.38×10^{-23} J/K; 8.62×10^{-5} eV/K)
m	Particle mass (kg; amu = 1.66×10^{-27} kg)
n	Neutral particle density (m^{-3}); exponent (dimensionless)
n^+	Ion density (m^{-3})
n^-	Electron density (m^{-3})
p_{ion}	Power density delivered to target (W/m^2)
q	Magnitude of the electronic charge (1.60×10^{-19} C)
t	Time (s)
v	Velocity (m/s)
v_a	Applied voltage (V)
v_{av}	Average speed (m/s)
x	Position coordinate, usually distance from cathode (m)
z	Impingement rate (m$^{-2} \cdot$ s^{-1})
A	Area of an electrode
B	Magnetic flux density (tesla; gauss = 10^{-4} T)
C	(Blocking) capacitance (F)
E	Energy (J; eV = 1.60×10^{-19} J)
E_i	Ionization potential (J; eV = 1.60×10^{-19} J)
I	Current (A)
KE_{av}	Average kinetic energy (J; eV = 1.60×10^{-19} J)
L_d	Debye thickness (m)
L	Sheath thickness (m)
Q	Electrical charge (C)
S_i	Ionization cross section (m^2)
T	Neutral particle characteristic temperature (K)
T^+	Ion characteristic temperature (K)
T^-	Electron characteristic temperature (K)
V	Electrostatic potential (V)
V_f	Floating potential (V)
V_p	Plasma potential (V)

W	Work function (J; eV = 1.60×10^{-19} J)
γ	Secondary electron emission coefficient (dimensionless)
ε_0	Permittivity of vacuum (8.85×10^{-12} F · m^{-1})
λ	Mean free path (m)
ρ	Volume charge density (C/m^3); gyro radius (m)
ω	Angular frequency (rad/s)

REFERENCES

Anderson, G. S., Mayer, W. N., and Wehner, G. K., 1962, "Sputtering of Dielectrics by High-Frequency Fields," *J. Appl. Phys.* **33**(10), 2991.

Barth, K., 1991, for constructing and operating the DC discharge apparatus.

Brown, S. C., 1966, *Introduction to Electrical Discharges in Gases*, Wiley, New York.

Butler, H. S., 1961, *Plasma Sheath Formation by RF Fields*, Stanford Univ. W. W. Hansen Laboratories of Physics, Microwave Lab. Report 820 (AD 260 088).

Chapman, B., 1980, *Glow Discharge Processes*, Wiley-Interscience, New York.

Chen, F. F., 1965, "Electric Probes," in R. H. Huddlestone and S. L. Leonard, eds., *Plasma Diagnostic Techniques*, Academic press, New York, Chapter 4.

Clements, R. M., 1978, "Plasma Diagnostics with Electric Probes," *J. Vac. Sci. Technol.* **15**(2), 193.

Cobine, J. D., 1941, *Gaseous Conductors*, McGraw-Hill, New York.

Godyak, V. A., Piejak, R. B., and Alexandrovich, B. M., 1991, "Electrical Characteristics of Parallel-Plate RF Discharges in Argon," *IEEE Trans. Plasma Sci.* **19**(4), 660.

Gu, L., and Lieberman, M. A., 1988, "Axial Distribution of Optical Emission in a Planar Magnetron Discharge," *J. Vac. Sci. Technol.* **A6**(5), 2960.

Hayes, E., 1996, private communication.

Holland, L., 1956, *Vacuum Deposition of Thin Films*, Wiley, New York.

Hoffman, D. W., 1985, "A Sputtering Wind," *J. Vac. Sci. Technol.* **A3**(3), 561.

Kaufman, H. R., and Robinson, R. S., 1987, *Operation of Broad-Beam Sources*, Commonwealth Scientific Corp., Alexandria, VA, p. 141.

Kaufman, H. R., and Rossnagel, S. M., 1988, "Analysis of Area-Ratio Effect for Radio-Frequency Diode," *J. Vac. Sci. Technol.* **A6**(4), 2572.

Knipp, L. J., and Wang, N., 1991, for constructing and operating the RF discharge apparatus.

Koenig, H. R., and Maissel, L. I., 1970, "Application of RF Discharges to Sputtering," *IBM J. Res. Devel.* **14**, 168.

Langmuir, I., 1929, "The Interaction of Electron and Positive Ion Space Charges in Cathode Sheaths," *Phys. Rev.* **33**(June) 954.

Lieberman, M. A., and Lichtenberg, A. J., 1994, *Principles of Plasma Discharges and Materials Processing*, Wiley, New York.

Llewellyn-Jones, F., 1966, *The Glow Discharge*, Methuen, London.

Maissel, L. I., and Glang, R., eds., 1970, *Handbook of Thin Film Technology*, McGraw-Hill, New York.

Manos, D. M., and Flamm, D. L., eds., 1989, *Plasma Etching — an Introduction*, Academic Press, New York.

Nasser, E., 1971, *Fundamentals of Gaseous Ionization and Plasma Electronics*, Wiley, New York.

Raizer, Y. P., 1991, *Gas Discharge Physics*, Springer, New York.

Rossnagel, S. M., 1989, "Magnetron Plasma Deposition Processes," *Thin Solid Films* **171**, 125.

Rossnagel, S. M., and Kaufman, H. F., 1988, "Current-Voltage Relations in Magnetrons," *J. Vac. Sci. Technol.* **A6**(2), 223.

Ruzic, D. N., 1994, *Electric Probes for Low Temperature Plasmas*, AVS Monograph Series, The American Vacuum Society, New York.

Turner, M. M., Doyle, R. A., and Hopkins, M. B., 1993, "Measured and Simulated Electron Energy Distribution Functions in a Low-Pressure Radio Frequency Discharge in Argon," *Appl. Phys. Lett.* **62**(25), 3247.

von Engel, A., 1965, *Ionized Gases*, Oxford Univ. Press, London.

Waits, R. K., 1978, "Planar Magnetron Sputtering," *J. Vac. Sci. Technol.* **15**(2), 179.

Walton, A. J., 1989, *Three Phases of Matter*, Oxford Univ. Press, Oxford.

VII

SPUTTERING

> Sputtering resembles most closely a three-dimensional billiard game with atoms.
>
> — Wehner and Anderson [1970]

VII.1 GENERAL CHARACTERISTICS AND BACKGROUND

Definition of Sputtering

The sputtering event, when utilizing the ions from a gaseous discharge, is portrayed schematically in Figure VII.1. It consists of

1. Acceleration of an ion across the cathode sheath
2. Penetration of the target, resulting in a series of atomic collisions
3. Backward ejection of one or more recoiling target atoms

Our main concern in this chapter is knowing the *sputter yield*, the average number of ejected target atoms per incident ion, and understanding this fundamental sputtering event.

The maximum possible projectile energy corresponds to the cathode fall, the difference between the plasma potential and that of the cathode:

$$E = q[V_p - v_{cat}] \qquad \text{(VII.1)}$$

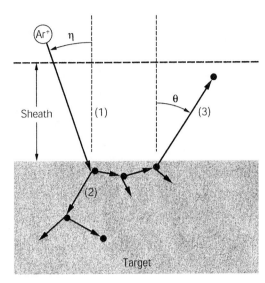

Figure VII.1 The sputtering event with a gaseous discharge consists of (1) acceleration of an ion across the cathode sheath, (2) an atomic collision cascade within the target, and (3) ejection of a target atom; η is the angle of incidence and θ, the emission angle.

(but is usually less than this because of charge exchange in the sheath), where q is the magnitude of the electronic charge.

A broad definition of sputtering is "the erosion of material surfaces by particle impact" [Sigmund, 1993]. Sigmund clarifies: "Sputtering is a phenomenon on the atomic scale. By this is meant that one can identify an individual sputter event, i.e., the emission of a number of atoms or molecules from a material surface initiated by a single bombarding particle" [Sigmund, 1993]. Ions are the projectiles employed for thin film deposition by sputtering, but photons, electrons, and neutral particles are known to cause sputtering, as well. It has been an active cosmological process for eons, eroding the surface of the moon and other celestial bodies [Wehner et al., 1963; Tombrello, 1993].

Behrisch [1981], in his introduction and overview to physical sputtering of single-element solids, observes that "The particles emitted are predominantly neutral atoms in the ground state." However, the target may emit ions and clusters (of surprising size) in the sputtered flux [Urbassek and Hofer, 1993].

Sputtering has a number of useful applications—above all, the ejected particles can be condensed on a substrate as a thin film. Low-energy sputtering has found application in the milling, etching, thinning, and polishing of microstructures [Spencer and Schmidt, 1971]. Also, the sputtered particles can be contaminants removed in sputter cleaning, can be mass-analyzed for a compositional determination of the sputtered object (secondary ion mass spectrometry), or can themselves be ionized to make a unique ion source of the target atoms.

GENERAL CHARACTERISTICS AND BACKGROUND   201

In some instances, sputter deposition offers significant advantages over evaporation. A representative kinetic energy of sputtered particles is 3–5 eV, orders of magnitude above that of typical evaporated particles. This excess kinetic energy imparts a higher surface mobility to condensing particles, a factor in obtaining smooth and conformal film morphologies. Sputtering sources are typically of relatively large area, which aids film thickness uniformity and conformality of coverage. Finally, under similar conditions the sputter yields of metals rarely differ by more than a factor of 10; in evaporation, however, the vapor pressures of common metals can differ by many orders of magnitude at the same source temperature.

The Mechanisms of Sputtering

The authorities like to call the process of Figure VII.1 "collisional sputtering." It is also *backward*, rather than *forward*, sputtering. (Forward emission can occur in the sputtering of very thin foils.) Sigmund [1981] described the mechanism of collisional sputtering in this way:

> The elementary event is an atomic collision cascade wherein the incident ion knocks atoms off their equilibrium sites in the target, thus causing these atoms to move through the material, to undergo further collisions, and eventually to cause the ejection of atoms through the target surface. The general consensus is that this sputtering mechanism is the most universal one, being active for ionic bombardment of all types of solids at appropriate ion energies.

Figure VII.2 shows diagrams by Sigmund [1981] of the collision cascade in different energy regimes. In the *linear cascade regime*, the sputter yield is found, both experimentally and theoretically, to be proportional to the first power of the *deposited energy density* at the surface. The density of recoils is sufficiently low that most collisions involve one moving and one stationary particle. Linear, collisional, sputtering is believed to describe what occurs with the most common combinations of projectile, energy, and target. Kelly [1984] stated that

> In general, slow collisional [i.e. linear cascade; see discussion of Kelly's timescale, below] sputtering should be regarded as the dominant mechanism with many (though not all) systems.
>
> Because of its dominant role, slow collisional sputtering establishes the baseline according to which all sputtering experiments tend to be compared. If excessive yields are found ... then mechanisms other than slow collisional are inferred to be important.

For film deposition, the alternative to collisional sputtering from a linear cascade is *thermal sputtering*. At higher recoil densities, when the majority of atoms within the cascade volume are simultaneously in motion, the cascade is said to have entered the *spike regime*. The atoms within the spike temporarily

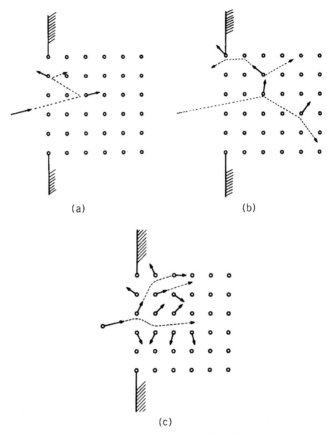

Figure VII.2 Sputtering, according to Sigmund [1981], has three energy regimes: (*a*) the low-energy, *single-knockon* regime; (*b*) the *linear cascade* regime, where "a recoil cascade occurs but collisions between moving atoms are infrequent"; and (*c*) the *spike* regime, a high-energy regime where "the majority of atoms within the 'spike volume' are in motion." (Reprinted with permission from P. Sigmund, in R. Behrish, ed., *Sputtering by Particle Bombardment I*. Copyright 1981 by Springer-Verlag.)

resemble a vapor and the ejection mechanism is similar to their vaporization — a thermal, rather than a collisional, mechanism [Sigmund and Claussen, 1981]. For such a spike, theory predicts and experiment shows that the sputter yield rises superlinearly with deposited energy, and thus such a cascade has been termed *nonlinear*. When the spike occurs, the yield often is surprisingly high, even termed *excess yield*. Perusing Andersen and Bay [1981] will show that these cases are not uncommon; sputtering of the noble- and near-noble-metal targets are prime examples.

When do the nonlinear cascade, and the associated excess sputter yield, occur? Thompson [1981] suggested that the criterion is whether the nuclear stopping power "... *is large enough to provide the binding energy to all atoms neighbouring the track* ..." This criterion is most easily met with a projectile-target combination having a high projectile nuclear stopping power (typically a heavy projectile; see Part 2), or with a target having a low binding energy, but may also be satisfied for lighter projectiles and normal targets at sufficiently high projectile energies. The above criterion effectively becomes whether the sputter yield exceeds some critical value. Thompson [1981] suggested that the critical yield is about 20. Sigmund and Claussen [1981] put the value of Y_{crit} at 10, while Zalm and Beckers [1984] suggested that its value is 7. Mahan and Vantomme [1999] suggest that excess yield becomes just discernable whenever the yield exceeds ~ 1.

A Brief History of Sputtering Theory and Simulation

One of the earliest theories was the momentum transfer theory of Lamar and Compton [1934], which ascribes the sputtering to a mechanical "bumping off" of surface atoms by ions which have penetrated below the surface and rebounded." It has led some to call sputtering a "momentum transfer" process, but it is not — it is fundamentally an *energy* transfer process in elastic collisions during which momentum must be conserved (as we will see in Section VII.3). Next came thermal hot-spot or "spike" theories [e.g., Seitz and Koehler, 1956; Thompson and Nelson, 1962]. In this picture, ejection of target atoms occurs by *evaporation from a transient microscopic hot spot* created by an energetic projectile. The lack of a general dependence of yield on target temperature [Bohdansky et al., 1987] is strong evidence against a universal thermal mechanism. However, the occurrence of spikes has been confirmed by the disappearance of lattice-correlated effects in yield (see Section VII.3), and by the observation of high yields that cannot be explained by linear collisional sputtering models. The collisional theories of Thompson [1968] and Sigmund [1969] are relatively recent developments. The energy spectra of emitted particles constitute one of the main confirmations of the collisional picture. The ideas of Sigmund have become widely accepted as standard sputtering theory [Falcone et al., 1987]. The yield expressions of several collisional theories are presented in Appendix C.

Kelly [1984] unified our general understanding of the several thermal and collisional mechanisms of sputtering by placing them on a timescale passing from *prompt collisional* (ejection of the first recoil) through *slow collisional* (ejection after a cascade develops) and *prompt thermal*, to *slow thermal*. The linear cascade, the main subject of this chapter, is "slow collisional." The spike is the prompt thermal regime in Kelly's framework.

Considerable effort has been devoted to developing semiempirical sputter yield expressions, such as those of Bohdansky [Bohdansky et al., 1980; Bohdansky, 1984], Matsunami et al. [1984], and García-Rosales et al. [1994]. They are based on expressions for the nuclear energy loss cross section as a function of reduced projectile energy (see discussion in Section VII.3). These semiempirical expressions have been quite useful as sources for yield predictions and estimates; that of Matsunami et al. is used extensively in this chapter to represent experimental yield values.

Much insight into the mechanisms of sputtering has come from computer simulations, where the individual events are followed by solving the equations of motion numerically, and many compromises associated with averaging are not made. (See Andersen [1987] and Robinson [1993] for very informative reviews of these simulations.) Computer simulations may be divided into two broad classes: *molecular dynamics simulations*, where all the interactions among projectile and target atoms are calculated simultaneously; and *binary collision approximation simulations*, where only one binary collision is handled at a time, all the other particles of the system being temporarily frozen in place.

TRIM (transport of ions in matter) has become the most widely applied computer model. It is a binary collision approximation simulation of lattice damage and sputter yield that was developed for personal computers [Ziegler, 1992]. Figure VII.3 shows the results of a TRIM simulation of ion trajectories and lattice damage in an aluminum target after bombardment by 500 normally incident 1-keV argon ions. The average ion trajectory is serpentine, and terminates at a depth of 29 Å. This simulation gave a sputter yield of 0.56.

In the original publication of his model, Sigmund [1969] said that it "cannot in any respect compete with a careful computer simulation of the ejection

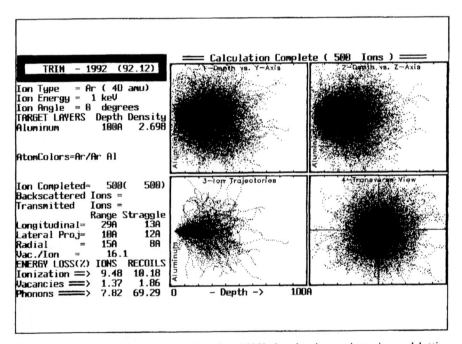

Figure VII.3 A TRIM simulation [Ziegler, 1992] showing ion trajectories and lattice damage after bombardment of aluminum by 500 normally incident 1-keV argon ions. (Used with permission from J. F. Ziegler, "The Transport of Ions in Matter," Version 92.xx. Copyright 1992 by the International Business Machines Corp.)

process." This position is too modest—an analytic model can compete with the best computer simulations in two senses. It can provide quick estimates of sputter yield that are of competitive (to the simulations) accuracy, and a well-constructed yield expression can convey important understandings of the sputtering mechanism. In Section VII.4, a simplified collisional model [Mahan and Vantomme, 1997], which was developed to meet these two objectives for conditions typical of the sputter deposition of thin films, is presented. It predicts the basic trends in sputter yield data and is quantitatively competitive with standard (Sigmund) sputtering theory, and with TRIM, in cases where the linear cascade is operative.

"Sputtering" is a uniquely unglamorous name—Sigmund [1993] stated that "Historically, the term came up early in this century to illustrate the elementary event, which was thought to resemble what happens when a stone falls on a water surface." There is reason to believe that the original English name was "spluttering."

Sources of Sputter Yield Data

Over the years, sputter yield data have been published in hundreds of scientific journal articles. Sputtering is a developing science and, as such, the data do not yet exist in a highly refined form equivalent to critically evaluated thermochemical data. In their review, Andersen and Bay [1981] stated that "a vast amount of data have been published... these early data may now be discarded as obsolete.... Over the years, sputtering-yield measurements have been marred by bad reproducibility."

In recent years, errors have been corrected and new phenomena have been discovered; the situation prompted Harrison [1983] to write that the effort of sputtering theoreticians is "aimed at a moving target." The most recent comprehensive compilation of this data for elemental targets is that of Matsunami et al. [1984]. They prepared graphs showing yield as a function of energy (several are reproduced in Section VII.2). Their database consisted of all the published sputter yield measurements that were available in early 1983 (little has been published since that time). The data were screened using criteria that increased their reliability.

In addition to the actual data points that they obtained from their literature search, Matsunami et al. showed in each graph an empirical curve for $Y(E)$, the yield as a function of projectile energy. The curve is calculated using three empirical parameters in a formula (see Appendix A) that resembles Sigmund's high-energy yield expression (see Appendix C). These parameters were obtained as best fits to all the screened yield data, encompassing many projectile–target combinations.

Example Calculate the sputter yield of 1-keV argon ions striking an aluminum target with the empirical formula of Matsunami et al.

This calculation is worked out in Appendix A. All the necessary equations are summarized there, together with the nonobvious constants. ∎

The empirical curves of Matsunami et al. [1984] are an attempt to address the problem of experimental error in the large body of yield data. Their empirical fitting procedure is based on the beliefs that the full set of data for all available projectile–target combinations is more reliable than any specific subset, and that the variations in yield between one combination of projectile, target, and energy, and a neighboring combination, are modest and smooth. Thus, the behavior of one elemental target can be inferred from that of its neighbors. In assessing the success of the empirical curves, Matsunami et al. stated that "Agreement between the solid curve and the data points for each ion-target combination is generally satisfactory.... For the most part ... we think that the sputtering yield for any ion-target combination evaluated ... is accurate to within ±20% near the maximum of the sputtering yield while deviating somewhat more at lower and higher energies."

The empirical curves are probably the truest estimates of sputter yield in the film deposition regime that one has today. We will use the Matsunami et al. empirical values (whenever possible) to show various data trends in Section VII.3. In addition, it is possible to utilize their empirical approach to predict yields for projectile–target combinations that have no experimental yield data, thus providing at least an educated guess of their yields.

To conclude this introduction, for this author the most helpful sputtering reviews have been those of Sigmund [1981], Thompson [1981], and Kelly [1984]. An older, landmark review may be found within Carter and Colligon [1968]. The single most influential paper in the sputtering field is Sigmund's [1969] publication of his theory. The most widely quoted single sources of data are probably Almén and Bruce [1961] and Laegreid and Wehner [1961]. Wide-ranging symposia were held in 1986 [Betz et al., 1987] and 1992 [Sigmund, 1993]. There is another comprehensive survey of yield data that should be mentioned, that of Andersen and Bay [1981]. This was based on a systematic search of the literature up to July 1977, and includes an authoritative critical analysis of the data.

VII.2 TRENDS IN SPUTTER YIELD DATA

The sputter yield in the case of amorphous or polycrystalline (fine-grained and randomly oriented) targets exhibits interesting and systematic dependences on (1) the bombarding projectile's kinetic energy, (2) the surface binding energy of the target, (3) the projectile's identity, and (4) the projectile's angle of incidence. The general energy (5) and angular distributions (6) of sputtered particles are well known. Single-crystal targets exhibit a sputter yield directional distribution (7) that can be strongly correlated with certain crystallographic directions. Finally, the experimental yield varies with the amount of target conditioning (8) by previous sputtering. Thus, the experimental characterization of sputtering includes a very broad group of phenomena. Harrison [1983] was "reminded of the fable of the blind man and the elephant."

We present below these trends for the sputtering of pure elements. Some of the plots that follow are a mixture of linear and nonlinear cascade yields. Still, the data selected are the best available (and sometimes only) examples of the broad trends of sputter yield for conditions typical of those used in film deposition.

Projectile Energy Dependence

Data for the sputtering of metal targets is represented by aluminum and tungsten in Figures VII.4 and VII.5, from the two broad data compilations [Andersen and Bay, 1981; Matsunami et al., 1984]. With respect to the incident ion's kinetic energy, there is a sharp yield *threshold*, followed by a rapid increase in yield with energy. There is a *maximum* at higher energies, and an eventual *decline* in yield at very high energy. For metallic targets, measured threshold energies are typically 20–40 eV. The yield maximum occurs at or above 10 keV in most situations. At 100 keV, the yield typically has diminished significantly.

The threshold exists because there is a minimum energy that must be provided for a particle to escape from the target, the *surface binding energy*. In Section VII.4 we show that the general shape of the $Y(E)$ curve is due to the nuclear stopping power of the projectile. The eventual decline is due to a decrease in the *nuclear energy loss* probability together with the onset of *electronic energy* loss, a projectile stopping mechanism that does not normally generate recoils. With the electronic energy loss mechanism, the projectile energy goes into stripping electrons from the target atoms, instead of dislodging the atoms from their lattice sites.

The data from Andersen and Bay in Figures VII.4a and VII.5a also contain theoretical yield curves of the Sigmund theory, which generally overestimates this data. Aluminum and tungsten usually exhibit yields typical of the linear cascade. Sigmund's model fits best the cases of *excess* yield [Mahan and Vantomme, 1997; Steinbrüchel, 1985]. The data from Matsunami in Figures VII.4b and VII.5b are accompanied by their own fitted empirical curves.

Many have proposed empirical or semiempirical formulas for the energy dependence of yield. Bohdansky et al. [1980] observed that the yield curves for many ion–target combinations are of nearly the same *shape* from near-threshold up to projectile energies of a few kiloelectronvolts, and they may be made to fall on a master empirical curve with proper normalization. Their empirical expression is

$$Y = (6.4 \times 10^{-3})m_r\gamma^{5/3}E^{0.25}\left(1 - \frac{E_{th}}{E}\right)^{3.5}, \quad \text{(VII.2)}$$

where γ is the energy transfer mass factor to be presented in Section VII.3, m_r is the recoil mass in amu, and E is the initial energy of the projectile in electronvolts. The normalizing factor for the projectile energy is the threshold

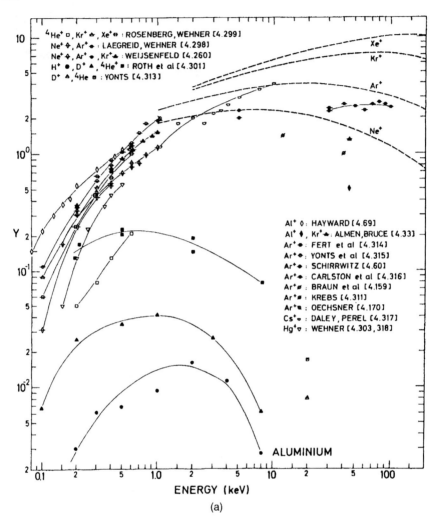

Figure VII.4 (a) The projectile energy dependence of yield for aluminum targets struck by various projectiles (data: symbols and solid lines). Calculated yields (dashed lines) for four noble gas projectiles are shown, according to the Sigmund theory. (Reprinted with permission from H. H. Andersen and H. L. Bay, "Sputtering Yield Measurements," in R. Behrisch, ed., *Sputtering by Particle Bombardment I*. Copyright 1981 by Springer-Verlag.) (b) Yield data for aluminum targets. Best-fit empirical curves are also shown. (Reprinted with permission from N. Matsunami, et al., "Energy Dependence of the Ion-induced Sputtering Yields of Monatomic Solids," *Atom. Data Nucl. Data Tables* **31**. Copyright 1984 by Academic Press.)

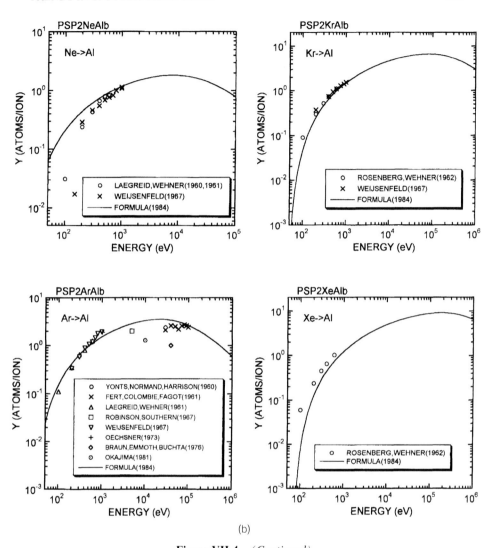

Figure VII.4 *(Continued)*

value, E_{th}, in electronvolts. This expression has been used to extract threshold energies from $Y(E)$ curves. Bohdansky et al.'s [1980] empirical expressions for the threshold energies have found widespread use:

$$E_{th} = \begin{cases} \dfrac{U_{sb}}{\gamma(1-\gamma)} & \left(\dfrac{m_p}{m_r} < 0.3\right) \\ 8U_{sb}\left(\dfrac{m_p}{m_r}\right)^{2/5} & \left(\dfrac{m_p}{m_r} > 0.3\right), \end{cases} \quad \text{(VII.3)}$$

where U_{sb} is the surface binding energy (see Section VII.3).

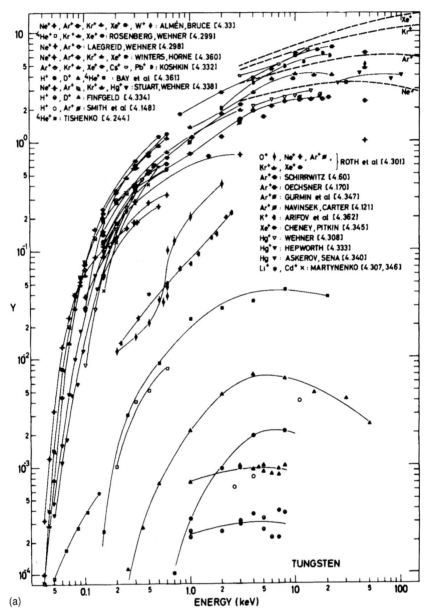

Figure VII.5 (*a*) The projectile energy-dependence of yield for tungsten targets struck by various projectiles (data: symbols and solid lines). Calculated yields (dashed lines) for four noble-gas projectiles are shown, according to the Sigmund theory. (Reprinted with permission from H. H. Andersen and H. L. Bay, "Sputtering Yield Measurements," in R. Behrisch, ed., *Sputtering by Particle Bombardment I*. Copyright 1981 by Springer-Verlag.) (*b*) Yield data for tungsten targets. Best-fit empirical curves are also shown. (Reprinted with permission from N. Matsunami, et al., "Energy Dependence of the Ion-induced Sputtering Yields of Monatomic Solids," *Atom. Data Nucl. Data Tables* **31**. Copyright 1984 by Academic Press.)

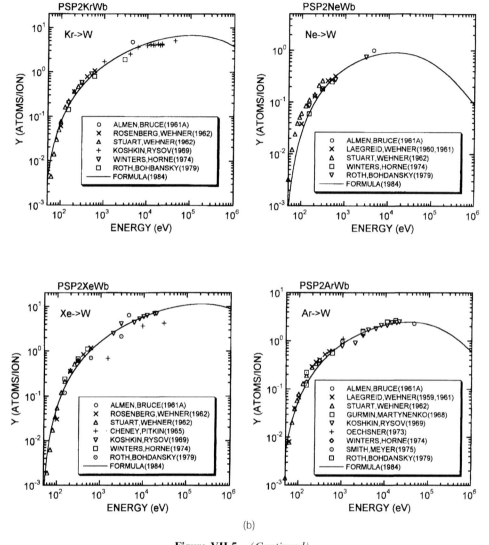

Figure VII.5 *(Continued)*

Example Estimate the threshold projectile energy for argon sputtering of gold, using Bohdansky et al.'s formula.

We need the following data: $m_p = 39.9$ amu, $m_r = 197$ amu, and $U_{sb} = 3.8$ eV (from Appendix B). Then $m_p/m_r = 0.203$, so the first expression is needed. We must first calculate the energy transfer mass factor:

$$\gamma = \frac{4 m_p m_r}{(m_p + m_r)^2} = \frac{4 \times 39.9 \times 197}{(39.9 + 197)^2} = 0.56.$$

Then

$$E_{th} = \frac{3.8\,\text{eV}}{0.56(1-0.56)} = 15.4\,\text{eV}.$$

Now estimate the sputter yield at 500 eV using the formula of Bohdansky et al.:

$$Y = 6.4 \times 10^{-3} \times 197 \times 0.56^{5/3} \times 500^{0.25} \times \left(1 - \frac{15.4}{500}\right)^{3.5} = 2.03. \quad \blacksquare$$

Steinbrüchel [1985] proposed a semiempirical formula for low energy yield in the range $0.1\,\text{keV} \le E \le 1\,\text{keV}$:

$$Y(E) = \frac{5.2}{U_{sb}} \cdot \frac{Z_r}{(Z_r^{2/3} + Z_p^{2/3})^{3/4}} \cdot \left(\frac{Z_p}{Z_r + Z_p}\right)^{0.67} \cdot E^{1/2}, \quad \text{(VII.4)}$$

where Z_p and Z_r are the atomic numbers of projectile and recoil, respectively; U_{sb} is assumed to be in electronvolts and E in kiloelectronvolts. The $E^{1/2}$ factor approximates the projectile energy dependence of the nuclear energy loss cross section, but the representative data in Figures VII.4 and VII.5 clearly show that while the exponent of E is around $\frac{1}{2}$, it decreases as E increases in this range. Steinbrüchel reported that "for $Y_{exp} > 1.3$ the agreement between theory Eq. (VII.4) and experiment is within $\sim 25\%$." The formula more seriously underestimates yields below 1.3.

For simply estimating the yield at other energies given the 1-keV yield, Mahan and Vantomme [1999] proposed

$$Y(E) = Y(1\,\text{keV}) \cdot \left(\frac{E}{1\,\text{keV}}\right)^{0.5} \quad \text{(VII.5)}$$

in the energy range $0.5\,\text{keV} \le E \le 2\,\text{keV}$. This energy dependence is identical to that of Steinbrüchel and is an approximation to the Matsunami et al. empirical formula, which is valid when the projectile energy is much greater than the threshold energy and when the reduced projectile energy is much less than one.

Dependence on Surface Binding Energy

With the projectile and its energy held constant, the yield increases with increasing volatility of the target material. Carbon is an outstanding (counter) example of this trend, having in the form of graphite the lowest vapor pressure and the lowest sputter yield of all the elements.

All the available yield data from Matsunami et al. [1984] for 1-keV argon ion sputtering are shown in Figure VII.6a [Mahan and Vantomme, 1999]. Why does the yield vary with target atomic number as it does? A strong correlation between yield and the binding energy of the target was identified many years ago [Carter and Colligon, 1968]. Kelly [1984] observed that "a

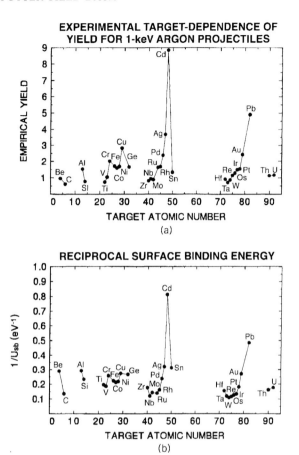

Figure VII.6 (*a*) Experimental 1-keV Ar$^+$ yields as a function of atomic number of the target. These are empirical values taken from Matsunami et al. [1984]. (*b*) Reciprocal of heat of sublimation as a function of atomic number of the target. The similarity between the two plots has long been recognized. (Reprinted with permission from J. Mahan and A. Vantomme, "Trends in Sputter Yield Data in the Film Deposition Regime," 1999, unpublished.)

highly characteristic aspect of slow collisional [i.e., linear cascade] sputtering is that, when values of Y are plotted against Z_r, there is a pronounced periodicity. The periodicity is that of the heat of sublimation."

Figure VII.6*b* shows the reciprocals of surface binding energy for the same set of targets [Mahan and Vantomme, 1999]. The similarity between the two plots is striking and one is tempted to suggest that $Y \propto 1/U_{sb}$. Indeed, the surface binding energy appears in the denominator of the yield expressions of several analytic models (see Appendix C), including the simplified collisional model presented in Section VII.4.

However, the dependence is not simply reciprocal [Mahan and Vantomme, 1999]. One may describe the relationship between these yields and surface binding energy with the following:

$$Y(1\text{-keV argon projectiles}) \propto \frac{1}{U_{sb}^{1.3}} \quad \text{(all empirical data)}. \qquad \text{(VII.6)}$$

This result comes from a nonlinear regression fit to all the empirical yields in Figure VII.7a. This strong U_{sb}-dependence is due to the fact that the nonlinear cascade is present in some of the experimental yield data. The nonlinear cascade accentuates the relative differences in yield in Figures VII.6a and VII.7a by enhancing the yields of materials having a low heat of sublimation. (See the discussion of Fig. VII.7b in Section VII.4.)

Dependence on Choice of Projectile

Argon is by far the most widely used projectile for sputter deposition of thin films. However, other noble gases frequently offer a higher sputter yield. We show in Figure VII.8a a qualitative contour plot of all the available 1-keV yields in the Matsunami et al. compilation (empirical yield values are plotted), as a function of both projectile atomic number and target atomic number. The ~90 data points underlying this plot have been smoothed and fitted with a three-dimensional surface. For heavy targets, the experimental yield increases with projectile in the order Ne → Ar → Kr → Xe. However, for light targets, the yield exhibits a maximum at an intermediate projectile mass, usually krypton. This behavior can be understood as being due principally to variations in projected range of the projectile [Mahan and Vantomme, 1999]. A small projected range leads to deposition of more of the projectile's energy near the surface, and thus to a higher sputter yield.

In Figure VII.9a we show all the available empirical data from Matsunami et al. for the 1-keV sputtering of column IVB targets [Mahan and Vantomme, 1999]. For a silicon target, the yield is maximized for the argon projectile. For carbon, the limited amount of data also is greatest for argon. For germanium and tin, however, the yield rises all the way to xenon.

Effect of Angle of Incidence

The yield *increases for off-normal incidence* — up to a point, and then it decreases for cases of glancing incidence. Some representative experimental data are given in Figure VII.10 [Yamamura et al., 1983].

It is found experimentally that the near-normal dependence is $1/\cos\eta$ in many instances, where η is the angle of incidence as measured from the surface normal (see Fig. VII.1). The increase may be understood to be due to an increase in the probability of escape of the recoils for off-normal incidence: The relationship between yield and a representative depth for the recoil distribution is $Y \sim 1/\text{depth}$, while a representative depth may be calculated from

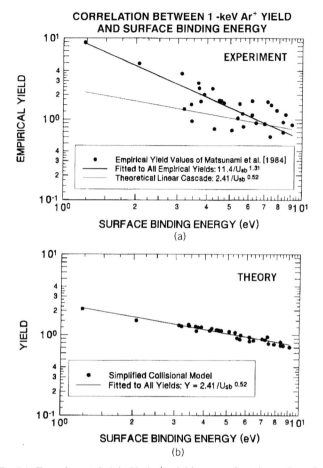

Figure VII.7 (*a*) Experimental 1-keV Ar^+ yields as a function of surface binding energy. (All the available empirical data from Matsunami et al. [1984] are shown.) A line is fitted to *all* the empirical yield data by nonlinear regression, and a second line representing the theoretical behavior of the linear cascade (obtained in *b*) is also shown. (*b*) Theoretical yield values calculated with the simplified collisional model (presented in Section VII.4) are plotted as a function of surface binding energy. The line is a nonlinear regression fit to all the theoretical yields. (Reprinted with permission from J. Mahan and A. Vantomme, "Trends in Sputter Yield Data in the Film Deposition Regime," 1999, unpublished.)

depth $= R_p \cos \eta$, where R_p is the projected range of the projectile within the target. (All these relationships will become apparent when the simplified collisional model is developed in Section VII.4.) Thus, the near-normal $1/\cos \eta$ dependence of yield is rationalized.

In the data of Figure VII.10, there is a departure from this ideal behavior at high angles of incidence. The decrease in yield above $\sim 60°$ may be explained as

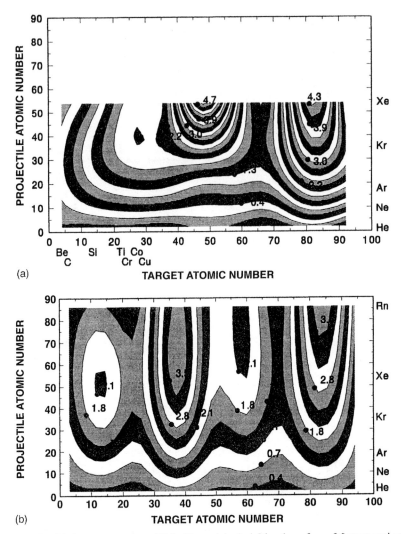

Figure VII.8 (*a*) A contour plot of 1-keV empirical yield values from Matsunami et al. [1984] as a function of both projectile and target atomic number. (*b*) A contour plot of (1-keV) theoretical yield values, according to the simplified collisional model. The heaviest inert gas, radon, has been added, even though no experimental yield data for radon was available. (Reprinted with permission from J. Mahan and A. Vantomme, "Trends in Sputter Yield Data in the Film Deposition Regime," 1999, unpublished.)

being due to an increased probability of reflection or escape of a still-energetic projectile, such that the number of recoils created is reduced. Yamamura et al. [1983] proposed an empirical expression to describe the overall dependence of yield on angle of incidence η:

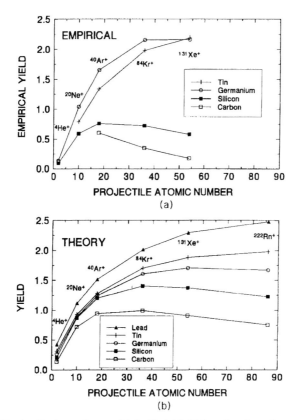

Figure VII.9 Projectile dependence of 1-keV yield. Values for the column IVB elements are plotted as a function of projectile atomic number. (*a*) Empirical yields from Matsunami et al. [1984]; (*b*) theoretical values calculated with the simplified collisional model of Section VII.4. (Reprinted with permission from J. Mahan and A. Vantomme, "Trends in Sputter Yield Data in the Film Deposition Regime," 1999, unpublished.)

$$Y(\eta) = \frac{Y(0)}{(\cos^f \eta)} \cdot \exp\left[f \cdot \cos \eta_{\text{opt}} \cdot \left(1 - \frac{1}{\cos \eta}\right)\right], \quad \text{(VII.7)}$$

where $Y(0)$ is the sputter yield at normal incidence and η_{opt} is the angle of incidence corresponding to the maximum in yield. Yamamura et al. obtained the following empirical expressions for f and η_{opt}:

$$f = \sqrt{U_{\text{sb}}}\left(0.94 - 1.33 \times 10^{-3} \times \frac{m_{\text{r}}}{m_{\text{p}}}\right)$$

$$\eta_{\text{opt}} = 90° - 286° \cdot \left((an_{\text{t}}^{1/3})^{3/2} \sqrt{\frac{Z_{\text{p}} Z_{\text{r}}}{(Z_{\text{p}}^{2/3} + Z_{\text{r}}^{2/3})^{1/2}} \cdot \frac{1}{E}}\right)^{0.45} \quad (U_{\text{sb}} \text{ and } E \text{ in eV}),$$

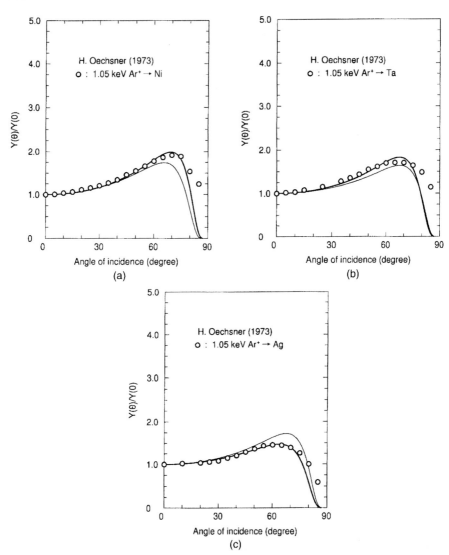

Figure VII.10 Yield versus angle of incidence. Actual experimental data is represented by the circles and is from Oechsner [1973]. The broad lines are best-fit curves to the data shown using the Yamamura et al. formula. The thin lines are calculated using expressions given in the text for f and η_{opt}. (Reprinted with permission from Y. Yamamura et al., *Angular Dependence of Sputtering Yields of Monatomic Solids*, Research Information Center, Institute of Plasma Physics, Nagoya University, Japan. Copyright 1983 by Research Information Center, Institute of Plasma Physics, Nagoya University, Japan.)

where a is the Thomas–Fermi screening length (see Section VII.3) and n_t is the number density of the target.

Example Calculate the Yamamura et al. exponent and the incidence angle of maximum yield for argon sputtering silver at 1.05 keV.

TRENDS IN SPUTTER YIELD DATA 219

For this system we have $Z_p = 18$, $Z_r = 47$, $m_p = 39.9$ amu, $m_r = 107.9$ amu, $U_{sb} = 3.0$ eV, and $n_t = 0.0586$ Å^{-3} (the latter two are from Appendix B). The exponent for the angle of incidence dependence is

$$f = (3.0)^{1/2}\left(0.94 - 1.33 \times 10^{-3} \times \frac{107.9}{39.9}\right) = 1.62.$$

In order to calculate η_{opt} we must first find the Thomas–Fermi screening length (from Section VII.3):

$$a = \frac{0.8853 \times 0.529 \text{ Å}}{(18^{2/3} + 47^{2/3})^{1/2}} = 0.105 \text{ Å}.$$

Finally

$$\eta_{opt} = 90° - 286° \times \left\{[0.105 \text{ Å} \times (0.0586 \text{ Å}^{-3})^{1/3}]^{3/2} \right.$$
$$\left. \times \left[\frac{18 \times 47}{(18^{2/3} + 47^{2/3})^{1/2} \times 1050}\right]^{1/2}\right\}^{0.45} = 67.5°.$$

This angle is close to the data shown in Figure VII.10. ∎

Energy Distribution of Sputtered Particles

A classic experimental study of the energy distribution of sputtered particles was published by Stuart et al. [1969]; a sample of their results is presented in Figure VII.11a. Behrisch [1981] observed that the energy distribution "generally has a maximum between half and the full surface binding energy. At high emerging energies the number of sputtered particles mostly decreases proportional to $1/(E_o)^2$." (The subscript "o" refers to *outside* the target, as opposed to the energy of recoils *inside* the target, which will have the subscript "i" in this chapter; relationships between internal and external flux spectra will be derived in Section VII.5.)

For the linear cascade, the following relative energy distribution, known as the *Thompson formula*, was derived from fundamental principles (see Section VII.5)

$$dY/dE_o \propto \frac{E_o U_{sb}}{(E_o + U_{sb})^3}. \tag{VII.8}$$

The Thompson formula fits the measured external energy spectra well. It has a maximum at $U_{sb}/2$, and a $(1/E_o)^2$ falloff at higher energies.

Figure VII.11b shows detailed flux spectra of sputtered gold atoms from Thompson [1981]. The $1/E_o^2$ dependence is clearly revealed. These spectra are some of the strongest evidence for the linear cascade mechanism, and were quite influential in the confirmation of the concept.

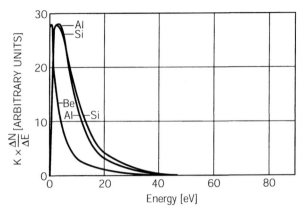

Energy distributions of atoms ejected in the normal direction from polycrystalline Be, Al, and Si targets under bombardment by normally incident Kr ions at 1200 eV.

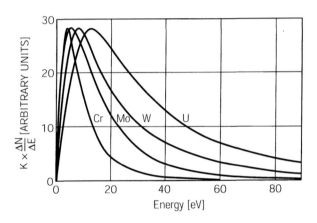

(a) Energy distributions of atoms ejected in the normal direction from polycrystalline Cr, Mo, W and U targets under bombardment by normally incident Kr ions at 1200 eV.

Figure VII.11 (*a*) Energy distributions of sputtered particles. (Reprinted with permission from R. V. Stuart et al., "Energy Distribution of Atoms Sputtered from Polycrystalline Metals," *J. Appl. Phys.* **40**(2). Copyright 1969 by the American Institute of Physics.) (*b*) detailed sputtered flux spectra. (Reprinted with permission from M. W. Thompson, "Physical Mechanisms of Sputtering," *Phys. Rep.* **69**(4). Copyright 1981 by Elsevier Science.)

Angular Distribution of Sputtered Particles

Behrisch [1981] states that *"For normal incidence the angular distributions may mostly be described in a first approximation by a cosine distribution."* This distribution is that of the ideal cosine emission law: $j_\Omega(\theta) = Yz^+ \cos\theta/\pi$, where

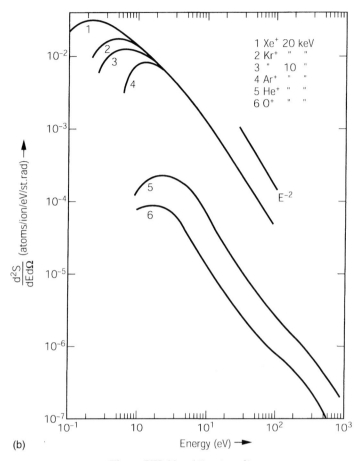

(b)

Figure VII.11 *(Continued)*

j_Ω is the emission flux angular distribution and z^+ is the impingement flux of projectiles. θ is the emission angle (measured from the vertical; see Fig. VII.1). The cosine law results from an isotropic angular distribution of recoil velocities, and is derived for the ideal sputtering source model in Section VII.5. It should be recognized that sputter deposition sources are often large; while even if an infinitesimal area of a large source emits particles according to the cosine law, the source as a whole is likely to be better described as an undercosine distribution, as described next.

There is some dependence of the angular distribution on the projectile mass and energy, and with angle of incidence [Behrish, 1981]. An "undercosine" distribution, in which less particles are ejected in the surface normal direction, has been observed experimentally for low ion energies. On the other hand, an "overcosine" distribution, peaked in the surface normal direction, is sometimes observed for light projectiles. One expression that has been used to describe the yield per unit solid angle of real sputtering sources is

$$\frac{dY}{d\Omega} = Y_{\text{total}} \cdot \frac{\cos\theta}{(\rho\sin^2\theta + \cos^2\theta)\pi}, \qquad (\text{VII.9})$$

where Y_{total} is the total integrated yield, ρ is an empirical parameter, and $d\Omega$ is the differential solid angle in the direction of emission [Blech and Van der Plas, 1983]; $\rho < 1$ models an undercosine distribution and $\rho > 1$, overcosine. [In Section VII.5 we show that $j_\Omega(\theta) = (dY/d\Omega)z^+$.]

For *oblique* incidence at relatively low energies, the maximum in the angular distribution of sputtered particles is typically shifted to the opposite side of the surface normal, an asymmetric distribution. This may be understood as being due to preferred forward ejection of target atoms.

Single-Crystal Targets

The angular distribution described above pertains to amorphous and to polycrystalline (fine-grained, randomly oriented) targets. However, some lattice-correlated effects are apparent in the sputtering of single crystals [Roosendaal, 1981; Winograd, 1993]. These include *channeling* of the projectile deep into the target, while doing little damage, *blocking* of recoils, which prevents their escape along certain crystallographic directions, and *focusing* of collision sequences, particularly along close-packed directions (such as $<110>$ directions in a face-centered cubic lattice), which leads to a lobed emission pattern. One of these patterns, having the so-called Wehner spots, is shown in Figure VII.12 [Wehner, 1962].

For the development of our understanding of the mechanisms of sputtering, these lattice-correlated effects were very important, because they contributed to the demise of the early thermal theories of sputtering. On the microscopic level, they remain very important because they are unavoidable in any crystalline target. Thompson [1981] wrote that "even in a randomly oriented grain in a polycrystalline target the probability of channeling is not negligible." Any theory that does not explicitly calculate lattice-correlated effects should be recognized for what it really is — an *amorphous* target theory. The simplified collisional model, and all of those summarized in the end-of-chapter Appendices, are such. TRIM [Ziegler, 1992] is an amorphous model.

Amorphous target theories are still quite relevant to sputter deposition, however; polycrystalline targets average out lattice-correlated effects; their $(1/E_o)^2$ emission spectra are in agreement with the amorphous target theories, proving the latter relevant. Furthermore, "Away from the main ejection spots. . . a crystal emits an amorphous type of [energy] spectrum" [Thompson, 1981].

Target Conditioning and Dose Effects

Past problems in reproducibility of yield data may sometimes be traced to variations in the targets. This includes

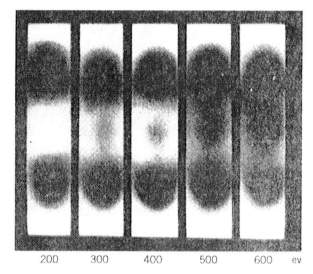

Figure VII.12 Wehner spots: emission patterns from a single crystal target. (Reprinted from G. K. Wehner, 1962, "Physical Sputtering," *Transact. 8th Natl. Vacuum Symp.*, Vol. 1. Copyright 1962 by Pergamon Press.)

- The possible existence of oxides on the surface
- The evolution of surface morphology during sputtering
- The accumulation of implanted ions

In the past, it was considered preferable to make yield measurements on a structurally perfect target, "Ideally... with one single ion fired at an undamaged clean target" [Carter and Colligon, 1968].

However, Townsend et al. [1976] prescribe the practice that yield measurements be made with atomically clean target surfaces, on targets that have been brought to a *stationary state* by previous conditioning; a steady-state damage level is typically achieved with an ion dose of $\sim 10^{16} \, \text{cm}^{-2}$. The acceptable stationary state is probably an amorphized surface, without *blistering*, a macroscopic erosion phenomenon caused by occluded gas [Scherzer, 1983]. Because of these different effects, the yield has been observed to both increase and decrease with dose, presenting a moving target for the theoreticians, indeed.

VII.3 BASIC CONCEPTS FOR MODELING

The Surface Binding Energy

An atom must possess at least a minimum kinetic energy to escape from the target. The height of the potential-energy barrier that it must overcome is the *surface binding energy*. While the *bulk binding energy* is the energy required to remove a target atom from an interior, bulk, lattice site (leaving a vacancy) to an infinite distance outside the crystal, the surface binding energy is the amount required to remove a target atom from a *surface* site (leaving no vacancy) to this same infinite separation.

Displacement, a concept more relevant to radiation damage than to sputtering, is the removal from its lattice site of a target atom that then remains within the target, specifically, the creation of a vacancy plus an interstitial atom. Because of the presence of the interstitial atom, the *displacement energy* (~ 15–30 eV [Carter and Colligon, 1968]) is typically much larger than even the bulk binding energy (~ 10 eV). The displacement energy is not relevant to our calculation of sputter yield, and our recoils need not be *displaced atoms*, as such. Why? In an actual displacement, the interstitial atom must be removed to a sufficient distance that recombination with the vacancy does not occur, while still remaining within the target. This recombination zone can extend over many crystalline unit cells, and the radius of this recombination zone is typically greater than the depth from which most sputtered atoms originate. Thus, *recoils need not have the energy of displaced particles in order to be sputtered* [Sigmund, 1981].

Sigmund [1993] stated that "There is clear evidence to support the notion that the effective surface binding energy does not differ significantly from the heat of sublimation." The heat of sublimation (per particle) is typically calculated as the sum of the heats of fusion and vaporization:

$$U_{sb} = \frac{\Delta_{fus}H + \Delta_{vap}H}{N_a} = \frac{\Delta_{sub}H}{N_a}, \quad (VII.10)$$

where N_a is Avagadro's number (these enthalpies are typically given in kilojoules per mole); $\Delta_{sub}H$ is temperature-dependent, but for calculations in this chapter we will assume room-temperature values only. When corrected to absolute zero, it is known as the *cohesive energy*.

It is sometimes difficult to select the best estimate of surface binding energy from the reference literature. Appendix B presents values for most of the elements from some of today's widely used sources. Using a single value for each target ignores the fact that the binding energy varies with crystallographic orientation of the surface, and with the number of nearest neighbors of a given surface site, but is consistent with an amorphous theory. The science of sputtering is not sufficiently accurate that these relatively small differences matter much.

In addition to its magnitude, the surface binding energy potential barrier must also be given some spatial orientation. The two predominant models have been the spherical binding model and the planar binding model. In the *spherical model*, the binding force holding the recoil within the target directly opposes its velocity vector during escape; it is isotropic and the escaping recoil is slowed, but undeflected. Physically, this potential might arise from the attraction between a recoil and a spherically symmetric feature on the surface, such as a single atom.

BASIC CONCEPTS FOR MODELING

For the *planar model*, the binding force is perpendicular to the surface of the target. Thus, in addition to the slowing, there is also a deflection, or *refraction*, of the escaping recoil, because all of the kinetic energy required to surmount this barrier must come from the normal component of the recoil's velocity. Physically, this potential might be due to the aggregate attraction between the recoil and all the atoms of a flat target surface.

Energy Transfer in Binary Elastic Collisions of Hard Spheres

The purpose of this analysis of hard-sphere collisions is to predict the amount of energy transferred from projectile to recoil atom, and their final directions of travel after a collision. These properties are determined by conservation of energy and momentum.

Consider the two particles represented as hard spheres that are shown in Figure VII.13. The projectile is traveling with initial velocity

$$v_{pi} = v_{pi}[\cos \delta_r\, a_x + \sin \delta_r\, a_y]. \tag{VII.11}$$

The recoil is initially stationary:

$$v_{ri} = 0. \tag{VII.12}$$

The place on the recoil atom where the projectile strikes is very important. It determines where the target particle goes, because in this "billiard ball" model the resulting force is always along the line connecting their centers at the moment of impact. Also, it determines the orientation of the coordinate system. The target particle, after the collision, moves exactly along the x axis:

$$v_{rf} = v_{rf}\, a_x. \tag{VII.13}$$

The projectile's velocity after collision is determined by conservation of momentum for the two-body system. Its y component remains unchanged, because the recoiling atom has no y component of momentum, either before or

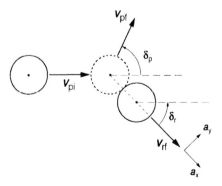

Figure VII.13 A binary elastic collision between two hard spheres. The force is along their line of centers at the moment of impact.

after the collision. The x component is calculated from conservation of momentum in the x direction: $m_p v_{pi} \cos \delta_r = m_p v_{pfx} + m_r v_{rf}$. The result is that

$$v_{pf} = v_{pi} \left(\cos \delta_r \frac{m_p - m_r}{m_p + m_r} a_x + \sin \delta_r a_y \right), \tag{VII.14}$$

and the final speed of the projectile is

$$v_{pf} = v_{pi} \left\{ 1 - \frac{4 m_p m_r}{(m_p + m_r)^2} (\cos \delta_r)^2 \right\}^{1/2}. \tag{VII.15}$$

To solve for v_{rf} in terms of v_{pi}, δ_r, and the mass ratio, we relate them with conservation of energy:

$$\frac{m_p (v_{pi})^2}{2} = \frac{m_p}{2} \left[\frac{m_p v_{pi} \cos \delta_r - m_r v_{rf}}{m_p} \right]^2 + \frac{m_p [v_{pi} \sin \delta_r]^2}{2} + \frac{m_r (v_{rf})^2}{2}. \tag{VII.16}$$

The result is that

$$v_{rf} = \frac{2 v_{pi} \cos \delta_r}{1 + m_r/m_p}. \tag{VII.17}$$

Note from Figure VII.13 that δ_r cannot exceed 90°. As δ_r approaches 90° (grazing impact), v_{rf} approaches zero and the incident particle is undeflected. For a head-on collision ($\delta_r = 0$), v_{rf} will be positive but v_{pfx} may be either positive or negative depending on the mass ratio. Thus, as one knows from practical experience, if the projectile is relatively massive it will continue forward after the collision; on the other hand, if it is relatively light, it may reverse direction. If the masses are equal, then in a head-on collision the incident particle stops and the struck particle takes up all the momentum.

Through additional algebra, one may obtain the final direction of travel of the projectile [Long, 1993]:

$$\delta_p = \tan^{-1} \left[\frac{\sin 2\delta_r}{(m_p/m_r) - \cos 2\delta_r} \right]. \tag{VII.18}$$

The sum $\delta_r + \delta_p$ equals 90° only for the cases of equal masses.

The *energy transfer function* specifies the fraction of energy that is transferred to the target ion. It is the ratio of the final kinetic energy of the target to the initial kinetic energy of the projectile:

$$T(\delta_r) \equiv \frac{m_r (v_{rf})^2}{m_p (v_{pi})^2} = \frac{4 m_p m_r}{(m_p + m_r)^2} (\cos \delta_r)^2. \tag{VII.19}$$

BASIC CONCEPTS FOR MODELING

The *energy transfer mass factor* (γ) is part of the energy transfer function:

$$\gamma \equiv \frac{4 m_p m_r}{(m_p + m_r)^2}. \quad \text{(VII.20)}$$

Transfer of energy is enhanced when the two masses are equal – $T(\delta_r) = (\cos \delta_r)^2$; when $m_p \neq m_r$ then T is less than this value. The energy transfer function approaches zero when the masses differ greatly.

The reader should be aware that the hard sphere model is very crude. More realistic interaction potentials include the Molière potential, the Born–Mayer potential, and the Thomas–Fermi potential [Harrison, 1983]. The hard-sphere model is most useful, however for visualizing the encounters, calculating final velocities, and obtaining an estimate of threshold energy.

Threshold Energy for Sputtering at Normal Incidence

We have obtained the surface binding energy and surveyed the kinematics of hard-sphere collisions, and will now consider the possibility of ejecting a target atom. The purpose of this analysis is to calculate an ideal threshold projectile energy for ejection of a recoil, using an idealized model. We assume normal incidence of the projectile, a single, isotropic surface binding energy, the spherical model for the surface potential-energy barrier, and zero kinetic energy for the expelled recoil.

It is important to realize that *it is not possible to eject a target atom from a planar surface, for normal incidence, as a direct result of the primary collision*. However, it is possible to expel a target atom with one of the secondary collisions. The effect is illustrated in Figure VII.14. There are two options

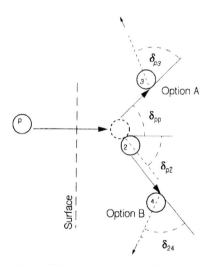

Figure VII.14 Two possible collision sequences leading to an ejected target atom. At threshold, the ejected atom is traveling exactly parallel to the target surface.

shown for the secondary collision: option A, where the projectile strikes and expels the second target atom 3; and option B, where the primary recoil atom 2 expels the second target atom 4. There could also be an option C, where the primary recoil is itself expelled after its second collision. For calculating the absolute threshold energy, options B and C are equivalent and are preferred to option A, which will have a higher threshold. We will analyze option B assuming that atoms 2 and 4 occupy *surface* sites characterized with the surface binding energy:

Referring to Figure VII.14, if atom 4 is to escape, the kinetic energy of atom 2 (the primary recoil) on colliding with 4 must be at least $U_{sb}/(\cos \delta_{24})^2$. (This would displace and minimally eject atom 4, which would have *no* kinetic energy outside the target in the limiting condition of threshold.) If atom 2 is to have this kinetic energy after being struck by the projectile, the projectile p must have had a primary kinetic energy within the target of at least $[U_{sb}/(\cos \delta_{24})^2]/\gamma(\cos \delta_{p2})^2$. Finally, if atom 4 is to escape the target, it must be true that $\delta_{p2} + \delta_{24} \geq 90°$. At threshold $\delta_{p2} = 90° - \delta_{24}$.

The threshold value for the primary kinetic energy of the projectile within the target is, therefore, $E_{th} = U_{sb}/[\gamma(\cos \delta \sin \delta)^2]$, where δ is either one of the scattering angles, since the one must be the complement of the other. The denominator is maximized at 45°. Thus

$$E_{th} = \frac{4U_{sb}}{\gamma}. \qquad (VII.21)$$

This is not the absolute theoretical threshold energy, but a practical estimate that is probably the one to use in comparison to experimental measurements. If *three* collisions result in the expulsion of a target atom, one may show that the minimum projectile energy required is $U_{sb}/\{\gamma[(\cos 30°)^2]^3\}$, which yields a threshold energy of $2.37 U_{sb}/\gamma$; if four collisions, the threshold projectile energy would be $U_{sb}/\{\gamma[(\cos 22.5°)^2]^4\}$. In the limit of a large number of collisions, the theoretical limit is simply U_{sb}/γ. However, getting even three precisely aligned collisions is quite improbable; a minimum of two collisions is required, so we will adopt the above value, $4U_{sb}/\gamma$, as a practical estimate.

Some experimental values of threshold energy for argon sputtering are given in Table VII.1. These experimental values were taken from yield–energy curves that followed approximately the empirical fit of Bohdansky et al. [1980] near E_{th}. The theoretical predictions in the table were calculated using the enthalpy of sublimation values of Weast [1975], which may be found in Appendix B. The theoretical threshold energies exhibit a wider range of values, and yet more internal consistency, than the experimental data. For example, within the context of the other experimental data the experimental value for aluminum seems abnormally high, considering its enthalpy values and mass. For the most part, the experimental values are greater than the theoretical values; this is certainly what one would expect for measurements of a threshold quantity. On the other hand, there are reasons to expect experimental thresholds to be below

BASIC CONCEPTS FOR MODELING

TABLE VII.1 Threshold Energy (in eV) for Argon Sputtering

	U_{sb}^a	γ	Theoretical E_{th}	Experimental[b] E_{th}
Al	3.37	0.964	14.0	47
Si	4.30	0.968	17.8	40
Ti	5.10	0.992	20.6	32
Fe	4.51	0.972	17.5	28
Ni	4.60	0.964	19.1	24
Cu	3.67	0.948	15.5	28
Mo	7.00	0.832	33.7	32
W	9.11	0.588	62.0	35

[a] Thermodynamic data, $\Delta_{sub}H(298)/N_a$, from Weast [1975].
[b] Andersen and Bay [1981].

the theoretical values. These include the presence of atom-scale surface features that contain weakly bonded target atoms (characterized by an exceptionally low U_{sb}), and rare occurrences of multiple interactions in which the ejected target atom is struck more than once by the projectile. It is not well established whether these experimental values are true, absolute threshold energies, or rather just the energy at which measurement of yield becomes impossible, practically speaking.

Nuclear Energy Loss Theory

An energetic ion traveling through a solid loses energy continually, because passage through a crystal is a continuous series of collisions, the next one starting before the last one is finished (like driving in Boston or Los Angeles). While the energy transfer function says that a particle gives up a certain *fraction* of its energy with each collision, the statistically based energy loss theory says that it gives up energy at a certain *rate* per unit distance traveled, the *stopping power*. For sputtering, the dominant ion energy loss mechanism for both the projectile and the recoils is *nuclear* energy loss. This is the transfer of energy to the atoms occupying lattice sites in the target crystal. It occurs as the projectile passes through the target, creating recoiling atoms, and as these energetic recoils (actually ions themselves) create other recoils in the collision cascade, perhaps making their way to the surface.

The average depth of penetration of the projectiles in sputtering is a key factor in determining the escape probability of the recoils. Penetration is determined by the stopping power. And stopping power, in turn, is determined by the nuclear energy loss cross section.

Wilson et al. [1977] showed that the nuclear energy loss cross section is well approximated by

$$S_n(E) = \frac{4\pi a Z_p Z_r q^2 m_p}{m_p + m_r} \cdot \frac{\ln(1+\varepsilon)}{2(\varepsilon + 0.14\,\varepsilon^{0.42})}, \qquad \text{(VII.22)}$$

where a is the Thomas–Fermi screening length, given by

$$a = \frac{0.8853\, a_0}{[Z_p^{2/3} + Z_r^{2/3}]^{1/2}}, \qquad \text{(VII.23)}$$

where a_0 is the Bohr radius, 0.529 Å. The electronic charge q in this above expression for S_n should be entered as $(14.4\,\text{eV}\text{Å})^{1/2}$ (this comes from representing interparticle potential energy as $q_1 q_2/r$). ε is the reduced projectile energy given by

$$\varepsilon(E) = \frac{a m_r}{Z_p Z_r q^2 (m_p + m_r)} E, \qquad \text{(VII.24)}$$

the projectile energy in the center-of-mass frame normalized by the potential energy of the projectile-target combination when separated by the screening length. A recent development, the nuclear energy loss cross section, which is based on the krypton–carbon potential, is preferred by some [Eckstein et al., 1993]; it could be substituted for Wilson's cross section, but the point of this chapter is understanding, not quantitative accuracy.

The nuclear stopping power is the product of the cross section and the number density of target atoms (n_t) in Å^{-3}:

$$\left.\frac{dE}{dR}\right|_n = S_n n_t. \qquad \text{(VII.25)}$$

The units of stopping power are $\text{eV} \cdot \text{Å}^{-1}$. It is the convention to give the stopping power positive values, although the derivative dE/dR is certainly negative. Using Equations (VII.22) and (VII.25), the nuclear stopping power for 1-keV Ar^+ ions in aluminum is 39.4 eV/Å. This value suggests that a modestly energetic argon ion is capable of creating numerous recoils as it passes through even one crystalline unit cell, and that a nonlinear cascade might well occur.

Example Calculate the nuclear stopping power for a 2-keV krypton ion in copper.

Data for the problem are as follows: $E = 2000\,\text{eV}$, $n_t = 0.0845\,\text{Å}^{-3}$, $Z_p = 36$, $Z_r = 29$, $m_p = 83.8\,\text{amu}$, and $m_r = 63.5\,\text{amu}$.

Solution The value of the Thomas–Fermi screening length is

$$a = \frac{0.8853 \times 0.529\,\text{Å}}{(36^{2/3} + 29^{2/3})^{1/2}} = 0.104\,\text{Å}.$$

The reduced projectile energy is

$$\varepsilon = \frac{0.104\,\text{Å} \times 63.5\,\text{amu} \times 2000\,\text{eV}}{36 \times 29 \times 14.4\,\text{eV} \cdot \text{Å} \times (83.8 + 63.5)\,\text{amu}} = 5.96 \times 10^{-3}.$$

The nuclear energy loss cross section is

$$S_n = \frac{4\pi \, 0.104 \text{ Å} \times 36 \times 29 \times 14.4 \text{ eV} \cdot \text{Å} \times 83.8 \text{ amu}}{(83.8 + 63.5) \text{ amu}} \times$$

$$\frac{\ln(1 + 0.00596)}{2(0.00596 + 0.14 \times 0.00596^{0.42})} = 1490 \text{ eV} \cdot \text{Å}^2.$$

Finally, the nuclear stopping power is

$$\left. \frac{dE}{dR} \right|_n = 1490 \text{ eV} \cdot \text{Å}^2 \times 0.0845 \text{ Å}^{-3} = 126 \text{ eV/Å}. \qquad \blacksquare$$

As shown in Figure VII.15, the *range* (R) is the total pathlength of the projectile within the target. The *mean projected range* (R^p), along the initial direction of the projectile, is also illustrated in the figure. It is equal to the average depth of penetration for normal incidence only. While the range is given simply by

$$R(E) = \int_0^E \frac{dE'}{dE'/dR|_n}, \qquad (\text{VII}.26)$$

a projected range expression is a bit more difficult to formulate. Because of scattering of the projectile, R^p is always less than R.

To calculate the projected range of projectiles, R_p^p, we utilize an empirical relationship that comes from a plot due to Sigmund [1981]: $(m_p/m_r) \cdot$

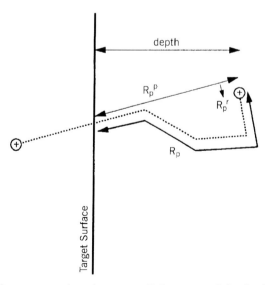

Figure VII.15 The range, projected range, radial range, and depth of penetration of a projectile are illustrated.

$\{R_p - R_p^p\}/R_p^p \approx 0.4$. This plot presents a calculation of the relationship between range and projected range as a function of mass ratio and reduced projectile energy. This empirical relationship approximates the calculated result for ion energies characteristic of sputter deposition. It may be solved for R_p^p to give the following:

$$R_p^p \approx \frac{R_p}{[1 + 0.4(m_r/m_p)]}. \tag{VII.27}$$

Other attempts to find expressions for R_p^p in terms of R_p are discussed in García-Rosales et al. [1994]. As one might expect from the preceding analysis of energy transfer in binary collisions, R_p^p/R_p is $\ll 1$ for light projectiles and nearly unity for very heavy ones.

Example Estimate the range and projected range for a 2-keV krypton ion in copper.

Using the results of the previous example, the range is

$$R = \frac{2\,\text{keV}}{126\,\text{eV}/\text{Å}} = 15.9\,\text{Å}.$$

The projected range is

$$R^p = \frac{15.9\,\text{Å}}{1 + 0.4(63.5/83.8)} = 12.2\,\text{Å}. \qquad \blacksquare$$

This summary of the fundamentals of nuclear energy loss theory gives a picture of the projectile's behavior. We need, finally, to understand and model the behavior of the recoiling target atoms within the collision cascade:

Linear Cascade Theory

To model the behavior of recoils, we will obtain their number, their probability distribution function in energy, the directional distribution of their velocities, and their depth distribution. Some recoils may escape from the surface as sputtered particles, but we will ignore this small loss until we calculate sputter yield in Section VII.4.

The collision cascade is a rapid-fire sequence of collisions that are caused first by the projectile and then by primary, secondary, tertiary (etc.) *knockons*. Thompson provided a schematic illustration of these events, which is reproduced in Figure VII.16. As the cascade progresses the number of recoils increases. The energy of any given recoil decreases monotonically (because a moving atom is assumed to collide only with stationary atoms), as does the average energy of the recoils, and the minimum energy of the recoil population. The disturbance ultimately merges smoothly into the lattice vibrations of the target.

Thus, the energy distribution is a function of time. However, during sputtering there is a *steady-state* population of recoils, due to an enormous rate

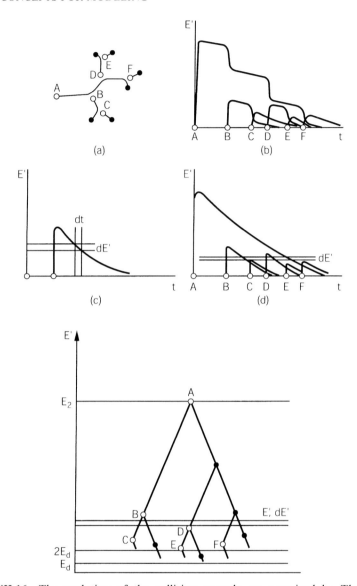

Figure VII.16 The evolution of the collision cascade, as conceived by Thompson [1981]. (Reprinted with permission from M.W. Thompson, "Physical Mechanisms of Sputtering," *Phys. Rep.* **69**(4). Copyright 1981 by Elsevier Science.)

($\gtrsim 10^{15}\,\text{cm}^{-2}\cdot\text{s}^{-1}$ under practical film deposition conditions) of ion bombardment and subsequent collision cascades. This population of moving atoms may be characterized with a steady-state kinetic energy distribution. We estimate this distribution by considering a single cascade, letting it evolve, and then terminating it appropriately. It doesn't matter whether one single cascade has only a few recoils and may not, by itself, possess statistical validity — we are representing many such cascades with these statistics. The derivation itself resembles one presented by Sigmund [1981] for the distribution of the *initial* energies of recoils; something equivalent to the final result was also derived by Thompson [1981] for the case of the hard sphere interaction potential. Kelly [1984] called this approach to deriving the energy distribution "oversimplified but instructive" and that is why we like it.

A linear cascade is one for which the number of recoils is proportional to the initial projectile energy. For a single cascade, the greatest possible number of recoils that could have been created is simply $E/E_{i,min}$, where $E_{i,min}$ is the minimum energy of the recoil population. The *actual* number, on the other hand, is a certain fraction (K) of this, which has been estimated from transport theory. (It is a classical problem in radiation damage; as an order of magnitude estimate, this fraction is 0.5 [Sigmund, 1981], but it doesn't really matter because K will cancel out of the normalized probability distribution function.) We shall shortly rationalize this basic result with a simplified model of the cascade.

First, however, with $E_{i,min}$ as the lower limit to the energy distribution of the recoils in a single cascade, the total number of recoils in a single cascade is

$$N \equiv \frac{KE}{E_{i,min}} = \int_{E_{i,min}}^{\gamma E} f(E_i) dE_i \qquad \text{(VII.28)}$$

(γE is the largest energy a recoil could have). This equation represents a dynamic process; as the cascade matures, $E_{i,min}$ decreases and N increases. The equation also defines the distribution in energy of the recoils, $f(E_i)$, where E_i is the recoil energy *inside* the target. By differentiation we arrive at the actual expression for $f(E_i)$:

$$f(E) = \begin{cases} \dfrac{KE}{E_i^2} & E_i \geq E_{i,min} \\ 0 & E_i < E_{i,min} \end{cases} \qquad \text{(VII.29)}$$

It is worthwhile to understand the physical origin of this unique energy distribution. We will use a simplistic picture of the collision cascade to derive the starting concept, which was that the total number of recoils is proportional to $E/E_{i,min}$. Suppose that when the projectile enters the target, collisions occur every λÅ of distance traveled, and that every collision involves one moving and one stationary particle. The projectile loses energy continually, and it creates energetic recoils that then go on to strike stationary target atoms every λÅ of distance they travel, producing additional recoils just as the original projectile does. The energetic recoils become, in effect, the projectiles of later collisions. This and other details of the cascade are summarized in Table VII.2, where it is shown that after the projectile has traveled $n\lambda$Å, the number of recoils is 2^n.

BASIC CONCEPTS FOR MODELING   235

TABLE VII.2 Evolution of an Idealized Linear Collision Cascade

Distance Traveled by Projectile (in Units of λ)	Number of Moving Atoms (Including Original Projectile)	$E_{i,min}$ (Assuming $b > \frac{1}{2}{}^a$)
0	1	E
1	2	$(1-b)E$
2	4	$(1-b)^2 E$
3	8	$(1-b)^3 E$
4	16	$(1-b)^4 E$
...		
n	2^n	$(1-b)^n E$

^{a}If b were $< \frac{1}{2}$, the minimum energies would be bE, $b^2 E$, $b^3 E$, and so on; the choice of b is arbitrary.

Now suppose that the fraction of energy *retained* by the projectile of each collision (whether by the original projectile or by an energetic recoil) is a constant b; the fraction lost is $1 - b$. After $n\lambda$ Å of distance, the minimum energy of the population of moving atoms is $(1 - b)^n E$, as also shown in Table VII.2. Combining these results, the relationship between the total number of recoils and the minimum energy of their distribution is $N \equiv (E/E_{i,min})^{[-\log 2/\log(1-b)]}$. In this simplified model of the cascade, if $b = \frac{1}{2}$, then N is proportional to (and in fact equal to) $E/E_{i,min}$, which was the basis of the derivation of the energy distribution of the recoils [Eg. (VII.29)].

In fact, b may be estimated as the average value of the energy transfer function in binary collision theory, $\gamma[(\cos \delta_r)^2]_{av}$ ($= \gamma/2$ if all values of δ_r are equally probable); γ is precisely one for a target atom–target atom collision, and rather close to one for argon ions and most targets (see Table VII.1). Thus, $b = \frac{1}{2}$ is not an unreasonable assumption for this simplistic model. Of course, a range of b values will occur in a real cascade.

We mentioned earlier that as the cascade evolves, $E_{i,min}$ decreases. As Harrison [1983] elegantly explains, *termination* (the choice of $E_{i,min}$) is always a problem for both simulation and modeling. A practical lower limit to consider for $E_{i,min}$ as representing the endpoint of a cascade is U_{sb}, because no recoils having an energy below this value can escape the target, and U_{sb} is the average binding energy of surface atoms.

A normalized probability distribution function for the recoils may be derived. Assuming that no recoil has an energy less than some $E_{i,min}$ nor greater than γE (the mass transfer function times the initial kinetic energy of the projectile), the probability distribution function in energy is

$$F_i(E_i) \equiv \begin{cases} \dfrac{KE/E_i^2}{\int_{E_{i,min}}^{\gamma E}(KE/E_i^2)dE_i} & E_i \geq E_{i,min} \\ 0 & E_i < E_{i,min}. \end{cases} \quad (VII.30)$$

After the obvious cancellations of K and E, the denominator is $1/E_{i,\min} - 1/\gamma E$. Using a termination value for $E_{i,\min}$ of U_{sb}, the denominator becomes $1/U_{sb} - 1/\gamma E$.

We simplify the denominator to simply $1/U_{sb}$, since in most practical instances the initial projectile energy times the mass function is more than an order of magnitude greater than the surface binding energy. Thus, the (limiting) probability distribution function in energy is

$$F_i(E_i) \approx \begin{cases} \dfrac{U_{sb}}{E_i^2} & E_i \geq E_{i,\min} \\ 0 & E_i < E_{i,\min} \end{cases} \qquad \text{(VII.31)}$$

It is interesting that $F_i(E_i)$ does not contain the projectile energy. Indeed, the experimental data for the external energy distribution of a given projectile–target combination do not change much even when the projectile energy varies widely (say, from 10^2 to 10^5 eV) [Oechsner, 1975].

This distribution function is quite different from the Maxwell–Boltzmann distribution, since it has different origins. It results from the way in which the projectile's kinetic energy is shared among the recoils. This energy distribution function has been indirectly verified with measurements of the energy spectra of *ejected* particles. At recoil energies $\gg U_{sb}$, the distribution of ejected particles varies as $1/(E_o)^2$, where E_o is the recoil energy *outside* the target. Kelly [1984] wrote that "The most explicit evidence for slow collisional sputtering [i.e. the linear cascade] probably lies in the existence of sputtered fluxes which approach $(1/E_o)^2$ behavior at high energies and which have been observed repeatedly both in experiments and in simulations." The effect of the surface potential barrier—how it converts the internal energy distribution of recoils into the external energy distribution of sputtered particles—is explored in Section VII.5.

The average energy of recoils for this limiting distribution is

$$E_{i,av} \approx \int_{U_{sb}}^{\gamma E} E_i \cdot \left(\frac{U_{sb}}{E_i^2}\right) dE_i = U_{sb} \ln\left(\frac{\gamma E}{U_{sb}}\right) \qquad \text{(VII.32)}$$

Thus, under normal sputtering conditions the average recoil energy is no more than a few times U_{sb} after the cascade has run its practical course.

Regarding the *directional distribution* of the recoil velocities, the initial direction of the projectile becomes unimportant when $E \gg U_{sb}$; many collisions occur, and the direction of travel becomes random—this distribution is isotropic. Of course, for energies near threshold, the distribution is very strongly directed into the target.

The next issue to address in this summary of linear cascade theory is "the *spatial distribution* of recoils. On the average, the recoils will be distributed with cylindrical symmetry about the path of the typical projectile, which is given by the projected range (as shown in Fig. VII.17). For normal incidence, the lateral

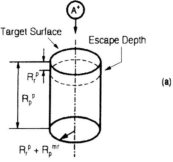

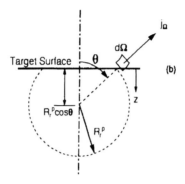

Geometry for Angular Yield Spectrum

Figure VII.17 Escape probability and angular yield spectrum. (*a*) It is assumed that the recoils are uniformly distributed in a cylindrical volume whose depth is the projected range of the projectile; (*b*) θ is the emission angle for recoils having a projected range R_r^p. The maximum depth from which they may come is $R_r^p \cos\theta$. (Reprinted with permission from J. E. Mahan and A. Vantomme, "A Simplified Collisional Model of Sputtering in the Linear Cascade Regime," *J. Vac. Sci. Technol.* **A15**(4), 1976 (1997). Copyright 1997 by the American Vacuum Society.)

part of the distribution is immaterial to the sputter yield — what counts is the *depth* distribution.

The recoil depth profile is not the lattice damage profile, because, as we have already seen, the recoils are not displaced atoms as such. Neither is it the implanted ion profile, because recoils are created along the entire length of the projectile's path, not just at its final resting place. Sigmund [1981] asserts that the recoil depth profile is that of the *deposited energy*: "when being shared among the atoms down to the very end of linear energy dissipation.... This is different from the stopping-power concept that focuses on the energy lost by one moving particle, disregarding further relocation of energy by means of recoil atoms."

This has become one of the most central, and durable, ideas of collisional sputtering theory. Sigmund approached the calculation of this deposited energy profile using the Boltzmann transport theory. We approximate it with some particularly simplifying assumptions; in fact, we use the stopping power concept in spite of Sigmund's distinction, because in Section VII.4 we develop a *simplified* collisional model of sputtering.

VII.4 A SIMPLIFIED COLLISIONAL MODEL FOR SPUTTER YIELD

A Yield Expression

The purpose of this section is to present a simplified collisional model of yield, for the linear cascade regime of sputtering. It is simplified in that we make particularly simple assumptions about the distribution functions (spatial, directional, energetic) which characterize the collision cascade. One goal is to construct an uncluttered, closed-form yield expression whose physical meaning is transparently clear, embodying the essential concepts of linear cascade sputtering. Another is to explain, and qualitatively predict, sputter yield data trends that were summarized in Section VII.2. A third is to be as accurate as possible, while remembering that the primary intent of the model is explanation, rather than quantitative accuracy. This model was first published by Mahan and Vantomme [1997].

The major concepts of the collisional picture of sputtering might be summarized as follows:

- Stopping of the projectile via nuclear energy loss to target atoms
- Development of a collision cascade
- Escape of certain recoils through the surface potential energy barrier

We model the spatial distribution of recoils with the geometry shown in Figure VII.17. We are concerned only with projectiles at normal incidence. The recoiling particles are assumed to be distributed uniformly in depth within a *disturbed volume*, which is a circular cylinder with flat ends. The length of the cylinder is assumed to be the projected range of the projectile (R_p^p). Now, just like the projectile, the recoils have a range within the target. The radius of the cylindrical disturbed volume, then, is assumed to be equal to $R_r^p + R_p^{mr}$, where the former is the projected range of recoils and the latter is the mean *radial* range of projectiles. (By "projected range of a recoil" we mean how far it travels along its *initial* direction, whatever that may be, after having been struck by an energetic atom.) While it is important for a physical picture of the cascade, the lateral distribution of target recoils is actually immaterial to the calculation of yield in the case of normal incidence; what counts is their depth distribution.

Only those recoils within the *escape depth* may escape, typically a small fraction of the recoils. We will take the escape depth to be R_r^p (this is justified in

Section VII.5). For all the calculations presented in this chapter, the largest value of $R_\mathrm{r}^\mathrm{p}/R_\mathrm{p}^\mathrm{p}$ is 0.29 — for 40 eV – $\mathrm{Ar}^+ \to \mathrm{Al}$ — and virtually all values are less than 0.2. Thus, the shape of the bottom of the disturbed volume is also immaterial for typical deposition cases, because no ejected particles come from such depths. (A rounded lower end for the disturbed volume would probably be more realistic, but this would introduce algebraic complications that are unnecessary for present purposes.)

The projectile range is estimated simply from its *initial* energy:

$$R_\mathrm{p} \approx \frac{E}{dE/dR|_\mathrm{np}}. \tag{VII.33}$$

As noted at the conclusion of Section VII.3, one of the main ideas of the Sigmund theory is that the depth profile of recoiling atoms is that of the deposited energy. We are in effect representing the recoil depth profile with a deposited energy profile. This profile is simplified first by approximating it with the energy lost by the projectile, neglecting further redistribution by recoils. It is simplified secondly by assuming that this projectile energy loss is uniform down to depth R_p^p. Similarly, Sigmund [1981] and Thompson [1981] have in their final yield expressions the stopping power of the projectile *evaluated at its initial energy*. (Of course, one of the main points of Sigmund's [1969] model was to calculate the redistribution of energy lost by the projectile, using Boltzmann transport theory.) Kelly's yield expression is based on calculating the fraction of the projectile's energy which is *deposited in the outer atomic layer* [Kelly, 1984].

The *effective* range of the recoils is estimated using their average energy at the practical endpoint of the cascade, $E_\mathrm{i,av}$:

$$R_\mathrm{r,eff} \approx \frac{E_\mathrm{i,av} - E_\mathrm{i,th}}{dE/dR|_\mathrm{nr}}. \tag{VII.34}$$

We have inserted into this equation the effect of diminishing probability of escape as projectile energy approaches our practical estimate of threshold energy from Section VII.3, $4U_\mathrm{sb}/\gamma$. The corresponding average energy of the recoils at threshold is

$$E_\mathrm{i,th} = \frac{U_\mathrm{sb} \ln \gamma E_\mathrm{th}}{U_\mathrm{sb}} = U_\mathrm{sb} \ln(4) \approx 1.39 U_\mathrm{sb}. \tag{VII.35}$$

Thus, the effective range of the recoils goes to zero when $E \to E_\mathrm{th}$, because none of them are able to escape the target. (This point is slightly different than in the original publication of the simplified collisional model, where the factor of 1.39 did not appear [Mahan and Vantomme, 1997].)

Of course, near threshold, the expression for average recoil energy, $U_\mathrm{sb}\ln(\gamma E/U_\mathrm{sb})$, is not very accurate, since it was derived assuming $\gamma E \gg U_\mathrm{sb}$. A more realistic way to incorporate threshold physics into the yield expression would be to require that $E_{i,av} > U_{sb}$, and to estimate the increasingly anisotropic directional distribution of recoil velocities as $E \to U_{sb}$. However, remember that the simplified collisional model is only a qualitative one; undertaking these complex tasks would defeat the purpose of the model.

The yield will be the number of recoiling target atoms per incident ion at termination of the cascade, multiplied by the probability that they are created close enough to the surface of the target to escape, and by the probability that they are traveling in the right direction.

These three factors are formulated as follows. The effective number of recoils resulting from a single penetration is given by

$$N = \frac{E}{E_{i,av}} \tag{VII.36}$$

The fraction of the total number of recoils that are close enough to the surface to escape is given by the ratio of volumes shown in Figure VII.17a: $\pi(R_r^p + R_p^{mr})^2 \cdot R_{r,\text{eff}}^p / \{\pi(R_r^p + R_p^{mr})^2 \cdot R_p^p\}$, which reduces to

$$f = \frac{R_{r,\text{eff}}^p}{R_p^p}. \tag{VII.37}$$

Now, even among those that are close enough to the surface to escape, only a fraction are traveling in the right direction to do so with the energy they possess. This escape probability is a function of the depth at which they are created. We assume an isotropic velocity distribution, and thus the probability of traveling in the right direction is calculated simply as a ratio of solid angles, as suggested in Figure VII.17b. (We again assume the spherical surface binding model. This means that regardless of the exact angle with which it approaches the surface, if a recoil of the terminated cascade reaches the surface, it will escape.) The mean value of the escape probability for all the recoils within the escape volume is $\frac{1}{4}$, which is demonstrated in Section VII.5. This factor of $\frac{1}{4}$ has the same geometric origin as that which is found in the impingement rate expression of kinetic theory $(nv_{av}/4)$.

The complete sputter yield expression for normal incidence of the projectile is constructed by combining the three parts mentioned above:

$$Y = \frac{E}{E_{i,av}} \cdot \frac{R_{r,\text{eff}}^p}{R_p^p(E)} \cdot \frac{1}{4}. \tag{VII.38}$$

Example Calculate the sputter yield of 1-keV argon ions striking aluminum according to the simplified collisional model.

This solution is worked out in Appendix D. All of the necessary equations of the simplified model are summarized there, together with the nonobvious constants. ∎

The assumptions of the simplified collisional model may be summarized as follows:

- A projectile range calculated from the initial energy of the projectile using the nuclear stopping power corresponding to that energy
- A uniform depth distribution of recoils, down to R_p^p
- An isotropic distribution of recoil velocities
- Termination of the cascade when the minimum recoil energy is equal to U_{sb}
- An escape depth $R_{r,eff}^p$ calculated from the average recoil energy at termination and limited by threshold;
- The first-order estimate of threshold energy from Section VII.3.

Predictions

We show in Figure VII.18 the yield as a function of argon projectile energy for aluminum, tungsten, and copper targets. For comparison, the empirical yields of Matsunami et al. [1984] are also shown.

The theoretical curves owe their basic shape to the energy dependence of the projectile range, $R_p(E)$, but superposed on this is the low energy cutoff which was built into the effective recoil range. Clearly, Figure VII.18 shows that the simplified collisional model can aspire to be only a qualitative model. Near threshold, it greatly overestimates the yields of two of the three targets. When only a few recoils are produced, one might expect their velocity distribution *not* to be isotropic, but rather directed into the target. Assuming an isotropic distribution, as the model does, would therefore lead to overestimating the yield.

Above $\sim 100\,\text{eV}$, the model basically agrees with the tungsten empirical yields, if one accepts that "within a factor of 2" means "agreement" in the sputtering field. Above $\sim 100\,\text{eV}$ the model definitely underestimates the copper empirical yields. This may be because copper is susceptible to the nonlinear cascade mechanism, with its "excess" yield, and of course, the nonlinear cascade is not included in the model. Aluminum may also be susceptible to the nonlinear cascade mechanism; its highest empirical yields are above 3. No theoretical yield model has done particularly well, quantitatively. For some perspective, we show in Table VII.3 the empirical and some theoretical 1-keV argon ion yields for the same targets.

The simplified collisional model does exhibit important qualitative features of the empirical data, and is useful for understanding these. The calculated yields of six 1-keV inert gas projectiles, for 462 projectile–target combinations, are shown in Figure VII.19 [Mahan and Vantomme, 1999]. These are all the elemental targets for which density and surface binding energy values could be found (see Appendix B). The calculated yield variations from one target to the next are less extreme than in the data (Fig. VII.6a). However, the basic structure of the experimental curves is reproduced in the theoretical curves.

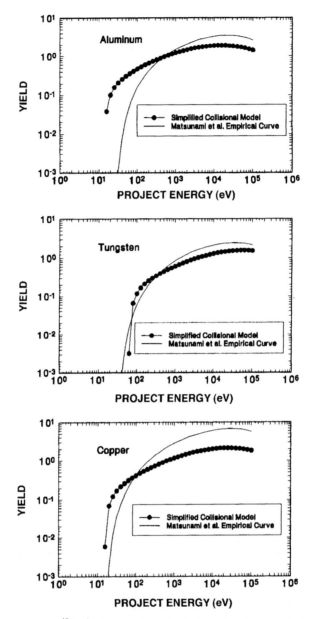

Figure VII.18 Argon (^{40}Ar$^+$) ion sputter yields as a function projectile energy are shown for aluminum, tungsten, and copper targets. The points are the theoretical yields of the simplified collisional model; the solid lines are the empirical yields of Matsunami et al. [1984].

TABLE VII.3 1-keV Ar$^+$ Yields

	Empirical [Matsunami et al.]	Simplified Collisional Model	Sigmund's Theory[a]	TRIM[b]
Aluminum	1.53	1.23	2.2	0.56
Tungsten	0.86	0.62	~2	1.5
Copper	2.82	1.16	3.5	3.9

[a]From Andersen and Bay [1981].
[b]Obtained by running the TRIM program [Ziegler, 1992] as in Figure VII.3.

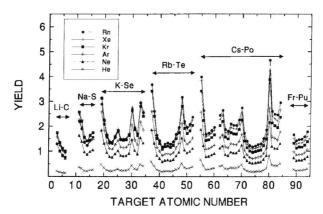

Figure VII.19 Yields of simplified collisional model. The calculated sputter yields of the simplified collisional model are shown as a function of target atomic number; 1-keV inert gas projectiles were assumed. (Reprinted with permission from J. Mahan and A. Vantomme, "Trends in Sputter Yield Data in the Film Deposition Regime," 1999, unpublished.)

There is a periodicity in yield as a function of target atomic number, which repeats with each row of the Periodic Table. For a given row, the curve is basically U-shaped with one or two internal maxima. The principal factor here is the surface binding energy variation among the targets. The theoretical U_{sb} dependence of yield is shown in Figure VII.7b, where the theoretical yields are plotted against surface binding energy for the same set of targets as in Figure VII.7a (adding all the possible targets of Fig. VII.19 does not significantly change the plot). The nonlinear regression best-fit to the theoretical yields gave a binding energy dependence of

$$Y_{\text{theor}}(1\text{-keVAr}^+) \approx \frac{2.41}{U_{sb}^{0.52}}. \quad (\text{VII.39})$$

On the other hand, for the experimental 1-keV Ar$^+$ yields, the nonlinear and linear cascade mechanisms are both present, accentuating the differences between the low and high yields. Consequently, the experimental U_{sb} dependence is stronger ($\propto U_{sb}^{1.31}$ in Fig. VII.7a.)

The calculated 1-keV yields as a function of both (inert gas) projectile and target atomic number are portrayed in Figure VII.8b. The effect of choice of projectile is interesting, and varied. For the heaviest targets, the highest yield occurs with the radon projectile. However, for lighter targets, the highest yields occur for krypton or xenon projectiles. As a detailed example, the behavior for the column IVB targets is presented in Figure VII.9b. The yield maximum for carbon and silicon occurs for the krypton projectile, while for germanium the maximum occurs for xenon. For tin and lead, the yield exhibits a terminal maximum at radon. Thus, the experimental projectile dependence is basically reproduced by the simplified collisional model. Mahan and Vantomme [1999] showed that this projectile dependence of yield is due primarily to variations in the projected range of the projectile.

Summary

Basic trends in the sputter yield data of Section VII.2 are predicted by the simplified collisional model as follows:

- The projectile energy dependence of yield has a threshold at a few times U_{sb}, followed by a rapid increase and then a broad maximum within the 10–100-keV range.
- For the target dependence of yield there is a periodicity in yield values, repeating with each row of the Periodic Table.
- There is a strong correlation between yield and surface binding energy, which is the most important target parameter. For the linear cascade only, the yield is approximately proportional to $1/U_{sb}^{0.5}$.
- The projectile dependence of yield is as follows. For target atomic numbers above ~ 35, the 1-keV yield increases monotonically with inert gas projectile mass; for lighter targets the yield exhibits a maximum at an intermediate projectile mass.
- The ideal cosine law of emission *will be predicted* in Section VII.5.
- The Thompson formula for the energy distribution of emitted particles *will be predicted* in Section VII.5, when the planar surface binding model is substituted for the spherical model.

(In Section VII.5, the energy distribution of emitted particles for the spherical binding model is found to be slightly different from the Thompson formula.) Finally, the effect of angle of incidence of the projectile *could* be modeled, by tipping the disturbed volume and adjusting the calculations for probability of escape (but this has not been done).

Many simplifying assumptions have been used in constructing the yield expression of the simplified collisional model. These were discussed by Mahan and Vantomme [1997] in their original article and include (1) calculating the projectile range from the *initial* projectile energy, (2) calculating the recoil range from the *average energy at the termination* of the cascade, (3) assuming a *uniform* depth distribution of recoils within the disturbed volume, (4) assuming an *isotropic* velocity distribution of recoils, (5) using *Wilson's* expression for the nuclear energy loss cross section, (6) estimating the surface binding energy from the enthalpy of sublimation, (7) neglecting electron energy loss, (8) neglecting the nonlinear cascade, and (9) neglecting lattice-correlated effects. They stated that the model "has to be viewed as a case of compensating errors, since there are many simplifications and they do not all cause the yield to deviate in the same direction."

At this stage in the development of sputtering theory and experiment, it is still true that "within a factor of 2" frequently sums up the state of the art for quantitative accuracy of yield predictions. Clearly there is an additional need for a *non*linear cascade model, because in many film deposition situations this mechanism is operative.

VII.5 AN IDEAL SPUTTER DEPOSITION SOURCE

The goal here is to model an "ideal" sputter deposition source, one whose emission flux, in analogy to the ideal evaporation source, exhibits the cosine law of emission. Sputter deposition sources are typically large, however, in that the size of the sputtering target is not negligible compared to the *throw distance*, the distance from the target to the substrate. It is only the local emission flux, from an infinitesimal area element dA of the target surface, that exhibits the cosine law. A second aspect of the ideal source is that the external energy spectrum of emitted particles is the Thompson formula.

The Cosine Law of Emission

This is an exact derivation of escape probability, whose result confirms the expressions that were adopted a priori in Equations (VII.37) and (VII.38) for the fractions of recoils that were close enough to the surface to escape, and that were traveling in the right direction. We will retain the assumption of the spherical surface potential barrier, so that all the recoils that reach the surface with energy U_{sb} will escape, undeflected. In a later section we will actually calculate the transformations that link the energy distributions of recoils inside and outside the target.

The derivation is based on the geometry shown in Figure VII.17b. To perform this calculation, all the recoils are translated laterally back to the axis of the disturbed volume, and are assumed to be distributed uniformly along this axis *within* the disturbed volume, so that their linear density is a uniform N/R_p^p. The total number of particles originating at depths between z and $z + dz$, and emitted into the differential solid angle $d\Omega$ about θ and φ, is given by $N \cdot dz/R_p^p \cdot d\Omega/4\pi$. The total emission into $d\Omega(=\sin\theta\,d\theta\,d\varphi)$ is summed over depths starting at the surface and going down to a maximum depth $R_{r,\text{eff}}^p \cos\theta$, at which escape into $d\Omega$ becomes impossible:

$$dY = N \cdot \int_0^{R_{r,\text{eff}}^p \cos\theta} \frac{dz}{R_p^p} \cdot \frac{d\Omega}{4\pi}. \qquad \text{(VII.40)}$$

After evaluating this simple integral, we obtain

$$\frac{dY}{d\Omega} = N \cdot \frac{R_{r,\text{eff}}^p}{R_p^p} \cdot \frac{\cos\theta}{4\pi}. \qquad \text{(VII.41)}$$

It may be easily seen that the integral of dY over all possible values of θ and φ ($0 \leq \theta \leq \pi/2$ and $0 \leq \varphi \leq 2\pi$) gives the previous value of the total yield, Equation (VII.38), and that

$$\frac{dY}{d\Omega} = Y_{\text{total}} \frac{\cos\theta}{\pi}. \qquad \text{(VII.42)}$$

Example Starting from the expression for the angular yield spectrum, Equation (VII.42), integrate over all possible emission directions to show that the total yield is simply Y_{total}.

The value of $d\Omega$ is given by $\sin\theta \, d\theta \, d\varphi$. The total possible range of emission directions is $0 \leq \theta \leq \pi/2$ and $0 \leq \varphi \leq 2\pi$. Then

$$\int_0^{2\pi} \int_0^{\pi/2} Y_{\text{total}} \frac{\cos\theta}{\pi} \sin\theta \, d\theta \, d\varphi =$$

$$\int_0^{2\pi} Y_{\text{total}} d\varphi \cdot \frac{1}{2\pi} =$$

$$Y_{\text{total}} \cdot 2\pi \cdot \frac{1}{2\pi} = Y_{\text{total}}. \qquad \blacksquare$$

Thus, the ideal cosine distribution for the angular yield spectrum is derived. Of course, in the above derivation we ignored the effect of the surface potential barrier on the angular spectrum of emitted particles. This might seem a serious omission, particularly if the planar binding model holds. In that model there is a deflection of the escaping particles away from the surface normal direction. Despite such a refraction effect, calculations in a later section show that a cosine directional distribution *within* the target (which we just derived) is retained as an *external* cosine distribution, although the particle energies and their total number are reduced.

Now for a sputtering source, if the ion flux locally striking the substrate is z^+, then the flux of sputtered particles emitted from this same place on the target surface is Yz^+. In analogy to the expression for evaporation in Chapter V, the *emission flux angular distribution* of the ideal sputtering source is defined as

$$j_\Omega(\theta) = Yz^+ \frac{\cos\theta}{\pi}, \qquad \text{(VII.43)}$$

AN IDEAL SPUTTER DEPOSITION SOURCE

the cosine law of emission for a sputtering source. Thus the relationship between emission flux and yield angular distribution is

$$j_\Omega(\theta) = \frac{dY}{d\Omega} z^+. \tag{VII.44}$$

The Beam Intensity of a Sputtering Source

As was emphasized in Chapter V, a physical vapor deposition source is fundamentally characterized by its *beam intensity*. For a small, flat sputtering source of area δA, the beam intensity is given by

$$J_\Omega = j_\Omega \delta A = Y z^+ \delta A \frac{\cos \theta}{\pi}. \tag{VII.45}$$

Normally, sputtering targets are *large sources*, in the sense that their physical dimensions are not much smaller than the distance to the substrate. Thus, the deposition flux onto the substrate must be calculated from the contributions from different portions of the target, as suggested in Figure VII.20. To accomplish such a calculation, the surface of a large sputtering target is characterized with a *differential* beam intensity (dJ_Ω):

$$dJ_\Omega = j_\Omega dA, \tag{VII.46}$$

where dA is a differential area of the source. In Chapter VIII we consider the quantitative calculation of the deposition flux (at the substrate) of large sources.

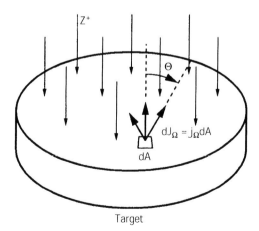

Figure VII.20 The deposition flux of a large sputtering source may be calculated from the contributions of many infinitesimal area elements dA, each of which have their own differential beam intensity.

Next, we extend the simplified collisional model, first by deriving the combined internal energy and angular spectra of recoils reaching the surface. Then we calculate the external spectra (for particles actually emitted) corresponding to the two surface binding energy models (spherical and planar).

Combined Internal Flux Spectra for the Simplified Collisional Model

By an approach similar to that of the cosine law derivation, the combined internal flux spectra (energy and angular, for recoils internal to the target) are

$$\frac{dj_\Omega}{dE_i} = z^+ N \frac{U_{sb}}{E_i^2} \cdot \frac{R_r^p(E_i - U_{sb})}{R_p^p} \cdot \frac{\cos \theta_i}{4\pi}. \qquad (\text{VII}.47)$$

The first factor on the right gives the total internal rate of creation of recoils (per unit surface area per second), having energy between E_i and $E_i + dE_i$. The second two factors result from integrating over the volume of the collision cascade from the surface down to $R_r^p(E_i - U_{sb}) \cos \theta_i$, the maximum depth from which a recoil of energy E_i can escape. On replacing $R_r^p(E_i - U_{sb})$ with $R_r^p(E_{i,av} - 1.39 U_{sb})$ as in the simplified collisional model's derivation of total yield, and then integrating over all possible values of E_i, the cosine law [Eq. (VII.43)] is obtained.

This is not the general expression for the combined spectra of all the recoils created, but rather for those that reach the surface with internal energy greater than or equal to U_{sb}. These combined spectra are modified as the recoils pass through the surface. In both the planar and spherical binding models there is an energy shift as particles escape:

$$E_o = E_i - U_{sb}. \qquad (\text{VII}.48)$$

Combined External Spectra Assuming the Spherical Surface Binding Model

For the spherical model, the particles are slowed but undeflected. The angular transformation is trivial: $\theta_i = \theta_o$. Equation (VII.47) becomes

$$\frac{dj_\Omega}{dE_o} = z^+ N \frac{U_{sb}}{(E_o + U_{sb})^2} \cdot \frac{R_r^p(E_o)}{R_p^p} \cdot \frac{\cos \theta_o}{4\pi} \qquad (\text{VII}.49)$$

The cosine emission law still holds, as one might have expected. Since in the simplified collisional model $R_r^p(E_o)$ is proportional to E_o, the energy distribution is *nearly* the Thompson formula, except that the denominator is squared rather than cubed. There is a maximum at U_{sb} and the distribution falls off as $1/E_o$. It takes the planar surface binding model to get the Thompson formula.

Combined External Spectra Assuming the Planar Surface Binding Model

With the planar binding model, there is a refraction effect as a sputtered particle passes through the surface. Two relations must hold–its kinetic energy must be reduced by U_{sb}, and its momentum in the direction parallel to the surface must be conserved. Thus, some of the internal flux of recoils is cut off, and for the rest, those traveling in direction θ_i are shifted into a new direction θ_o. Using Figure VII.21 as a guide, it may easily be shown that

and
$$\sin^2 \theta_i = \frac{E_i - U_{sb}}{E_i} \sin^2 \theta_o \qquad (\text{VII.50})$$

$$\cos^2 \theta_i = \cos^2 \theta_o + \frac{U_{sb}}{E_i}[1 - \cos^2 \theta_i]. \qquad (\text{VII.51})$$

The former shows that as $E_i \to U_{sb}$, $\theta_i \to 0$. This means that for low-energy recoils, only those traveling in directions near to the perpendicular can escape. The latter shows that

$$\theta_o \geq \theta_i, \qquad (\text{VII.52})$$

implying that all particles are deflected away from the direction normal to the target surface.

To obtain the external spectra, we use

$$\frac{dj_{\Omega_o}}{dE_o} = \frac{dj_{\Omega_i}}{dE_i} \cdot \frac{dE_i}{dE_o} \cdot \frac{d\Omega_i}{d\Omega_o} \qquad (\text{VII.53})$$

with

$$\frac{dE_i}{dE_o} = 1 \qquad (\text{VII.54})$$

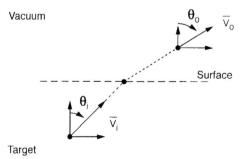

Figure VII.21 Effects of a planar surface potential barrier. Refraction of an escaping particle involves a slowing, and a deflection away from the surface normal direction.

and

$$\frac{d\Omega_i}{d\Omega_o} = \frac{E_o \cos\theta_o}{E_i \cos\theta_i}. \quad \text{(VII.55)}$$

The latter derivative is obtainable from Figure VII.21. The external combined flux spectra become

$$\frac{dj_{\Omega_o}}{dE_o} = z^+ N \frac{U_{sb} \cdot E_o}{(E_o + U_{sb})^3} \cdot \frac{R_r^p(E_o)}{R_p^p} \cdot \frac{\cos\theta_o}{4\pi}. \quad \text{(VII.56)}$$

Again, the cosine distribution holds. Ignoring the energy dependence of $R_r^p(E_o)$, the energy spectrum would be exactly that of the Thompson formula.

On replacing the function $R_r^p(E_o)$ with the expression $R_r^p(E_{av} - 1.39U_{sb})$ as in the simplified collisional model, and then integrating over all possible values of E_o, the cosine law is again obtained, except that there is an additional factor of $\frac{1}{2}$. (There *must* be a reduction in the yield as compared to that of the spherical model, since escaping particles are deflected away from the surface normal direction and yet the cosine distribution is retained.) Computer simulations predict a reduction by a factor of $\frac{1}{2}$ to $\frac{2}{3}$ [Andersen, 1987].

How to decide which surface potential model holds? The reduction in yield for the planar potential model would cause it not to fit most of the data of Section VII.2, as well as the spherical model does. On the other hand, the external energy spectrum of the planar model is closer to the Thompson formula.

VII.6 SUMMARY OF PRINCIPAL EQUATIONS NOT FOUND IN SAMPLE CALCULATION OF YIELD

Maximum ion energy	$E = q[V_p - v_{cat}]$
Bohdansky et al. empirical yield expression	$Y = (6.4 \times 10^{-3}) m_r \gamma^{5/3} E^{0.25} (1 - E_{th}/E)^{3.5}$
Bohdansky et al. empirical threshold energy	$E_{th} = \begin{cases} [U_{sb}/\gamma(1-\gamma)], (m_p/m_r) < 0.3 \\ U_{sb} 8(m_p/m_r)^{2/5}, (m_p/m_r) > 0.3 \end{cases}$
Steinbrüchel's semiempirical formula	$Y(E) = (5.2/U_{sb}) \cdot [Z_r/(Z_r^{2/3} + Z_p^{2/3})^{3/4}] \cdot [Z_p/(Z_r + Z_p)]^{0.67} \cdot E^{1/2}$
Approximate energy dependence of yield	$Y(E) = Y(1\text{ keV}) \cdot (E/1\text{ keV})^{0.5}$
Yield versus surface binding energy	$Y(1\text{-keV argon projectiles}) \propto (1/U_{sb}^{1.3})$ (all empirical data).
Yamamura's formula for angle of incidence	$Y(\eta) = Y(0)/(\cos^f \eta) \cdot \exp[f \cdot \cos\eta_{opt} \cdot [1 - (1/\cos\eta)]]$
Thompson formula	$dY/dE_o \sim [E_o U_{sb}/(E_o + U_{sb})^3]$
Blech and Van der Plas' angular distribution	$dY/d\Omega = Y_{total} \cdot [\cos\theta/(\rho\sin^2\theta + \cos^2\theta)\pi]$

APPENDIXES

Energy transfer function	$T(\delta_r) \equiv [4m_p m_r/(m_p + m_r)^2](\cos \delta_r)^2$
Energy transfer mass factor	$\gamma \equiv 4m_p m_r/(m_p + m_r)^2$
Projectile threshold energy	$E_{th} \equiv 4U_{sb}/\gamma$
Probability distribution function	$F_i(E_i) \approx U_{sb}/E_i^2$
Threshold average energy of recoils	$E_{i,th} \equiv 1.39 U_{sb}$
Theoretical binding energy dependence of yield	$Y_{theor}(1\text{-keVAr}^+) \equiv 2.41/U_{sb}^{0.52}$
Angular distribution of yield	$dY/d\Omega = Y_{total}(\cos\theta/\pi)$
Emission flux angular distribution	$j_\Omega(\theta) = Yz^+(\cos\theta/\pi)$
Combined internal flux spectra	$dj_{\Omega_o}/dE_i = z^+ N(U_{sb}/E_i^2)$ $\cdot [R_r^p(E_i - U_{sb})/R_p^p]\cdot(\cos\theta_i/4\pi)$
Energy transformation	$E_o = E_i - U_{sb}$
Emission angle transformation for spherical binding model	$\theta_o = \theta_i$
Combined External flux spectra for spherical binding model	$dj_{\Omega_o}/dE_o = z^+ N[U_{sb}/(E_o + U_{sb})^2]\cdot$ $[R_r^p(E_o)/R_p^p]\cdot(\cos\theta_o/4\pi)$
Emission angle transformation for planar binding model	$d\Omega_i/d\Omega_o = (E_o/E_i)(\cos\theta_o/\cos\theta_i)$
Combined external flux spectra for planar binding model	$dj_{\Omega_o}/dE_o = z^+ N[(U_{sb}\cdot E_o)/(E_o + U_{sb})^3]\cdot[R_r^p(E_o)/R_p^p]\cdot(\cos\theta_o/4\pi)$

VII.7 APPENDIXES

Appendix A: The Empirical Yield Formula of Matsunami et al. [1984]

The purpose of Appendix A is to present the empirical yield formula of Matsunami et al. [1984], and give an example of its application, to the 1-keV argon ion sputtering of aluminum:

$$Y(E) = 0.42 \frac{\alpha^* Q s_n(\varepsilon)}{U_s[1 + 0.35 U_s s_e(\varepsilon)]} \cdot \frac{8.478 Z_p Z_r m_p}{(Z_p^{2/3} + Z_r^{2/3})^{1/2}(m_p + m_r)} \cdot \left[1 - \left(\frac{E_{th}}{E}\right)^{1/2}\right]^{2.8} \quad \text{(VII.57)}$$

where U_s is the sublimation energy. The three empirical parameters are α^*, Q, and E_{th} (whose physical meanings were not stated); α^* and E_{th} are calculated from the empirical formulas

$$\alpha^*\left(\frac{m_r}{m_p}\right) = 0.08 + 0.164 \cdot \left(\frac{m_r}{m_p}\right)^{0.4} + 0.0145 \cdot \left(\frac{m_r}{m_p}\right)^{1.29} \quad \text{(VII.58)}$$

and

$$E_{th}\left(U_s, \frac{m_r}{m_p}\right) = U_s \cdot \left[1.9 + 3.8 \cdot \left(\frac{m_r}{m_p}\right)^{-1} + 0.134 \cdot \left(\frac{m_r}{m_p}\right)^{1.24}\right]. \quad \text{(VII.59)}$$

The value of Q is tabulated in Appendix B.
The reduced energy is given by

$$\varepsilon(E) = \frac{0.03255}{Z_p Z_r (Z_p^{2/3} + Z_r^{2/3})^{1/2}} \cdot \frac{m_r}{m_p + m_r} E. \qquad (VII.60)$$

Lindhard's elastic (nuclear) reduced stopping cross section is approximated with

$$s_n(\varepsilon) = \frac{3.441\sqrt{\varepsilon}\ln(\varepsilon + 2.718)}{1 + 6.355\sqrt{\varepsilon} + \varepsilon(-1.708 + 6.882\sqrt{\varepsilon})}. \qquad (VII.61)$$

The inelastic (electronic) reduced stopping cross section is approximated with

$$s_e(\varepsilon) = 0.079 \cdot \frac{(m_p + m_r)^{3/2}}{m_p^{3/2} m_r^{1/2}} \cdot \frac{Z_p^{2/3} Z_r^{1/2}}{(Z_p^{2/3} + Z_r^{2/3})^{3/4}} \varepsilon^{1/2}. \qquad (VII.62)$$

Calculation of Empirical Yield for 1-keV Ar^+ Sputtering of Aluminum
Data:

$m_p = 39.95$ $Z_p = 18$ $Q = 1.09$
$m_r = 26.98$ $Z_r = 13$ $U_s = 3.39$

Reduced projectile energy from Equation (VII.60):

$$\varepsilon = 0.0159.$$

Nuclear and electronic reduced stopping cross sections, from Equations (VII.61) and (VII.62):

$$s_n(\varepsilon) = 0.244$$
$$s_e(\varepsilon) = 0.0156$$

Values of α^* and E_{th} from Equations (VII.58) and (VII.59):

$$\alpha^* = 0.229, \quad E_{th} = 25.8\,\text{eV}$$

Yield from Equation (VII.57): $Y = 1.53$

Appendix B: A Summary of Target Parameters

In this Appendix we provide density data, and thermodynamic data that may be used to estimate the surface binding energy. Weast's [1975] values are "the standard heat of vaporization at room temperature." Gschneidner's [1964] values are "the heat of sublimation at 298 K." Matsunami et al.'s [1984] values referenced Kittel [1976]; while Matsunami et al. calls them "the sublimation energy," they are actually tabulated as cohesive energies in Kittel's book, (M denotes a missing value):

Target	Number Density (atom/Å³) Weast [1975]	Number Density (atom/Å³) Barrett and Massalski [1980]	Surface Binding Energy Estimate (eV/atom) Weast [1975]	Surface Binding Energy Estimate (eV/atom) Gschneider [1964]	Surface Binding Energy Estimate (eV/atom) Matsunami et al. [1984]	Matsunami et al. [1984] Q values
Lithium	M	0.0463	M	1.68	M	M
Beryllium	0.1210	0.1230	3.45	3.38	3.32	2.17 ± 0.82
Boron	M	0.1300	M	5.73	5.77	4.6 ± 1.5
Carbon	0.1760	0.1760	M	7.42	7.37	3.1 ± 0.9
Sodium	0.0265	0.0253	1.15	1.12	M	M
Magnesium	0.0430	0.0430	1.61	1.55	M	M
Aluminum	0.0602	0.0602	3.37	3.36	3.39	1.09 ± 0.14
Silicon	0.0500	0.0500	4.30	4.69	4.63	0.78 ± 0.17
Phosphorus	M	0.0603	M	3.27	M	M
Sulfur	M	0.0392	M	2.88	M	M
Potassium	M	0.0133	M	0.93	M	M
Calcium	M	0.0230	M	1.83	M	M
Scandium	M	0.0427	M	3.49	M	M
Titanium	0.0566	0.0567	5.10	4.90	4.85	0.58 ± 0.10
Vanadium	0.0722	0.0720	5.38	5.34	5.31	0.9 ± 0.3
Chromium	0.0833	0.0833	3.39	4.12	4.10	1.23 ± 0.21
Manganese	0.0818	0.0819	2.96	2.92	2.92	M
Iron	0.0850	0.0850	4.51	4.34	4.28	1.06 ± 0.18
Cobalt	0.0897	0.0898	4.72	4.43	4.39	1.0 ± 0.32
Nickel	0.0914	0.0914	4.60	4.47	4.44	1.06 ± 0.26
Copper	0.0845	0.0847	3.67	3.52	3.49	1.30 ± 0.22
Zinc	0.0655	0.0656	1.42	1.35	1.35	M
Gallium	M	0.0511	M	2.82	M	M
Germanium	0.0442	0.0442	3.77	3.88	3.85	0.83 ± 0.10
Arsenic	M	0.0464	M	1.26	M	M

	Number Density (atom/Å³)		Surface Binding Energy Estimate (eV/atom)				Matsunatmi et al. [1984] Q values
Target	Weast [1975]	Barrett and Massalski [1980]	Weast [1975]	Gschneider [1964]	Matsunami et al. [1984]	Matsunatmi [1984]	
Selenium	M	0.0367	M	1.71	M	M	
Rubidium	M	0.0108	M	0.86	M	M	
Strontium	M	0.0178	M	1.70	M	M	
Yttrium	M	0.0303	M	4.25	M	M	
Zirconium	0.0429	0.0430	5.68	6.34	6.25	0.70 ± 0.16	
Niobium	0.0556	0.0556	8.37	7.60	7.57	1.02 ± 0.09	
Molybdenum	0.0642	0.0642	7.00	6.86	6.82	0.84 ± 0.24	
Technetium	0.0704	0.0704	M	6.60	M	M	
Ruthenium	0.0736	0.0737	7.22	6.68	6.74	1.52 ± 0.20	
Rhodium	0.0726	0.0726	6.21	5.77	5.75	1.26 ± 0.18	
Palladium	0.0680	0.0679	4.22	3.91	3.89	1.10 ± 0.25	
Silver	0.0585	0.0586	3.12	2.97	2.95	1.21 ± 0.19	
Cadmium	0.0464	0.0463	1.23	1.16	1.16	M	
Indium	0.0383	0.0382	2.56	2.49	M	M	
Tin	0.0362	0.0362	3.20	3.12	3.14	0.47 ± 0.14	
Antimony	0.0331	0.0331	2.85	2.72	M	M	
Tellurium	0.0294	0.0294	2.26	2.02	M	M	
Iodine	M	0.0234	M	M	M	M	
Lanthanum	M	0.0269	M	4.42	M	M	
Cerium	M	0.0291	M	4.24	M	M	
Praeseodynium	M	0.0293	M	3.71	M	M	
Neodymium	M	0.0293	M	3.28	M	M	
Promethium	M	M	M	2.78	M	M	
Samarium	M	0.0303	M	2.17	M	M	
Europium	M	0.0205	M	1.85	M	M	

Gadolinium	M	0.0302	M	3.58	M	M
Terbium	M	0.0321	M	3.89	M	M
Dysprosium	M	0.0317	M	2.89	M	M
Holmium	M	0.0321	M	3.05	M	M
Erbium	M	0.0326	M	3.06	M	M
Thulium	M	0.0332	M	2.52	M	M
Ytterbium	M	0.0303	M	1.74	M	M
Lutetium	M	0.0339	M	4.29	M	M
Hafnium	0.0452	0.0451	M	6.34	6.44	0.75 ± 0.08
Tantalum	0.0555	0.0555	8.35	8.12	8.10	0.78 ± 0.19
Tungsten	0.0632	0.0631	9.11	8.68	8.90	1.10 ± 0.18
Rhenium	0.0680	0.0680	8.55	8.07	8.03	1.27 ± 0.22
Osmium	M	0.0715	7.83	8.12	8.17	1.47 ± 0.19
Iridium	0.0706	0.0707	7.45	6.90	6.94	1.37 ± 0.22
Platinum	0.0662	0.0662	5.50	5.86	5.84	1.13 ± 0.17
Gold	0.0590	0.0590	3.70	3.80	3.81	1.04 ± 0.23
Mercury	M	0.0427	M	0.64	M	M
Thallium	M	0.0350	M	1.88	M	M
Lead	0.0330	0.0330	2.06	2.03	2.03	M
Bismuth	M	0.0283	M	2.17	M	M
Polonium	M	0.0267	M	1.50	M	M
Francium	M	M	M	0.79	M	M
Radium	M	M	M	1.82	M	M
Actinium	M	0.0267	M	4.52	M	M
Thorium	M	0.0304	M	M	6.20	0.9 ± 0.3
Protactinium	M	0.0401	M	5.73	M	M
Uranium	M	0.0481	5.63	M	5.55	0.81 ± 0.13
Neptunium	M	0.0520	M	4.91	M	M
Plutonium	M	0.0427	M	3.99	M	M

Appendix C: Some Collisional Sputtering Theories

We summarize in this Appendix the yield expressions of several well-known collisional sputtering theories. (Some of the authors' original notation has been altered to conform with this present chapter.) More detailed attention is given to that of Sigmund, because of his unique influence on the field.

A Collision Model of Sputtering A rather early radiation damage model was that of Pease [1959], who gave the yield as a function of the average energy of the displaced target atoms:

$$Y = (\sigma E_{i,av} n_t^{2/3}) \cdot \frac{1}{4E_d} \cdot \left[1 + \left(\frac{\log(E_{i,av}/U_{sb})}{\log 2}\right)^{1/2}\right], \quad \text{(VII.63)}$$

where E_d is the displacement energy and σ is the cross-section for imparting energy greater than E_d. This expression is the product of the mean energy deposited in each atomic layer in primary displacement collisions, times half the mean total number of displacements per unit energy deposited, times the mean number of atomic layers that contribute to the sputtering.

A Transport Theory of Sputtering According to Sigmund [1969],

> A yield calculation consists of a number of steps: (1) to determine the amount of energy deposited by energetic particles (ion and recoil atoms) near the surface; (2) to convert this energy into a number of low-energy recoil atoms; (3) to determine how many of these atoms come to the very surface; and (4) to select those atoms that have sufficient energy to overcome the surface binding forces.

His general yield expression is

$$Y(E) = \Lambda \alpha n_t S_n(E), \quad \text{(VII.64)}$$

where Λ is a *material constant* that includes the range of a displaced target atom and the probability of ejection of an atom at the surface and α is a dimensionless *correction factor*. This yield expression takes low- and high-energy forms. For projectile energies smaller than 1 keV the result is

$$Y(E) = \left(\frac{3}{4\pi^2}\right) \alpha \frac{\gamma E}{U_{sb}}. \quad \text{(VII.65)}$$

Sigmund gives α in graphical form. This low-energy yield expression is linear in E, and does not predict a threshold.

For energies greater than 1 keV the result is

$$Y(E) = \frac{0.0420 \alpha S_n(E)}{U_{sb} \cdot \text{Å}^2}. \quad \text{(VII.66)}$$

Sigmund expressed the nuclear energy loss cross section as

$$S_n(E) = \frac{4\pi a_p Z_p Z_r q^2 m_p}{m_p + m_r} \cdot s_n(\varepsilon), \qquad (VII.67)$$

where s_n is a "universal function" which Sigmund gave in tabular form. As a general rule, Sigmund [1969] used thermodynamic data (heat of sublimation) for U_{sb}.

A Statistical Model of Sputtering This model defines a target volume in which atom displacements are probable [Schwarz and Helms, 1979]. The sputter yield is simply the area of intersection of this statistical volume with the target surface, times the surface density of displaced atoms, times the probability B that a displaced surface atom actually escapes:

$$Y = B\left(\frac{3\pi^{1/2}}{8E_d}\right)^{2/3} \frac{E^{2/3}}{1 + k\left(\frac{8}{3}\pi E_d A n_t\right)^{1/3} E^{2/3}} \left[1 - \left(\frac{k\left(\frac{8}{3}\pi E_d A n_t\right)^{1/3} E^{2/3}}{1 + k\left(\frac{8}{3}\pi E_d A n_t\right)^{1/3} E^{2/3}}\right)^2\right],$$

(VII.68)

where A is the fraction of atoms in the volume that are displaced. A fixed value for A of 0.0012 was found to give good results. Schwarz and Helms empirically determined B and published a table of its values. The term k is the inverse of twice the nuclear stopping power of the ion. The nuclear stopping power was calculated using an energy-independent expression.

Thompson's Theory Thompson [1981] calculated the yield due to a transient flux of recoils which constitute the collision cascade. His basic result is

$$Y \approx \frac{D \cdot k \cdot dE/dR|_n}{2U_{sb} \cos \eta}, \qquad (VII.69)$$

where D is the mean interatomic distance, given by $(n_t)^{-1/3}$ and k is a parameter in the radiation damage function used to calculate the number of displaced atoms per recoil ($kE_r/2E_d$). Thompson observes that this expression "shows the well known inverse dependence of yield on [surface] binding energy U_{sb} ... and is proportional to the collisional stopping power $[dE/dR|_n]$ at the surface, which determines the dependence of $[Y]$ on E."

Thompson also observed that his calculations are "qualitatively similar to the results of Sigmund's sputtering theory based on Boltzmann transport theory, incorporating anisotropy of recoil distribution, slowing down and reflection of ions. Its results, though more accurate, cannot be presented in such simple form."

Direct Derivation of Yield Referring to the yield expression of the Sigmund theory, Kelly [1984] stated that "A very direct and somewhat more correct derivation of S_{cascade} is also possible." It is based on the expression for impingement rate from the kinetic theory of gases: $nv_{\text{av}}/4$, where n is the density and v_{av} is the average speed. The $\frac{1}{4}$ gives "the fraction of the flux crossing an arbitrary plane in unit time assuming the flux to be isotropic."

The projectile deposits a fraction $[C_n(0)\lambda]$ of its energy in the outermost atomic layer, calculated as the energy distribution function (C_n) evaluated at the surface times the mean atomic spacing (λ). The number of recoils is given by $EC_n(0)\lambda/U_{\text{sb}}$. The yield is therefore

$$Y = \frac{1}{4} \frac{EC_n(0)\lambda}{U_{\text{sb}}}. \tag{VII.70}$$

Kelly observed that his and Sigmund's yield expressions "contain certain features in common. These are that the relevant energy deposition is confined to the outer atomic layer and that the motion is isotropic. The direct derivation has the advantage, however, of circumventing all assumptions relating to the binary atom-atom interactions, and this is why it can be regarded as more correct. The numerical difference is, of course, unimportant."

Kelly adds the following factor to Equation (VII.70) to account for the threshold: $1 - [E_{\text{th}}/\gamma E]^{1/2}$.

Appendix D: A Sample Calculation of Yield with the Simplified Collisional Model

The Sputtering of Aluminum by 1-keV Argon Ions Data:

$m_p = 39.9$ amu $Z_p = 18$ $\Delta_{\text{sub}}H = 3.37$ eV/atom $E = 1$ keV

$m_r = 27.0$ $Z_r = 13$ $n_t = 0.0602$ Å^{-3}

Constants:

$$a_0 = 0.529 \text{ Å}, \quad q = (14.4 \text{ eV/Å})^{1/2}$$

Model Equations*	Value
Surface binding energy: $U_{\text{sb}} = \Delta_{\text{sub}}H$	3.37 eV
Energy transfer mass factor: $\gamma = 4m_p m_r/(m_p + m_r)^2$	0.964
Average energy of recoils: $E_{i,\text{av}} \equiv U_{\text{sb}} \ln(\gamma E/U_{\text{sb}})$	19.1 eV
Thomas–Fermi screening length of projectile: $a_p = 0.8853 a_0/[Z_p^{2/3} + Z_r^{2/3}]^{1/2}$	0.133 Å
Thomas-Fermi screening length of recoil: $a_r = 0.8853 a_0/[2Z_r^{2/3}]^{1/2}$	0.141 Å

*With permission from J.E. Mahan and A. Vantomme, 1997, "A Simplified Collisional Model of Sputtering in the Linear Cascade Regime," *J. Vac. Sci. Technol.* **A15**(4), 1976. Copyright 1997 by the American Vacuum Society.

Reduced energy of projectile:
$\varepsilon_p(E) = [a_p m_r / Z_p Z_r q^2 (m_p + m_r)] E$ 0.0159

Reduced average energy of recoil:
$\varepsilon_{av}(E) = [a_r / [Z_r^2 q^2 \cdot 2] E_{i,av}$ 0.000551

Nuclear energy loss cross section of projectile:
$S_{np}(E) = [4\pi a_p Z_p Z_r q^2 m_p / (m_p + m_r)]$
$\cdot [\ln(1 + \varepsilon_p)] / [2(\varepsilon_p + 0.14 \varepsilon_p^{0.42})]$ 655 eV·Å²

Nuclear energy loss cross section of recoil:
$S_{nr}(E_{av}) = [4\pi a_r (Z_r)^2 q^2 / 2]$
$\cdot [\ln(1 + \varepsilon_{av})] / [2(\varepsilon_{av} + 0.14 \varepsilon_{av}^{0.42})]$ 90.7 eV·Å²

Stopping power for projectile: $dE/dR|_{np} = S_{np} n_t$ 39.4 eV/Å

Stopping power for recoil: $dE/dR|_{nr} = S_{nr} n_t$ 5.46 eV/Å

Range of projectile: $R_p(E) \approx E/(dE/dR|_{np})$ 25.4 Å

Effective range of recoil: $R_{r,eff} \approx (E_{i,av} - 1.39 U_{sb})/(dE/dR|_{nr})$ 2.63 Å

Projected range of projectile: $R_p^p \approx R_p / [1 + 0.4(m_r/m_p)]$ 20.0 Å

Effective projected range of recoil: $R_{r,eff}^p \approx 0.714 R_{r,eff}$ 1.88 Å

Number of recoils: $N = E/E_{i,av}$ 52

Yield: $Y = E/E_{i,ave} \cdot R_{r,eff}^p / R_p^p \cdot \frac{1}{4}$ 1.23

VII.8 MATHEMATICAL SYMBOLS, CONSTANTS, AND THEIR UNITS

SI units are given first, followed by other units in widespread use.

a	Screening length (m; Å = 10^{-10} m)
a_0	Bohr radius (0.529 Å)
a_i	Unit vector in the i direction
b	Fraction of kinetic energy retained by the projectile
f	Fraction of recoils close enough to the surface to escape
j_Ω	Emission flux distribution (m^{-2} · s^{-1} · steradian)
m_p	Projectile mass (kg)
m_r	Recoil ion mass (kg)
n_t	Number density of target atoms (m^{-3})
q	Magnitude of the electronic charge (1.60 × 10^{-19} C; 4.80 × 10^{10} (cm³ · g · s^{-2})$^{1/2}$; (14.4 eV/Å)$^{1/2}$)
s_e	Inelastic reduced stopping cross section (dimensionless)
s_n	Nuclear reduced stopping cross section (dimensionless)
v_{pi}	Projectile's initial velocity (m/s)
v_{pf}	Projectile's final velocity
v_{ti}	Target atom's initial velocity
v_{tf}	Target atom's final velocity
v_{cat}	Cathode potential (V)
z	Depth (m)
z^+	Impingement flux of projectiles (m^{-2} · s^{-1})
A	Sputtering target area (m²)

E	Projectile energy (J; $eV = 1.60 \times 10^{-19}$ J)
E_d	Displacement energy
E_i	Energy of recoil inside the target
E_o	Energy of recoil outside the target
E_R	Rydberg energy (Ry $= 2.18 \times 10^{-18}$ J $= 13.6$ eV)
E_{th}	Sputtering threshold energy
F_D	Energy deposited per unit depth (eV/Å)
F_i	Probability distribution function inside the target (eV^{-1})
F_o	Probability distribution function outside the target
J_Ω	Beam intensity (s^{-1} · steradian)
K	Constant used to estimate number of recoils produced (dimensionless)
N	Effective number of recoils produced
N_a	Avogadro's number (6.02×10^{23})
Q	A parameter in Matsunami et al.'s yield formula (dimensionless)
R	Range (m; Å $= 10^{-10}$ m)
R^p	Projected range
S_n	Nuclear energy loss cross section (eV · Å^2)
T	Energy transfer function (dimensionless)
U_s	Matsunami et al. sublimation energy (eV)
U_{sb}	Surface binding energy (eV)
V_p	Plasma potential (V)
Y	Sputter yield
Z	Atomic number
α^*	A parameter in Matsunami et al. yield formula (dimensionless)
γ	Energy transfer mass factor (dimensionless)
δ	Scattering angle (rad)
$\Delta_{fus}H$	Enthalpy of fusion (kJ/mol; eV/particle $= 96.3$ kJ/mol)
$\Delta_{sub}H$	Enthalpy of sublimation
$\Delta_{vap}H$	Enthalpy of vaporization
ε	Reduced energy (dimensionless)
η	Angle of incidence (rad)
θ	Emission angle (rad)
λ	Mean free path (m)
ρ	Empirical parameter for fitting sputter yield angular distribution (dimensionless)
φ	Polar angle (rad)
Ω	Solid angle (steradians)

REFERENCES

Almén, O., and Bruce, G., 1961, "Sputtering Experiments in the High Energy Region," *Nucl. Instr. Meth.* **11**, 279.

Andersen, H. H., and Bay, H. L., 1975, "Heavy-Ion Sputtering Yield of Silicon," *J. Appl. Phys.*, **46**(5), 5, 1919.

Andersen, H. H., and Bay, H. L., 1981, "Sputtering Yield Measurements," in R. Behrish, ed., *Sputtering by Particle Bombardment I*, Springer-Verlag, Berlin.

REFERENCES

Andersen, H. H., 1987, "Computer Simulations of Atomic Collisions in Solids with Special Emphasis on Sputtering," in G. Betz, W. Husinsky, P. Varga, and F. Viehböck, eds., *Symposium on Sputtering*, Proc. Symp. Sputtering, Spitz an der Donau, Austria, June, 2–6, 1986, North-Holland, Amsterdam.

Barrett, C., and Massalski, T. B., 1980, *Structure of Metals*, 3rd rev. ed., Pergamon Press, Oxford, pp. 626 ff.

Behrisch, R., 1981, "Introduction and Overview," in R. Behrish, ed., *Sputtering by Particle Bombardment I*, Springer-Verlag, Berlin.

Betz, G., Husinsky, W., Varga, P., and Viehböck, F., eds., 1987, *Symposium on Sputtering*, Proc. Symp. Sputtering, Spitz an der Donau, Austria, June, 2–6, 1986, North-Holland, Amsterdam.

Bird, J. R., Brown, R. A., Cohen, D. D., and Williams, J. S. 1989, "Data Lists," in J. R. Bird and J. S. Williams, *Ion Beams for Materials Analysis*, Academic Press, Australia, Marrickville, NSW, Chapter 14.

Blech, I. A., and Van der Plas, H. A., 1983, "Step Coverage Simulation and Measurement in a DC Planar Magnetron Sputtering System," *J. Appl. Phys.* **54**, 3489.

Bohdansky, J., Roth, J., and Bay, H. L., 1980, "An Analytical Formula and Important Parameters for Low-Energy Ion Sputtering," *J. Appl. Phys.* **51**(5), 2861.

Bohdansky, J., 1984, "A Universal Relation for the Sputtering Yield of Monatomic Solids at Normal Ion Incidence," *Nucl. Instr. Meth.* **B2**, 587.

Bohdansky, J., Lindner, H., Hechtl, E., Martinelli, A. P., and Roth, J., 1987, "Sputtering Yield of Cu and Ag at Target Temperatures Close to the Melting Point," *Nucl. Instr. Meth.* **B18**, 509.

Carter, G., and Colligon, J. S., 1968, *Ion Bombardment of Solids*, American Elsevier, New York.

Eckstein, W., García-Rosales, C., Roth, J., and Ottenberger, W., 1993, *Sputtering Data*, Max-Planck Institut für Plasmaphysik, Report IPP 9/82.

Falcone, G., Kelly, R., and Oliva, A. 1987, "Corrections to the Collisional Sputtering Yield," *Nucl. Instr. Meth.* **B18**, 399.

García-Rosales, C., Eckstein, W., and Roth, J., 1994, "Revised Formulae for Sputtering Data," *J. Nucl. Mat.* **218**, 8.

Gschneider, K. A., 1964, "Physical Properties and Interrelationships of Metallic and Semimetallic Elements," *Sol. St. Phys.* **16**, 275.

Harrison, D. E., 1983, "Sputtering Models—a Synoptic View," *Rad. Effects* **70**, 1.

Kelly, R., 1984, "The Mechanisms of Sputtering—Part I. Prompt and Slow Collisional Sputtering," *Rad. Effects* **80**, 273.

Kelly, R., 1987, "The Surface Binding Energy in Slow Collisional Sputtering," *Nucl. Instr. Meth.* **B18**, 388.

Kittel, C., 1976, *Introduction to Solid State Physics*, Wiley, New York, p. 74.

Laegreid, N., and Wehner, G. K., 1961, "Sputtering Yields of Metals for Ar+ and Ne+ Ions with Energies from 50 to 600 eV," *J. Appl. Phys.* **32**(3), 365.

Lamar, E. S., and Compton, K. T., 1934, "A Special Theory of Sputtering," *Science* **80**, 541.

Long, R. G., 1993, private communication.

Mahan, J. E., and Vantomme, A., 1997, "A Simplified Collisional Model of Sputtering in the Linear Cascade Regime," *J. Vac. Sci. Technol.* **A15**(4), 1976.

Mahan, J. E., and Vantomme, A., 1999, "Trends in Sputter Yield Data in the Film Deposition Regime," unpublished.

Matsunami, N., Yamamura, Y., Itikawa, Y., Itoh, N., Kazumata, Y., Miyagawa, S., Morita, K., Shimizu, R., and Tawara, H., 1984, "Energy Dependence of the Ion-Induced Sputtering Yields of Monatomic Solids," *Atom. Data Nucl. Data Tables* **31**, 1.

Oechsner, H., 1973, "Sputtering of Polycrystalline Metal Surfaces at Oblique Ion Bombardment in the 1 keV Range" (in German), *Z. Phys.* **261**, 37.

Oechsner, H., 1975, "Sputtering – a Review of Some Recent Experimental and Theoretical Aspects," *Appl. Phys.* **8**, 185.

Pease, R. S., 1959, "Sputtering of Solids by Penetrating Ions," *Proc. Internatl. School of Physics "Enrico Fermi"*, Academic Press, New York, p. 158.

Rauhala, E., 1995, "Energy Loss," in J. R. Tesmer and M. Nastasi, eds., *Handbook of Modern Ion Beam Materials Analysis*, Materials Research Society, Pittsburgh, PA, Chapter 2.

Robinson, M. T., 1993, "Computer Simulation of Sputtering," in P. Sigmund, ed., *Fundamental Processes in Sputtering of Atoms and Molecules*, Symp. on Occasion of 250th Anniversary of the Royal Danish Academy of Sciences and Letters, Copenhagen, Aug. 30–Sept. 4, 1992, The Royal Danish Academy of Sciences and Letters, Copenhagen.

Roosendaal, H. E., 1981, "Sputtering Yields of Single Crystalline Targets," in R. Behrisch, ed., *Sputtering by Particle Bombardment I*, Springer-Verlag, Berlin.

Roth, J., 1983, "Chemical Sputtering," in R. Behrisch, ed., *Sputtering by Particle Bombardment II*, Springer-Verlag, Berlin.

Scherzer, B. M. U., 1983, "Development of Surface Topography Due to Gas Ion Incorporation," in R. Behrisch, ed., *Sputtering by Particle Bombardment II*, Springer-Verlag, Berlin.

Schwarz, S. A., and Helms, C. R., 1979, "A Statistical Model of Sputtering," *J. Appl. Phys.* **50**(8), 5492.

Seitz, F., and Koehler, J. S., 1956, "Displacement of Atoms During Irradiation," in F. Seitz and D. Turnbull, eds., *Solid State Physics*, Vol. 2, Academic Press, New York, p. 305.

Sigmund, P., 1969, "Theory of Sputtering. I. Sputtering Yield of Amorphous and Polycrystalline Targets," *Phys. Rev.* **184**(2), 383.

Sigmund, P., 1981, "Sputtering by Ion Bombardment: Theoretical Concepts," in R. Behrisch, ed., *Sputtering by Particle Bombardment I*, Springer-Verlag, Berlin.

Sigmund, P., 1993, "Introduction to Sputtering," in P. Sigmund, ed., *Fundamental Processes in Sputtering of Atoms and Molecules*, Symp. on Occasion of 250th Anniversary of the Royal Danish Academy of Sciences and Letters, Copenhagen, Aug. 30– Sept. 4, 1992, The Royal Danish Academy of Sciences and Letters, Copenhagen.

Sigmund, P., and Claussen, C., 1981, "Sputtering from Elastic Collision Spikes in Heavy-Ion-Bombarded Metals," *J. Appl. Phys.* **52**, 990.

Spencer, E. G., and Schmidt, P. H., 1971, "Ion-Beam Techniques for Device Fabrication," *J. Vac. Sci. Technol.* **8**(5), S52.

Steinbrüchel, Ch., 1985, "A Simple Formula for Low-Energy Sputtering Yields," *Appl. Phys.* **A36**, 37.

Strydom, H. J., and Gries, W. H., 1984, "A Comparison of Three Versions of Sigmund's Model of Sputtering Using Experimental Results," *Rad. Effects Lett.* **86**, 145.

Stuart, R. V., Wehner, G. K., and Anderson, G. S., 1969, "Energy Distribution of Atoms Sputtered from Polycrystalline Metals," *J. Appl. Phys.* **40**(2), 803.

Thompson, M. W., 1968, "The Energy Spectrum of Ejected Atoms during the High Energy Sputtering of Gold," *Phil. Mag.* **18**, 377.

Thompson, M. W., 1981, "Physical Mechanisms of Sputtering," *Phys. Rep.* **69**(4), 335.

Thompson, M. W., and Nelson, R. S., 1962, "Evidence for Heated Spikes in Bombarded Gold from the Energy Spectrum of Atoms Ejected by 43 keV Ar^+ and Xe^+ Ions," *Phil. Mag.* **7**, 2015.

Tombrello, T. A., 1993, "Sputtering in Planetary Science," in P. Sigmund, ed., *Fundamental Processes in Sputtering of Atoms and Molecules*, Symp. on Occasion of 250th Anniversary of the Royal Danish Academy of Sciences and Letters, Copenhagen, Aug. 30–Sept. 4, 1992, The Royal Danish Academy of Sciences and Letters, Copenhagen.

Townsend, P. D., Kelly, J. C., and Hartley, N. E. W., 1976, *Ion Implantation, Sputtering, and their Applications*, Academic press, London.

Urbassek, H. M., and Hofer, W. O., 1993, "Sputtering of Molecules and Clusters: Basic Experiments and Theory," in P. Sigmund, ed., *Fundamental Processes in Sputtering of Atoms and Molecules*, Symp. on Occasion of 250th Anniversary of the Royal Danish Academy of Sciences and Letters, Copenhagen, Aug. 30 – Sept. 4, 1992, The Royal Danish Academy of Sciences and Letters, Copenhagen.

Weast, R. C., ed., 1975, *Handbook of Chemistry and Physics*, 56th, CRC Press, Cleveland, OH.

Wehner, G. K., and Anderson, G. S., 1970, "The Nature of Physical Sputtering," in L. I. Maissel and R. Glang, eds., *Handbook of Thin Film Technology*, McGraw-Hill, New York.

Wehner, G. K., 1962, "Physical Sputtering," 1961 *Transact. 8th Natl. Vacuum Symp.* Vol. 1, Pergamon Press, Oxford, p. 239.

Wehner, G. K., Kenknight, C. E., and Rosenberg, D. L., 1963, "Sputtering Rates Under Solar-Wind Bombardment," *Planet. Space Sci.* **11**, 885.

Wilson, W. D., Haggmark, L. G., and Biersack, J. P., 1977, "Calculations of Nuclear Stopping, Ranges, and Straggling in the Low-Energy Region," *Phys. Rev. B*, **15**(5), 2458.

Winograd, N., 1993, "Sputtering of Single Crystals: Experimental Evidence of the Ejection Process," in P. Sigmund, ed., *Fundamental Processes in Sputtering of Atoms and Molecules*, Symp. on Occasion of 250th Anniversary of the Royal Danish Academy of Sciences and Letters, Copenhagen, Aug. 30–Sept. 4, 1992, The Royal Danish Academy of Sciences and Letters, Copenhagen.

Yamamura, Y., Itikawa, Y., and Itoh, N., 1983, *Angular Dependence of Sputtering Yields of Monatomic Solids*, Research Information Center, Institute of Plasma Physics, Nagoya University, Japan.

Zalm, P. C., 1983, "Energy Dependence of the Sputtering Yield of Silicon Bombarded with Neon, Argon, Krypton, and Xenon Ions," *J. Appl. Phys.* **54**(5), 2660.

Zalm, P. C., and Beckers, L. J., 1984, "In Quest of the Spike: Energy Dependence of the Sputtering Yield of Zinc Bombarded with Neon and Xenon Ions," *Philips J. Res.* **39**, 11.

Ziegler, J. F., Biersack, J. P., and Littmark, U., 1985, *Stopping Powers and Ranges in All Elements*, Vol. 1, Pergamon Press, New York.

Ziegler, J. F., 1992, *The Transport of Ions in Matter*, Version 92.xx, copyright International Business Machines Corporation.

VIII

FILM DEPOSITION

Physical vapor deposition sources are fundamentally characterized with a beam intensity. This is the number of particles per second being emitted by the source in directions within $d\Omega$ ($\sin\theta\, d\theta\, d\varphi$) about the direction specified by θ (the polar angle) and φ (the azimuthal angle), as illustrated in Figure VIII.1a.

For most real sources, it is a good assumption that the beam intensity is independent of φ. For ideal small evaporation and sputtering sources,

$$J_\Omega = \begin{cases} \dfrac{z\,\delta A \cos\theta}{\pi} & \text{(ideal small evaporation source)} \\ \dfrac{Yz^+\delta A \cos\theta}{\pi} & \text{(ideal small sputtering source),} \end{cases} \quad \text{(VIII.1)}$$

where z is the thermal equilibrium impingement rate of the evaporant at the temperature of the source and Yz^+ is the total flux coming off a sputtering source, the product of the yield, and the incident ion flux (this latter formulation ignores the possibility of charge exchange in the cathode sheath, which creates a mixture of ions and neutrals that bombard the sputtering target); δA is the source area. "Ideal" encompasses both a cosine angular distribution and the intensity value contained in the above theoretical expressions.

For this cosine law [Eq. (VIII.1)] to hold, the source area must be both flat and small. "Flat" means that δA is planar. What makes a source "small" is simply that the linear extent of its source area (say, $\delta A^{1/2}$) must be sufficiently small compared to the distance to the substrate.

It is not uncommon for real sources to deviate to a practical degree from the ideal cosine law, even when they may be considered small. The polar angle

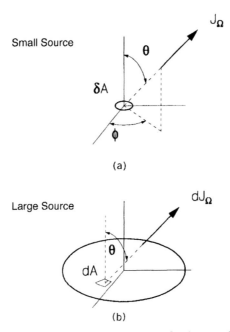

Figure VIII.1 Physical vapor deposition sources are fundamentally characterized with (a) a beam intensity or (b) a differential beam intensity.

dependence of beam intensity for an E-gun evaporator has been modeled with the following empirical expression [Graper, 1973]:

$$J_\Omega \sim (1 - C)\cos^n\theta + C. \quad \text{(VIII.2)}$$

For rather high-rate evaporation of aluminum (100–1000 Å/s), n was found to be in the range 2–6 and a typical value of C was 0.1.

Many practical applications of physical vapor deposition require the use of large substrate, and a high uniformity of the deposit. A high deposition rate is often required. These circumstances demand a "large" source (one for which the lateral size is not negligible relative to the distance to the substrate). For a large source, a differential beam intensity is assigned to each infinitesimal area element of the source (dA, as illustrated in Fig. VIII.1b). We will call large sources "ideal" if the differential source areas have the differential beam intensity of the ideal small source and if dJ_Ω is uniform across the source area:

$$dJ_\Omega = \begin{cases} \dfrac{z\,dA\cos\theta}{\pi} & \text{(ideal large evaporation source)} \\[2mm] \dfrac{Yz^+dA\cos\theta}{\pi} & \text{(ideal large sputtering source)}. \end{cases} \quad \text{(VIII.3)}$$

When combined with specific substrate configurations, a given beam intensity leads to a specific incident flux and lateral film thickness profile across the substrate. Film thickness control and uniformity can be crucial — applications in optics come to mind. Here the successful performance of the thin film may depend on optical interference effects, where the film thickness must be controlled to within a small fraction of the wavelength of visible light. We will see that the utilization of a large source primarily increases the deposition rate, while substrate movement during deposition can especially increase the film thickness uniformity.

In the historical development of the field of film deposition modeling, the name of Holland comes to the fore. However, one of the earliest contributions was (1) Knudsen's [1915] "Das Cosinusgesets in der kinetischen Gastheorie," where the cosine law was first proposed. In Fisher and Platt's [1937] "The Deposit of Films of Uniform Thickness for Interferometer Mirrors," a tilted substrate with off-axis substrate rotation was analyzed. The theory was systematized with (3) Holland and Steckelmacher's [1952] "The Distribution of Thin Films Condensed on Surfaces by the Vacuum Evaporation Method," where point sources, small flat sources, wire sources, ring sources, and planar and spherical receiving surfaces were considered, and with (4) Holland's [1963] book chapter, "The Emission Characteristics of Vapour Sources and Film Thickness Distribution," where a "thin plane strip source" and a ring source were analyzed.

VIII.1 INCIDENT FLUX AND FILM DEPOSITION RATE

The Incident Flux at the Substrate

We now calculate the local *incident flux* of particles onto a receiving surface for a given beam intensity. It depends on the position and orientation of the receiving surface. It will be calculated as a *flux*, meaning *rate of arrival of particles per unit area of the receiving surface*.

From solid geometry, a surface element dA_s subtends a solid angle $d\Omega$ with respect to a point O which is given by

$$d\Omega = \frac{(\vec{R}\cdot\vec{n})dA_s}{R^3}. \qquad (\text{VIII.4})$$

$\vec{R}$ is the vector from O to the element and $\vec{n}$ is the unit vector inwardly normal to the element, as shown in Figure VIII.2. Multiplying J_Ω by this $d\Omega$ and then dividing by dA_s gives a general expression for the local incident flux onto the receiving surface:

$$j_i = \frac{J_\Omega(\vec{R}\cdot\vec{n})}{R^3}. \qquad (\text{VIII.5})$$

If the angle between $\vec{R}$ and $\vec{n}$ is β, then the expression for the incident flux reduces to

$$j_i = \frac{J_\Omega \cos \beta}{R^2}. \tag{VIII.6}$$

This depends on the beam intensity, β, and the distance to the point of deposition in a quite reasonable fashion; β will be called the *deposition angle*.

Example Suppose that an aluminum film is being deposited with an effusion cell whose temperature is 1500 K. The orifice area is $1\,\text{cm}^2$ and the distance to the substrate is 10 cm. Calculate the impingement flux of aluminum atoms onto the substrate directly above the source (i.e., at $\theta = 0°$), considering the effusion cell to be an ideal small source.

First, we calculate the thermal equilibrium vapor pressure within the effusion cell. The standard molar entropy and enthalpy of vaporization are 118 J/K and 314 kJ, respectively, at the melting point of aluminum (from the Appendix of Chapter III; we ignore differences between the melting point and the source temperature). Then the vapor pressure of aluminum is given by

$$P_{eq}(1500\,\text{K}) = 10^5\,\text{Pa} \times \exp\frac{118\,\text{J/K}}{8.31\,\text{J/K}} \times \exp\frac{-314\,\text{kJ}}{8.31\,\text{J/K} \times 1500\,\text{K}}$$
$$= 1.26 \times 10^{-2}\,\text{torr}.$$

The thermal equilibrium impingement rate within the effusion cell is

$$z_{Al}(1500\,\text{K}) = \frac{1.26 \times 10^{-2}\,\text{torr}}{[2\pi \times 26.98 \times 1.66 \times 10^{-27}\,\text{kg} \times (1.38 \times 10^{-23}\,\text{J/K}) \times 1500\,\text{K}]^{1/2}}$$
$$= 221\,\text{Å}^{-2} \cdot \text{s}^{-1}.$$

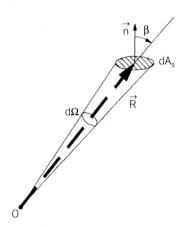

Figure VIII.2 A differential area element subtends a differential solid angle with respect to a point O that is given by $d\Omega = \cos \beta\, dA_s/R^2$.

This impingement rate corresponds to a beam intensity directly above the orifice of

$$J_\Omega(0°) = \frac{221\,\text{Å}^{-2}\cdot\text{s}^{-1} \times 1\,\text{cm}^2 \times \cos(0°)}{\pi} = 7.03 \times 10^{17}\,\text{s}^{-1}.$$

Finally, the incident flux at the point on the substrate under consideration is

$$j_i = \frac{7.03 \times 10^{17}\,\text{s}^{-1} \times \cos(0°)}{(10\,\text{cm})^2} = 0.703\,\text{Å}^{-2}\cdot\text{s}^{-1}. \qquad\blacksquare$$

Film Deposition Rate

The *condensation coefficient* (α_c) is the fraction of the incident flux that actually condenses:

$$j_c = \alpha_c j_i, \tag{VIII.7}$$

where j_c is the *condensation flux* (ignoring re-evaporation here, see page 63). For most strongly bonded elements, such as the refractory metals, α_c is very close to unity. However, in molecular, ionic, or other complex and anisotropic solids, α_c can be very much less than unity; α_c should always be considered to have the potential to vary, depending on the temperature and on the structure of the surface as it may have been affected by previous condensation or adsorption [Rosenblatt, 1976].

Resputtering reduces the net condensation flux of sputter-deposited films, by an amount varying from a few percent; to values approaching 100%. Hoffman [1990] defines resputtering as follows: "Resputtering refers to the concurrent removal of sputter deposited material by intentional or inadvertent bombardment during deposition." If f is the fraction of condensed particles that are subsequently resputtered, then the net condensation flux in the above equation would contain the additional factor $(1 - f)$.

Bias sputter deposition is a well-known instance of resputtering. The substrate is given a negative potential with respect to the plasma, so that it is intentionally bombarded by ions from the plasma during film growth. The amount and severity of the resputtering depends on the applied substrate bias. In addition, some resputtering mechanisms are *intrinsic* to sputter deposition. These include sputtering of the condensing film by energetic neutral projectiles, which have been reflected from the target, and self-sputtering by the condensing particles themselves (at least, those that have kinetic energies above the sputtering threshold). These occur with no substrate bias, and are difficult to remove from the sputter deposition process; hence they have been termed *residual* mechanisms.

Hoffman [1990] measured the intrinsic resputtering of a range of projectile–target combinations, finding that f falls on a universal curve, which is reproduced in Figure VIII.3. f is seen to be a function of the atomic mass

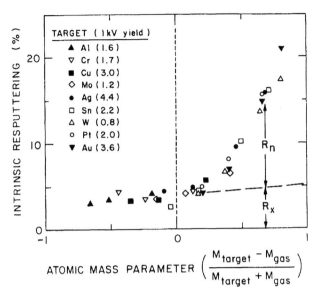

Figure VIII.3 Experimental data for the fraction resputtered from the substrate surface, as a function of atomic mass parameter. Neon, argon, krypton, and xenon projectiles, accelerated to 1.2 keV, were used; R_x is the nearly constant component and R_n, the steeply increasing component. (Reprinted with permission from D. W. Hoffman, "Intrinsic Resputtering—Theory and Experiment," *J. Vac. Sci. Technol.* **A8**(5), 3707. Copyright 1990 by the American Vacuum Society.)

parameter defined in the figure. For relatively *heavy* sputtering gases (a negative atomic mass parameter), the resputtered fraction is nearly a constant ~3.5%. This could be due to the self-sputtering yield of condensing film atoms themselves, as mentioned above. For relatively *light* sputtering gases, the resputtered fraction increases approximately linearly with atomic mass parameter. It was suggested that this additional resputtering (beyond the 3.5%) is due to energetic neutrals reflected from the sputtering target.

Given a net condensation flux, we would like to calculate the *deposition rate* as the rate of change of the film thickness (in the direction perpendicular to the local surface, along $\vec{n}$). This may be calculated with the continuity equation, which expresses the principle of conservation of matter:

$$\nabla \cdot j_c = \frac{-\partial n_f}{\partial t}. \qquad (\text{VIII.8})$$

Referring to Figure VIII.4, j_c goes to zero over a distance equal to the thickness of one monolayer of the condensing film ($a_{m\ell}$). Thus, the divergence of the condensation flux may be approximated with $-j_c/a_{m\ell}$. Now, at this growth front, the density of film which has condensed increases at a rate given by $n_f/t_{m\ell}$, where $t_{m\ell}$ is the monolayer deposition time, given by $a_{m\ell}/v_n$; v_n is the ultimate quantity of interest, the speed of the growth front in the normal

INCIDENT FLUX AND FILM DEPOSITION RATE

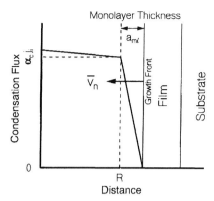

Figure VIII.4 Application of the continuity equation to calculation of film growth rate. The condensation flux falls to zero over a distance equal to the monolayer thickness, so that its divergence is $-\alpha_c j_i/a_{m\ell}$.

direction, or film deposition rate. The continuity equation becomes $j_c/a_{m\ell} = n_f v_n/a_{m\ell}$, which may be solved for deposition rate:

$$v_n = \frac{j_c}{n_f}. \quad\quad\quad (VIII.9)$$

The value of n_f may be calculated from the crystal structure, or from the bulk mass density. For example, silicon (with the diamond cubic structure) has eight atoms per conventional unit cell. With a cubic lattice parameter of 5.430 Å, the volume per particle is 20.0 Å^3 and the particle density is the reciprocal of this, 5.00×10^{28} m^{-3}. Similarly, for a bulk silicon mass density of 2.33×10^3 kg/m^{-3} and a silicon atomic mass of 28.09 amu (4.66×10^{-26} kg), the particle density is again 5.00×10^{28} m^{-3}. The silicon monolayer thickness, $a_{m\ell}$, is 1.35 Å in a $\langle 100 \rangle$ direction. (This is one-fourth of the length of the side of the cubic unit cell.) The atomic densities of many elements are tabu- lated in Appendix B of Chapter VII.

Example For the above example concerning the evaporation of an aluminum film, calculate the deposition rate at the point on the substrate which is directly above the source.

From Appendix B of Chapter VII, the atomic density of crystalline aluminum is 0.0602 Å^{-3}. For this example we assume a condensation coefficient of unity. The deposition rate then is

$$v_n = \frac{0.703 \text{ Å}^{-2} \cdot \text{s}^{-1}}{0.0602 \text{ Å}^{-3}} = 11.7 \text{ Å/s}. \quad \blacksquare$$

The deposition rate can be an important factor for film purity, because the film surface during growth continually adsorbs residual gas particles to some degree. The time required to completely cover an initially clean surface with adsorbed particles, the monolayer adsorption time, may be estimated from the fact that the surface atomic site density is approximately $(n_f)^{2/3}$. (It is exactly this only for a simple cubic crystal structure.) The monolayer adsorption time (t_{rg}) for residual gas particles is

$$t_{rg} \approx \frac{n_f^{2/3}}{\delta z_{rg}}, \tag{VIII.10}$$

where δ is the *trapping probability* of the adsorbing gas and z_{rg} is the impingement rate. On the other hand, the film monolayer formation time, as just discussed, is

$$t_{m\ell} \approx \frac{n_f^{2/3}}{j_c}. \tag{VIII.11}$$

At a chamber pressure of 10^{-6} torr, t_{rg} is on the order of one second. Thus, if the film deposition rate were also one monolayer per second, the film could easily contain ~ 50 at% impurities. This criterion for having an adequate deposition flux is that δz_{rg} be $\ll j_c$.

Associated Substrate Heating Mechanisms

Substrate heating during deposition is inevitable because the condensing particles bring with them both kinetic and potential energy. The potential energy is in the form of the thermodynamic heat of sublimation, which is released during condensation. In addition, physical vapor deposition sources are also sources of various kinds of radiation.

The thermodynamic heat of condensation is always present whenever a film is being deposited. The amount per condensing particle is given by $\Delta_{con}H°$ (equal in magnitude to the standard molar heat of sublimation) divided by Avogadro's number, N_a. The power density delivered to the substrate due to condensation is

$$j_{thermo} = \frac{\Delta_{con}H°}{N_a} \cdot \alpha_c j_i. \tag{VIII.12}$$

The condensation coefficient is included here because the particles must actually condense in order to release this energy.

With an evaporation source, the kinetic energy per impinging particle is $3kT_{source}/2$. The power flux delivered to the substrate by this means is

$$j_{evap} = \frac{3kT_{source}}{2} \cdot \delta j_i. \tag{VIII.13}$$

The trapping probability δ is inserted here because if the incident particle does not adsorb, its kinetic energy is not transmitted to the substrate.

As stated above, there is also radiation associated with both evaporation and sputtering sources. A hot effusion cell will emit *graybody radiation* from the orifice, and from any hot external surfaces. An E-gun source will emit radiation directly from the evaporant surface. We will model this radiant heating by considering the radiation coming from "δA" only, for an ideal source. This graybody radiation has a beam intensity analogous to that of the evaporant particles. The radiation beam intensity is given by [Anderson, 1981]

$$J_{\Omega\text{rad}} = \frac{\varepsilon_{\text{source}}\, \sigma T^4_{\text{source}} \cdot \delta A \cos\theta}{\pi}, \qquad \text{(VIII.14)}$$

where $\sigma = 5.67 \times 10^{-8}\, \text{W} \cdot \text{m}^{-2} \cdot \text{K}^{-4}$. For the E-gun, $\varepsilon_{\text{source}}$ is the emissivity of the evaporant surface averaged over all wavelengths; for the effusion cell, it is the emissivity of whatever surface within the effusion cell is seen by the substrate. The emissivity of a surface is equal to the absorbtivity, or one minus the reflectivity. Thus, clean metal surfaces often have emissivities in the vicinity of 0.1, while for porous nonmetallic materials the emissivity approaches unity.

The flux of this radiant power absorbed by the substrate is given by an expression similar to that of the particle condensation flux; the absorbtivity of the substrate is analogous to the condensation coefficient. As stated above, the absorbtivity is equal to the emissivity of the substrate. Thus

$$\dot{j}_{\text{rad}} = \frac{\varepsilon_{\text{sub}}\, J_{\Omega\text{rad}}}{R^2} \cos\beta. \qquad \text{(VIII.15)}$$

Thus, there is a one-to-one relationship between incident particle flux and power density delivered to the substrate by radiation:

$$\dot{j}_{\text{rad}} = \frac{\varepsilon_{\text{sub}}\varepsilon_{\text{source}}\, \sigma T^4_{\text{source}}}{z_{\text{source}}} \cdot j_i, \qquad \text{(VIII.16)}$$

where z_{source} is the thermal equilibrium impingement rate of the evaporant at the source temperature. We now present an example that gives a sense of the sizes of these associated substrate heating effects.

Example For the preceding example concerning the evaporation of an aluminum film, the impingement flux of aluminum atoms at the point on the substrate directly above the source was found to be $0.703\, \text{Å}^{-2} \cdot \text{s}^{-1}$. Estimate the substrate heating that is due to the release of the heat of condensation. Compare this to the heating effect of the kinetic energy of the condensing particles and to radiant heating of the substrate by the effusion cell. Assume a value of 0.08 for the emissivity of the aluminum source material and take the same value for the emissivity of the substrate.

The heat of condensation at the melting point of aluminum (from Chapter III) is 314 kJ/mol. It is less at 1500 K, but we will use this value anyway because the purpose of this and the following example is to present order-of-magnitude estimates of the various substrate heating effects. We assume trapping and condensation coefficients of unity. The substrate heating which is due to release of the heat of condensation is

$$j_{\text{thermo}} = \left(\frac{314 \text{ kJ}}{6.02 \times 10^{23}}\right) \times 0.703 \text{ Å}^{-2} \cdot \text{s}^{-1} = 3.67 \text{ mW/cm}^2.$$

The heat flux onto the substrate that is due to the kinetic energy of the aluminum atoms is

$$j_{\text{evap}} = \frac{3 \times (1.38 \times 10^{-23} \text{ J/K}) \times 1500 \text{ K} \times 0.703 \text{ Å}^{-2} \cdot \text{s}^{-1}}{2} = 0.22 \text{ mW/cm}^2.$$

Finally, the radiant heat flux absorbed by the substrate at the point under consideration is

$$j_{\text{rad}} = \frac{0.08 \times 0.08 \times 5.67 \times 10^{-8} \text{ W} \cdot \text{m}^{-2} \cdot \text{K}^{-4} \times (1500 \text{ K})^4 \times 0.703 \text{ Å}^{-2} \cdot \text{s}^{-1}}{221 \text{ Å}^{-2} \cdot \text{s}^{-1}}$$

$$= 0.58 \text{ mW/cm}^2.$$

Heating due to release of the heat of condensation is the most important substrate heating mechanism for this example. The total substrate heating for evaporation is

$$j_{\text{total}} = (3.67 + 0.22 + 0.58) \text{ mW/cm}^2 = 4.47 \text{ mW/cm}^2. \quad \blacksquare$$

The term $U_{\text{sb}}/2$, which is the most probable energy of the Thompson formula (Chapter VII), is a representative kinetic energy of a sputtered atom. The power flux delivered to the substrate that is due to the kinetic energy of sputtered particles is then

$$j_{\text{sput}} \approx \frac{U_{\text{sb}}}{2} \cdot \delta j_{\text{i}}. \qquad (\text{VIII.17})$$

(We neglect here any thermalization of the sputtered particles — which usually occurs, as discussed in Section VIII.3 — and are therefore estimating the greatest possible heating effect by this mechanism.)

With sputtering, radiation consisting of both electrons and ions bombards the substrate, but radiant heating is typically negligible since sputtering targets are water-cooled. We will again ignore any possible sheath collisions and assume that bombarding electrons and ions deliver their maximum possible energies to the substrate. We will assume that the substrate is the sole anode of a DC sputtering discharge. Both types of charged particle bombard the substrate at once because current flow at this anode corresponds to the electron retardation regime of the Langmuir probe.

The average kinetic energy of the electrons may be estimated with $3kT^-/2$, where T^- is the electron temperature of the plasma. The maximum kinetic energy with which the ions bombard the substrate may be estimated as $q(V_{\text{p}} - v_{\text{sub}})$, where the potential difference is the plasma potential minus that of the substrate. In addition, there is a heat of neutralization that is released when an ion striking the substrate is neutralized by an electron from the substrate. This energy release per incident ion is equal to the ionization potential (E_i, 15.8 eV for argon) of the neutral projectile. Thus, the power flux delivered to the substrate by charged particles from a sputtering plasma is

$$j_{\text{part}} \approx \frac{3kT^-}{2(-q)} \cdot j_e + (V_p - v_{\text{sub}}) \cdot j_{\text{ion}} + \frac{E_i}{q} \cdot j_{\text{ion}}. \tag{VIII.18}$$

In fact, j_e and j_{ion} here are the electron and ion *current* densities to the substrate.

Example Suppose that the aluminum film considered in the preceding examples is being deposited with a sputtering source. Estimate the amount of substrate heating that may be expected to occur. Take the surface binding energy of aluminum to be 3.43 eV and assume the practical plasma of Chapter VI.

Heating due to release of the heat of condensation was calculated in the previous example:

$$j_{\text{thermo}} = 3.67 \, \text{mW/cm}^2.$$

Substrate heating due to the kinetic energy of sputtered particles is

$$j_{\text{sput}} = \frac{3.43 \, \text{eV} \times 0.703 \, \text{Å}^{-2} \cdot \text{s}^{-1}}{2} = 1.93 \, \text{mW/cm}^2.$$

To make an order-of-magnitude estimate of the substrate heating due to the bombarding ions, we assume the enhanced ion current density of the practical plasma (Chapter VI) to be 0.35 mA/cm². We also assume that the substrate is the anode of the discharge and of an area equal to the cathode. In Chapter VI this gave a plasma potential that is 7.9 V above the anode potential. The substrate heating due to the bombarding ions is then

$$0.35 \, \text{mA/cm}^2 \times 7.9 \, \text{V} = 2.77 \, \text{mW/cm}^2.$$

Calculating the heating due to bombarding electrons is a little more involved. The random impingement current density for the electrons of the practical plasma is $-37.6 \, \text{mA/cm}^2$. However, as with the Langmuir probe analysis, this current density is reduced by a Boltzmann factor, $\exp[-q(V_p - v_{\text{sub}})/kT^-]$. Thus

$$j_e = -37.6 \, \text{mA/cm}^2 \times \exp\frac{-1.6 \times 10^{-19} \, \text{C} \times 7.9 \, \text{V}}{(1.38 \times 10^{-23} \, \text{J/K}) \times 23{,}000 \, \text{K}} = -0.70 \, \text{mA/cm}^2.$$

The substrate heating due to bombarding electrons is then

$$\frac{3 \times (1.38 \times 10^{-23} \, \text{J/K}) \times 23{,}000 \, \text{K} \times (-0.70 \, \text{mA/cm}^2)}{2 \times (-1.6 \times 10^{-19} \, \text{C})} = 2.09 \, \text{mW/cm}^2.$$

Finally, the substrate heating due to the neutralization of bombarding ions is

$$15.8\,\text{V} \times 0.35\,\text{mA/cm}^2 = 5.53\,\text{mW/cm}^2.$$

The sum of all the radiation heating in sputter deposition of the aluminum is

$$j_{\text{part}} = (2.77 + 2.09 + 5.53)\,\text{mW/cm}^2 = 10.4\,\text{mW/cm}^2.$$

This radiation heating associated with sputter deposition is substantially greater than that due to all other mechanisms considered in this chapter. The total substrate heating for sputter deposition is

$$j_{\text{total}} = (3.67 + 1.93 + 10.4)\,\text{mW/cm}^2 = 16.0\,\text{mW/cm}^2. \qquad \blacksquare$$

There are still other mechanism that might well make a contribution to substrate heating. These include bombardment of the substrate by high-energy electrons from an E-gun evaporation source, or by fast neutrals that had been reflected from the sputtering target. Furthermore, the relative contributions of the various mechanisms may well vary from one situation to the next. The preceding examples do show that sputter deposition has the potential for considerably more substrate heating than evaporation. But how significant are these heat fluxes?

Example Suppose that the substrate is a silicon wafer that is 0.5 mm thick, and that there is *no* dissipation of the heat delivered to the substrate during deposition of the aluminum film under consideration. Predict the rate of rise of the substrate temperature with evaporation and with sputter deposition.

The substrate temperature will rise at a rate given by $j_{\text{total}}/\rho w C$, where ρ is the mass density of the substrate, w is the substrate thickness, and C is the specific heat. For silicon, $\rho = 2.33\,\text{g/cm}^3$ and $C = 0.168\,\text{cal}\cdot\text{g}^{-1}\cdot\text{K}^{-1}$.
With our heat estimates for evaporation, the substrate temperature will rise at a rate given by

$$\left.\frac{dT}{dt}\right|_{\text{evap}} = \frac{4.47\,\text{mW/cm}^2}{(2.33\,\text{g/cm}^3) \times 0.5\,\text{mm} \times 0.168\,\text{cal}\cdot\text{g}^{-1}\cdot\text{K}^{-1}} = 3.3\,\text{K/min}.$$

On the other hand, for sputter deposition of the aluminum film, the substrate temperature will rise at a rate given by

$$\left.\frac{dT}{dt}\right|_{\text{sput}} = \frac{16.0\,\text{mW/cm}^2}{(2.33\,\text{g/cm}^3) \times 0.5\,\text{mm} \times 0.168\,\text{cal}\cdot\text{g}^{-1}\cdot\text{K}^{-1}} = 11.7\,\text{K/min}. \qquad \blacksquare$$

These examples, although approximate, show that there is a potential for significant substrate heating in physical vapor deposition, particularly with sputtering. Similar conclusions were reached by Pargellis [1989] and Class and Hieronymi [1982].

VIII.2 FILM THICKNESS PROFILES OF THE IDEAL SMALL SOURCE

Three Fundamental Receiving Surfaces

Figure VIII.5 shows three fundamental receiving surfaces:-a hemisphere, a plane, and a tangential sphere. The plane models a flat substrate. The hemisphere and the sphere, on the other hand, are common shapes of large fixtures that hold flat substrates and rotate during deposition in a simple or a planetary fashion. They are all characterized with the same *throw distance* (h), which is the distance from the source to the point on the receiving surface directly above it. For these receiving surfaces, we calculate the film thickness profiles of the *ideal small source*.

By ideal small source, we mean one for which the beam intensity is given by Eq. VIII.1, where the beam intensity varies as the cosine of the emission angle. It is rarely ideal in terms of the film thickness profile it creates, but is ideal in principle, representing the theoretical Knudsen cell, the ideal free evaporation source, and the ideal small sputtering source.

Using the expression for the beam intensity of a small evaporation source, a general expression for the local incident flux for any type of receiving surface is

$$j_i = \frac{z\,\delta A \cos\theta \cos\beta}{\pi R^2}. \quad (VIII.19)$$

For the geometry of the hemispherical receiving surface as shown in Figure VIII.5a, $\beta = 0°$ always and $R = h$. For the planar receiving surface in Figure

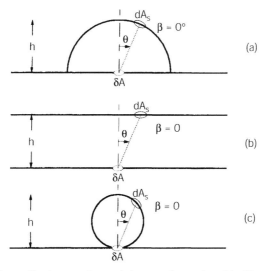

Figure VIII.5 Three fundamental receiving surfaces in thin-film physical vapor deposition are shown: (*a*) hemisphere, (*b*) plane, and (*c*) tangential sphere.

VIII.5b, $\beta = \theta$ and $R = h/\cos\theta$. Finally, for the tangential sphere geometry in Figure VIII.5c, $\beta = -\theta$ and $R = h\cos\theta$. The incident fluxes for each type of surface are as follows.

1. For the hemispherical surface:

$$j_{ih}(\theta) = \frac{z\,\delta A\,\cos\theta}{\pi h^2}. \quad (VIII.20)$$

Thus, for this surface, the cosine law of emission is itself reproduced by the film thickness distribution on the receiving surface.

2. For the planar receiving surface:

$$j_{ip}(\theta) = \frac{z\,\delta A(\cos\theta)^4}{\pi h^2}. \quad (VIII.21)$$

The angular variation in film thickness is rather strong for this surface.

3. For the spherical receiving surface:

$$j_{is}(\theta) = z\,\delta A/\pi h^2. \quad (VIII.22)$$

The spherical surface is optimal with respect to the uniformity of the film obtained, for the ideal small source. The incident flux onto this surface is uniform.

The incident flux is independent of position on the surface of this tangential sphere, which leads to a film of uniform thickness. The three film thickness profiles are compared in Figure VIII.6.

It is unfortunate that the planar substrate gives such a nonuniform thickness profile, because substrates are practically always planar. In recognition of the advantage of curved receiving surfaces, it is common when coating numerous small flat substrates together to configure them, with fixturing, into an assembly that approximates the hemispherical receiving surface. An example of such commercially available equipment is shown in Figure VIII.7. Good uniformity is achievable when this fixturing is given planetary rotations.

We note that the point source, with its fixed beam intensity, gives a modestly more uniform film on the planar substrate than does the ideal small source. The incident flux is $z\,\delta A/4\pi \cdot \cos\beta/R^2 = z\,\delta A/4\pi \cdot (\cos\theta)^3/h^2 = z\,\delta A/\pi h^2 \cdot (\cos\theta)^3/4$.

The Moving-Shutter Technique

It is possible to increase the uniformity of film thickness by changing the relative *exposure* of different parts of the substrate, to the source. This principle

FILM THICKNESS PROFILES OF THE IDEAL SMALL SOURCE 279

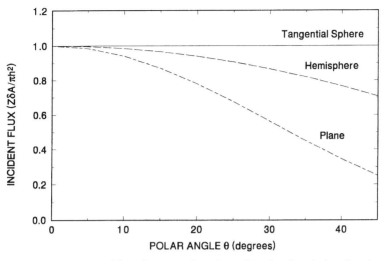

Figure VIII.6 Relative incident flux as a function of angle of emission for the three common receiving surfaces.

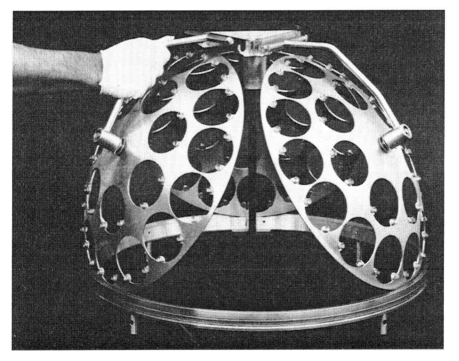

Figure VIII.7 A commercial substrate holder offering planetary rotation (photo provided by the Kurt J. Lesker Co.).

is most simply applied to the planar receiving surface placed directly above a source, through the use of a closely placed shutter or aperture, rotating about the axis of symmetry. We illustrate this approach in Figure VIII.8.

Equation (VIII.21) may be restated to give the relative film thickness profile as a function of radial position on the substrate (r), when an ideal small source is used with the planar receiving surface:

$$\frac{j_{\text{ip}}(r)}{j_{\text{ip}}(0)} = \frac{h^4}{[r^2 + h^2]^2}. \tag{VIII.23}$$

A rotating shutter to counteract this nonuniformity, shown conceptually in Figure VIII.8, may be devised. The required exposure is given by

$$\frac{\varphi(r)}{\varphi(0)} = \frac{[r^2 + h^2]^2}{h^4}, \tag{VIII.24}$$

where φ is the azimuthal angle.

The term $\varphi(r)/\varphi(0)$ expresses the relative exposure (relative time exposed to the source) of a point at position r on the substrate as compared to a point at the center; $\varphi(r)$ itself defines the shape of the shutter, which is the angular gap of the shutter opening as a function of r.

The preceding case was worked out by Behrndt [1962]. Experimentally, Behrndt achieved with a rotating shutter a thickness uniformity of ±0.35% across a 4 × 4-in. square substrate, at a throw distance of 10 in. using an E-gun evaporator. Ramsay et al. [1974] have demonstrated an *offset* shutter (one whose axis of rotation is offset from the axis of the source). This latter design eliminates any nonuniform spot at the center of the substrate.

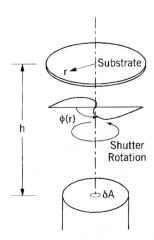

Figure VIII.8 The moving-shutter technique increases film thickness uniformity by altering the relative exposure, of different regions of the substrate, to the source. The shutter's shape is specified with $\varphi(r)$.

Example Design a moving shutter that will achieve uniformity for the planar substrate in combination with the ideal small source.

We will assume that the substrate is a disk whose radius is equal to the throw distance, h. Using Equation (VIII.24), we can determine the ratio of maximum to minimum exposure angles:

$$\frac{\varphi(h)}{\varphi(0)} = \frac{[h^2 + h^2]^2}{h^4} = 4.$$

We will design a double-lobed shutter, which means that $\varphi(h) = 180°$. Then $\varphi(0) = 180°/4 = 45°$. The equation for the edge of one lobe of the shutter is then

$$\varphi(r) = \frac{45°[r^2 + h^2]^2}{h^4}.$$

The design is sketched below. ∎

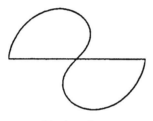

Moving shutter.

VIII.3 THERMALIZATION AND IONIZATION OF THE SPUTTERED BEAM

Evaporation sources and some sputtering systems can be modeled well with the assumption of ballistic trajectories for the emitted particles, as we have done in Sections VIII.1 and VIII.2 and will do again in Section VIII.4. This requires that the mean free path be large compared to the throw distance, so that scattering of the source beam is negligible. For example, at a background pressure of 10^{-6} torr, easily achieved in high-vacuum evaporators, the mean free path is on the order of 30 m. An ion beam sputtering system (a sputter deposition system design known for its ability to work at relatively low pressures) typically operates at a pressure in the low 10^{-4}-torr range, giving a mean free path of 30 cm.

On the other hand, at 1 mtorr (a working pressure where a magnetron sputtering source might operate) the mean free path is much smaller (~ 3 cm), and it would be a rare case where the throw distance is much less than this. Nonmagnetron sputtering sources operate at even higher pressures. Thus, an appreciable scattering of the sputtered beam is common. The directed beam is partially converted into a large, diffuse source, as suggested schematically in Figure VIII.9. The randomizing of directions of travel can be significant — with such scattering it is possible to coat the *back*side of an object placed in the sputtering chamber, under conditions that are not uncharacteristic of sputter deposition [Upadhye and Hsieh, 1990].

This scattering also causes energy transfer from the sputtered particles to the sputtering gas. With the characteristic temperature of the sputtered particles being reduced from thousands of degrees (5 eV = 60,000 K) to that of the gas (say, $\lesssim$ 500 K), this process is called *thermalization*. Thermalization also occurs to another class of particles present in the sputtering chamber. These are the *reflected neutrals*, still-energetic ions which have been reflected by the cathode and simultaneously neutralized. These particles can play an important role in the film growth processes when they bombard the surface of the growing film.

Several practical effects of thermalization have been observed:

- Reduction of the incident flux at the substrate
- An increase in film thickness uniformity
- Modification of the microstructure, and related properties, of the film that are due to the action of reflected neutrals striking the film surface during growth
- Reduction of the resputtering rate (removal of the growing film by sputtering)

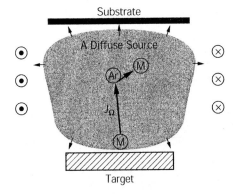

Figure VIII.9 With thermalization, part of the sputtered beam is converted to a distributed, diffuse source. One effect is a reduction of the incident flux at the surface. Also shown at the sides of the plasma are the turns of a RF induction coil, which is used with ionized physical vapor deposition (discussed later — plays no part in thermalization as such).

The Thermalization Distance

When does thermalization occur, to a practical degree? The fraction of sputtered particles that have been thermalized has been modeled with the following expression:

$$F = 1 - e^{-y/k\lambda}, \qquad (\text{VIII.25})$$

where y is the distance from the sputtering target, λ is the mean free path of the sputtered particle within the working gas, and k is an empirical parameter (often on the order of 10), expressing the fact that the thermalization distance is proportional to, but somewhat greater than, the mean free path.

Westwood [1978] presented a pioneering quantitative analysis of thermalization, assuming rigid-sphere scattering. For argon as the sputtering gas, and sputtered atoms having a mass within an order of magnitude of argon's 40 amu, Westwood calculated several interesting results: (1) typically, 20–60% of the sputtered particle's kinetic energy is lost with each collision; (2) the average scattering angle of the sputtered atom in a collision can be quite large (45° when the masses are equal); (3) as a representative estimate, 10 is the number of collisions required to thermalize a 5-eV sputtered atom to room-temperature argon; (4) at typical sputtering pressures of 0.5–10 Pa (3.8–75 mtorr), the thermalization distance is in the range 8–0.4 cm (see Westwood's graph, reproduced in Fig. VIII.10). Westwood concluded that "Apart from very heavy ... energetic ... atoms, essentially all sputtered material is thermalized before reaching the substrates in usual diode sputtering configurations."

Somekh [1984] computer-simulated the thermalization of sputtered atoms, assuming the Thompson formula for the energy spectrum of ejected particles and a more realistic, energy-dependent scattering cross section. The primary interest was not the reduction in incident flux caused by thermalization, but the reduction in the amount of energy delivered to the substrate by sputtered atoms and reflected neutrals. Somekh expressed his results in terms of a *PD* (pressure–distance) product, the gas pressure times the distance traveled from the target by the sputtered atom.

One of Somekh's chief results is shown in Figure VIII.11 — the variation of average sputtered atom energy with distance from the target, for three noble gas reflected neutrals and several sputtered metals. As expected, the energy falls approximately exponentially with distance. Somekh generalizes: "The main result is that for light atoms $PD_{0.1}$ [the pressure times an estimate of the thermalization distance] is about 200 Pa · mm and for heavy atoms less than 100 Pa · mm."

Reduction of the Incident Flux

Turner et al. [1992] calculated the fraction of sputtered atoms arriving at the substrate for a magnetron discharge. The sputtering gas was argon and the throw distance was 5 cm. The Thompson formula for the energy spectra of

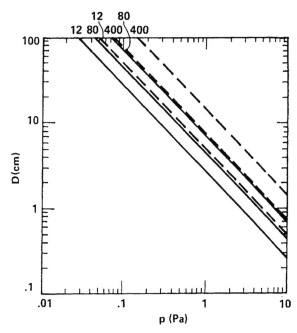

Figure VIII.10 The thermalization distance, as a function of argon pressure, for sputtered particle masses of 12, 80, and 400 amu. The solid lines are for 5 eV initial energies and the dashed, 1000 eV. (Reprinted with permission from W. D. Westwood, "Calculation of Deposition Rates in Diode Sputtering Systems," *J. Vac. Sci. Technol.* **15**(1), 1. Copyright 1978 by the American Vacuum Society.)

sputtered particles was assumed. Their results are summarized in Figure VIII.12. For a variety of common target materials, over a realistic range of sputtering pressures, the reduction of incident flux at the substrate is substantial.

Petrov et al. [1993] modeled the conversion of the directed sputtered flux of vanadium to a diffuse flux, for a DC planar magnetron sputter deposition system having a throw distance of 10 cm and a ring-shaped erosion track (of unspecified size). They used the same energy-dependent scattering cross section as Somekh [1984]. Some of their results are shown in Figure VIII.13a, where it is seen that the thermal fraction of the incident flux reaches 50% at a pressure of ~0.6 Pa in argon. We wish to mention, also, their calculated and experimentally measured reductions in the incident flux at the substrate. Results as a function of argon pressure are shown in Figure VIII.13b, for a constant discharge current of 0.3 A. The incident flux falls substantially, over a practical range of working pressure.

Although these various simulations of thermalization have some different details, they are unanimous in predicting that thermalization is significant under most practical sputtering conditions.

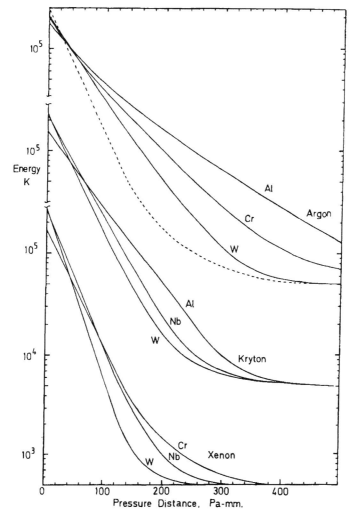

Figure VIII.11 The average energies of sputtered particles as a function of the product of working gas pressure and distance from the target. An initial energy spectrum according to the Thompson formula was assumed (1 eV = 11,600 K.) (Reprinted with permission from R. E. Somekh, "The Thermalization of Energetic Atoms during the Sputtering Process," *J. Vac. Sci. Technol.* **A2**(3), 1285. Copyright 1984 by the American Vacuum Society.)

One effect that accompanies thermalization of sputtered atoms is the concomitant heating of the sputtering gas. It has been shown that this heating can lead to an 80% rarefaction of the gas (density reduction to 20% of the nominal value) near the sputtering cathode, under conditions that are representative of magnetron sputter deposition [Rossnagel, 1989].

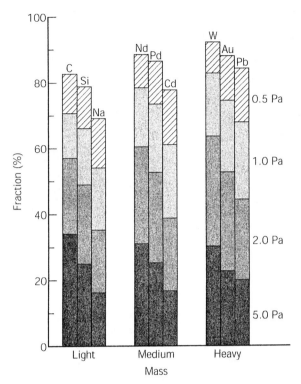

Figure VIII.12 Turner et al. calculated the fraction of sputtered atoms which reach the substrate plane—that is, are not prevented from doing so by scattering by the sputtering gas (throw distance 5 cm). Over the practical range of sputtering pressures shown, the reduction of the incident flux is substantial. (Reprinted with permission from G. M. Turner et al., "Monte Carlo Calculations of the Properties of Sputtered Atoms at a Substrate Surface in a Magnetron Discharge," *J. Vac. Sci. Technol.* **A10**(3), 455. Copyright 1992 by the American Vacuum Society.)

Ionized Physical Vapor Deposition

Ionized physical vapor deposition (I-PVD) is a technique that capitalizes on benefits of gas-phase collisions. Rossnagel [1998] defines the technique as follows: "the physical sputtering of metal atoms into a dense, inert gas plasma, ionization of the sputtered metal atoms, and subsequent deposition of the film from these metal ions."

The intent of ionizing the sputtered metal atoms is to be able to control their kinetic energy, and to be able to change their directional distribution from "diffuse" to "collimated." Such effects are achieved by biasing the substrate, which accelerates the metal ions toward the substrate with speeds that can greatly exceed their random thermal speed. The successful applications of

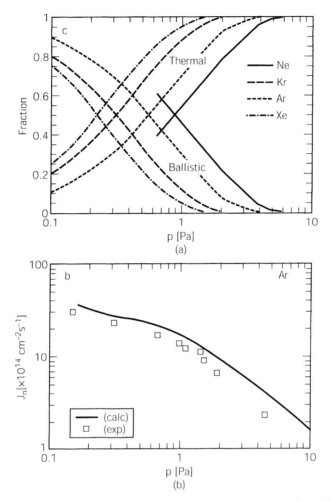

Figure VIII.13 (*a*) The fraction of sputtered particles which are "thermalized," and which are not, as calculated by Petrov et al. (throw distance 10 cm). (*b*) The incident flux of sputtered vanadium at the substrate, as calculated and experimentally measured by the same authors, for argon. (Reprinted with permission from I. Petrov et al., "Comparison of Magnetron Sputter Deposition Conditions in Neon, Argon, Krypton, and Xenon Discharges," *J. Vac. Sci. Technol.* **A11**(5), 2733. Copyright 1993 by the American Vacuum Society.)

I-PVD include the filling of high-aspect-ratio features on integrated circuit wafers, and the deposition of conformal coatings on topologically difficult substrates.

With the I-PVD technique, the metal atoms are typically sputtered from a magnetron cathode. They are then ionized in transit to the substrate via collisions with energetic electrons of the plasma. The relatively dense plasma that is required is created typically with an RF induction coil, as suggested in Figure VIII.9 (this inductively coupled plasma is, of course, auxiliary to the primary magnetron sputtering plasma).

Dickson et al. [1998] have characterized the I-PVD technique as an aluminum source. We present in Figure VIII.14 a sampling of their results.

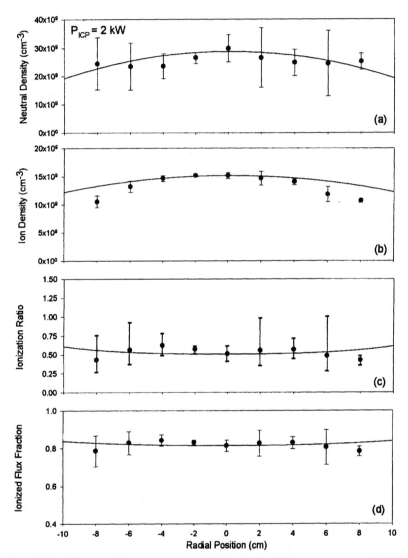

Figure VIII.14 An experimental characterization of aluminum neutral and ion profiles in ionized physical vapor deposition. The radial positions are for a deposition geometry similar to that of Figure VIII.9. (Reprinted with permission from M. Dickson et al., "Radial Uniformity of an External-Coil Ionized Physical Vapor Deposition Source," *J. Vac. Sci. Technol.* **B16**(2), 523. Copyright 1998 by the American Vacuum Society.)

They measured the densities of sputtered aluminum atoms and ions as functions of radial position in a deposition geometry similar to that of Figure VIII.9. For the experimental conditions that they selected, the ionization fraction is around 10%. Rossnagel [1998] reports, however, that metal ionization as high as 90% can be attained. Also shown in Figure VIII.14 are theoretical atom and ion density profiles that they calculated from solutions of the diffusion boundary value problem and from the "Klyarfeld parabolic approximation," respectively. The results shown in Figure VIII.14 were obtained at a sputtering gas pressure of 30 mtorr.

VIII.4 DEPOSITION WITH SUBSTRATE ROTATION AND WITH IDEAL LARGE SOURCES

A "large" source is sometimes desirable for increasing film deposition rate (and, of course, thickness uniformity to some extent). Other objectives include minimizing microscopic shadowing effects on the substrate surface, which cause roughness, and improving the uniformity of coverage of substrate steps. A Knudsen cell becomes a large source when the throw distance (h) is not a lot greater than the size of the orifice. This is not a likely occurrence in practice with effusion cells, but it is quite common to have large sputtering sources, and sometimes E-gun evaporators do not strictly qualify as small sources.

The utilization of substrate movement during deposition is one way by which a source of effectively large size may be obtained. The usual forms which substrate movement takes are *simple* and *planetary rotations*; the latter is a combination of simultaneous rotations about two different axes. A small source may be thought of as *zero-dimensional*; using several small-area sources gives a superposition of zero-dimensional sources. Simple substrate movement converts it to *one-dimensional*, a line as seen by a selected point on the substrate. For example, the average value of the incident flux at a certain point on a rotating substrate is obtained from an expression based upon

$$j_{i,\text{av}} = (1/T) \int_0^T J_\Omega(t) \frac{\cos \beta(t)}{[R(t)]^2} dt, \qquad (\text{VIII.26})$$

where T is the period of the substrate rotation. This single integral signifies the effectively one-dimensional nature of the source. Planetary movement would effectively convert it to a *two-dimensional* source, replacing the single integral in (VIII.26) by a double integral. Thus, large sources may be simulated to a limited extent by moving the substrate during deposition.

We will first consider simple, off-axis substrate rotation because it is in such widespread use, and is remarkably effective in increasing film thickness uniformity on planar substrates.

Off-Axis Substrate Rotation

Figure VIII.15 portrays schematically a rotating planar substrate situated off the axis of an ideal small source. At moving point $P'(r', \varphi')$ on the rotating substrate, the contribution to the incident flux during an incremental substrate rotation $d\varphi'$ is

$$dj_i = \frac{z \, \delta A \cos \theta \cos \beta}{\pi R^2} \cdot \frac{d\varphi'}{2\pi} \quad \text{(VIII.27)}$$

with $\theta = \beta$. Their cosine is h/R, where $R^2 = h^2 + (W - r' \cos \varphi')^2 + (r' \sin \varphi')^2$. ($R$, h, W, r', and φ' are defined in Fig. VIII.15.) Thus, we may calculate the average incident flux at P' during one revolution by integrating

$$dj_i = z \cdot \delta A \cdot \frac{h^2}{\pi \{h^2 + (W - r' \cos \varphi')^2 + (r' \sin \varphi')^2\}^2} \cdot \frac{d\varphi'}{2\pi}. \quad \text{(VIII.28)}$$

The average flux (after rearranging the denominator) is

$$j_{i,\text{av}} = \frac{z \, \delta A \, h^2}{2\pi^2} \cdot \int_0^{2\pi} \frac{d\varphi'}{[h^2 + W^2 + r'^2 - 2r'W \cos \varphi']^2}. \quad \text{(VIII.29)}$$

The result of the integration is (see Appendix)

$$j_{i,\text{av}}\left(\frac{r'}{h}, \frac{W}{h}\right) = \frac{z \, \delta A}{\pi h^2} \cdot \frac{1 + (r'/h)^2 + (W/h)^2}{[1 + (r'/h)^4 + (W/h)^4 + 2(r'/h)^2 + 2(W/h)^2 - 2(r'W/h^2)^2]^{3/2}}. \quad \text{(VIII.30)}$$

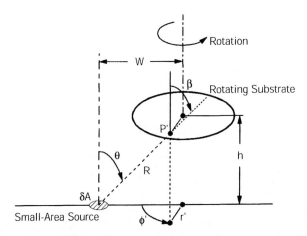

Figure VIII.15 Film thickness uniformity with a small source and a flat substrate is drastically improved with off-axis substrate rotation.

Relative film thickness profiles are portrayed in Figure VIII.16 for a range of dimensions of practical interest. It is clear that off-axis substrate rotation can drastically improve film thickness uniformity, but there is a tradeoff with deposition rate. It has been applied to the preparation of optical coatings, where very high film thickness uniformity is required [Macleod, 1989].

A Large Disk Source with a Planar Substrate

A large disk-shaped source might represent a nonmagnetron diode sputtering target, an ion-beam sputtering target, or perhaps a large E-gun evaporator. For a large two-dimensional source, the total incident flux at a fixed point on the substrate is obtained from an expression based on

$$j_{i,\text{total}} = \int_A \int dJ_\Omega \frac{\cos \beta}{R^2}, \qquad \text{(VIII.31)}$$

where A is the total source area. The double integral signifies the true two-dimensional nature of a large area source.

We will consider a large disk source of radius s with a stationary planar substrate as shown in Figure VIII.17. For a point (P) on this fixed substrate, the incremental incident flux due to an infinitesimal source area is

$$d^2 j_i(r, \varphi) = \frac{z \, dA \cos \theta \cos \beta}{\pi R^2} \qquad \text{(VIII.32)}$$

with $\theta = \beta$ and $dA = r \, dr \, d\varphi$. Their cosine is again h/R, and $R^2 = h^2 + (X - r \cos \varphi)^2 + (r \sin \varphi)^2$, a function of the same form as that considered in the

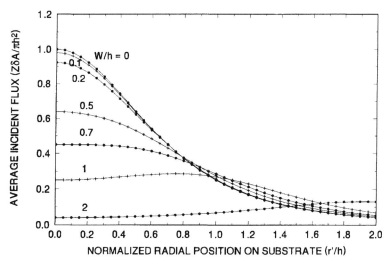

Figure VIII.16 Relative film thickness profiles for off-axis substrate rotation. The offset is W, and the throw distance is h.

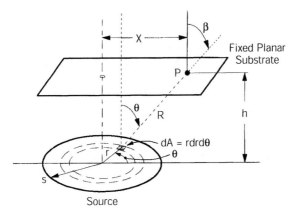

Figure VIII.17 Film thickness uniformity may be increased by using a large source; we show here a disk source with a fixed planar substrate.

analysis of substrate rotation (R, h, X, r, and φ are defined in Fig. VIII.17). Thus, we may calculate the total incident flux at P by integrating

$$d^2 j_i(r,\varphi) = z \cdot r\, dr\, d\varphi \cdot \frac{h^2}{\pi\{h^2 + (X - r\cos\varphi)^2 + (r\sin\varphi)^2\}^2}$$

$$= z \cdot 2\pi r\, dr \cdot \frac{h^2}{\pi\{h^2 + (X - r\cos\varphi)^2 + (r\sin\varphi)^2\}^2} \cdot \frac{d\varphi'}{2\pi} \quad \text{(VIII.33)}$$

around the ring. The double integral that must be evaluated is

$$j_{i,\text{total}} = (zh^2/\pi) \cdot \int_0^s \int_0^{2\pi} \frac{r\, dr\, d\varphi}{[h^2 + X^2 + r^2 - 2rX\cos\varphi]^2}. \quad \text{(VIII.34)}$$

The integral over φ has been already accomplished in the analysis of substrate rotation:

$$j_{i,\text{total}} = \frac{2z}{h^2} \cdot \int_0^s r\, dr \frac{1 + (r/h)^2 + (X/h)^2}{[1 + (r/h)^4 + (X/h)^4 + 2(r/h)^2 + 2(X/h)^2 - 2(rX/h^2)^2]^{3/2}}. \quad \text{(VIII.35)}$$

If we stop the analysis at this point, we have the film thickness profile due to a flat ring source of radius r, letting its area δA be $2\pi r\,\delta r$ and dropping the second integration. We write this out explicitly in the next section. Continuing, the integral over r gives (see Appendix) [Eisgruber, 1993]

$$j_{i,\text{total}}\left(\frac{s}{h}, \frac{X}{h}\right) = \frac{z}{2} \cdot \left\{1 - \frac{1 - (s/h)^2 + (X/h)^2}{[1 + (s/h)^4 + (X/h)^4 + 2(s/h)^2 + 2(X/h)^2 - 2(sX/h^2)^2]^{1/2}}\right\}. \quad \text{(VIII.36)}$$

Some representative film thickness profiles are plotted in Figure VIII.18. For high uniformity, the source must be larger than the substrate — the main benefit of a large source is usually the increase in deposition rate. For very large sources, the incident flux approaches the value of z itself, as expected from conservation of matter and as Equation (VIII.36) shows.

To summarize, off-axis substrate rotation is remarkably effective in achieving film thickness uniformity. Considering the practicalities of implementing a sufficiently large evaporation source, rotation can be an elegant solution to the need for uniformity. In the limit of a very large source, the incident flux at the substrate approaches z, the impingement rate of the evaporation source (or equivalently Yz^+ for a uniform sputtering source).

A Large Ring Source

A large ring-shaped source is a fairly common geometry in magnetron sputter deposition. An actual sputtering target is shown in Figure VIII.19. To analyze the film thickness profile due to a ring source, Equation (VIII.35) may be modified to give the total incident flux onto a planar substrate due to a thin ring source of average radius r and width δr:

$$j_{i,\text{total}} \approx \frac{z\,\delta A}{\pi h^2} \cdot \frac{1 + (r/h)^2 + (X/h)^2}{[1 + (r/h)^4 + (X/h)^4 + 2(r/h)^2 + 2(X/h)^2 - 2(rX/h^2)^2]^{3/2}}. \quad \text{(VIII.37)}$$

We assume here negligible thermalization and a uniform differential beam intensity across the source.

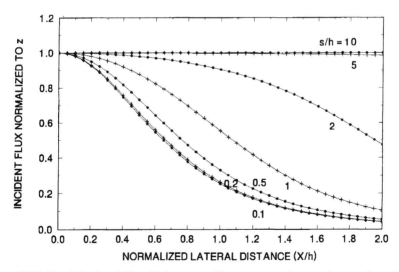

Figure VIII.18 Calculated film thickness profiles across a planar substrate for a large disk source. The disk radius is s, and the throw distance is h.

Figure VIII.19 A ring-shaped target from a planar magnetron sputtering system. Its diameter is on the order of 15 cm.

Some representative film thickness profiles are shown in Figure VIII.20. The novel aspect is that maxima in film thickness can occur at positions away from the center of the substrate.

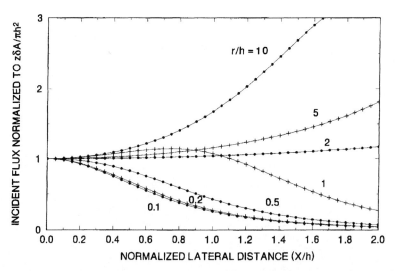

Figure VIII.20 Film thickness uniformity profiles on a planar substrate due to a large ring source (r is the ring radius, and h is the throw distance).

VIII.5 DEPOSITION MONITORS

In vacuo monitors of film deposition sense either the amount of material accumulated (the quartz oscillator) or the value of the incident flux (true flux sensors). They are often incorporated into a feedback control system in order to control the deposition rate. In the first case, to relate the film thickness on the sensor to that on the substrate, one makes use of a *tooling factor*, which is their ratio. The tooling factor is typically determined experimentally. The deposition rate is obtained by electronically differentiating the measurements of film thickness (a digital implementation of this is discussed in Long and MacLeod [1979]). In the second case, the incident flux itself is sampled by measuring some property of the effusive stream; again, a tooling factor is used to relate the incident flux at the sensor to that value somewhere on the substrate.

We present first a brief explanation of the operation of the ubiquitous quartz crystal microbalance, first proposed by Sauerbrey [1957]. Next, we give the basic operating principles of three technologies for measuring the deposition flux directly: the ionization gauge sensor, the quadrupole residual gas analyzer, and the EIES sensor (Electron Impact Emission Spectroscopy; a product of Inficon Leybold-Heraeus).

The Quartz Crystal Microbalance

A quartz crystal microbalance is shown schematically in Figure VIII.21. This monitor is reliable, relatively inexpensive, and capable of detecting a fraction of a monolayer of atoms deposited on its surface.

The crystal exhibits fundamental thickness-shear oscillations, as suggested in the Figure. These oscillations create a rapidly alternating electrostatic potential difference across the crystal, between top and bottom, a manifestation of the piezoelectric effect. (The effect also works in reverse — an applied potential difference *causes* such a shear distortion of the crystal.) In operation, the crystal can be incorporated into an oscillator circuit that drives these oscillations and determines the resonant frequency; at resonance, the AC

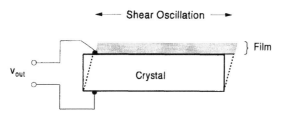

Figure VIII.21 The thickness–shear acoustic vibration mode is utilized in the quartz crystal oscillator. The resonant frequency drops as a film accumulates on the exposed surface of the crystal.

current delivered to the crystal is maximum. (There are other signal processing techniques, e.g., observing the oscillation period–film thickness relationship or acoustic impedance analysis [Lu, 1974; Maier-Komor, 1985].) The resonant frequency is typically in the range 5–10 MHz.

Film accumulated on the exposed face of the crystal causes the resonant frequency to change; the accumulated film thickness is thus indicated. The resonant frequency is given by

$$f_r = \frac{v_{tr}}{2x},\qquad\text{(VIII.38)}$$

where v_{tr} is the velocity of an elastic transverse wave, 3.340×10^3 m/s for "AT-cut" quartz crystals (the crystallographic orientation having the smallest temperature sensitivity [Glang, 1970] and x is the thickness of the crystal. The sensitivity of resonant frequency to thickness is thus $df_r/dx = -v_{tr}/2x^2$, and the change in resonant frequency due to a change in crystal thickness is

$$\Delta f_r = -f_r^2 \cdot \frac{2}{v_{tr}} \cdot \Delta x. \qquad\text{(VIII.39)}$$

Depositing a film onto the crystal surface simulates an increase in the crystal thickness, causing a drop in the resonant frequency. In the simplest analysis, the accumulating film is assumed to influence the resonant frequency only by virtue of the thickness it adds, and not by its elastic properties (this surprising result was proposed by Sauerbrey and justified by Miller and Bolef [1968]). The effective crystal thickness increase due to the addition of a coating of mass Δm_f is given by

$$\Delta x_{\text{eff}} = \frac{\Delta m_f}{\rho_{qz} A_{xl}}, \qquad\text{(VIII.40)}$$

where ρ_{qz} is the mass density of quartz and A_{xl} is the exposed surface area of the crystal. The change in film mass is related to the change in film thickness (t_f) through $\Delta t_f = \Delta m_f / \rho_f A_{xl}$ (ρ_f is the mass density of the film). For small changes in resonant frequency we find that the shift is proportional to film thickness:

$$\Delta f_r = -S \rho_f \, \Delta t, \qquad\text{(VIII.41)}$$

with the *sensitivity factor* given by

$$S \equiv \frac{-2 f_{ro}^2}{v_{tr} \rho_{qz}}. \qquad\text{(VIII.42)}$$

The S factor customarily has the units $\text{Hz} \cdot \text{cm}^2 \cdot \text{g}^{-1}$ in the literature.

Example For an AT-cut crystal, what thickness is required for a 6-MHz resonant frequency?

The crystal thickness is given by

$$x = \frac{3340 \text{ m/s}}{2 \times 6 \text{ MHz}} = 0.28 \text{ mm}.$$

If the minimum detectable frequency shift is 1 Hz, what is the minimum detectable thickness of aluminum film? Assume an aluminum density of 2.70 g/cm^3 and a quartz density of 2.65 g/cm^3.

The sensitivity factor of the starting crystal is given by

$$S = \frac{-2 \times (6 \text{ MHz})^2}{(3340 \text{ m/s}) \times 2.65 \text{ g/cm}^3} = -8.14 \times 10^7 \text{ Hz} \cdot \text{cm}^2 \cdot \text{g}^{-1}.$$

The minimum detectable thickness of aluminum would be

$$\Delta t = -\frac{\Delta f_r}{S \rho_f} = \frac{-1 \text{ Hz}}{(-8.14 \times 10^7 \text{ Hz} \cdot \text{cm}^2 \cdot \text{g}^{-1}) \times 2.70 \text{ g/cm}^3} = 0.46 \text{ Å}.$$

Does the sensitivity factor increase or decrease with accumulated film on the crystal?

We see in Equation (VIII.41) that $\Delta f_r / \Delta t < 0$. Since $S \propto f_r^2$, the magnitude of S decreases as t increases. Assuming that the crystal becomes unusable when the sensitivity factor changes by 10%, at what aluminum thickness does this occur for a 6-MHz crystal?

Since $S \propto -f_r^2$, $\Delta S \propto -2 f_r \Delta f_r$. The relative changes are then $\Delta f_r / f_r \approx -\Delta S / 2S$, so $\Delta f_r = -0.05 f_r$. The limiting thickness of aluminum would then be

$$\Delta t \approx \frac{-3 \times 10^5 \text{ Hz}}{(-8.14 \times 10^7 \text{ Hz} \cdot \text{cm}^2 \cdot \text{g}^{-1}) \times 2.70 \text{ g/cm}^3} = 14 \text{ μm}.$$

The temperature sensitivity of the natural frequency is about 0.5 ppm/°C. A change of temperature consequently gives a false indication of film deposition. A fluctuation of 1°C corresponds to what apparent thickness of aluminum?

This temperature change corresponds, first, to a frequency shift of $0.5 \times 10^{-6} \times 6 \text{ MHz} = 3 \text{ Hz}$. The corresponding aluminum thickness is then 3 times the minimum detectable thickness calculated above, or 1.37 Å. ∎

A change in resonant frequency of one part per million is routinely detectable; as we have just seen, this corresponds roughly to a ~ 1 Å increase in film thickness. A sensitivity of one part in 10^{10} has been achieved under special conditions [Thomas, 1979]. The greatest practical difficulty is spurious readings caused by the crystal's sensitivity to changes in temperature (for data, see Riegert [1968]). As the crystal becomes coated, it eventually ceases to function, as a result of damping and mode-hopping.

In some instances the required precision means that a more realistic analysis of resonance be utilized — one that includes the propagation of an elastic wave in the deposited film. Several authors have done this, treating the film/crystal system as a compound acoustical resonator [Miller and Bolef, 1968; Lu, 1974; Heyman and Miller, 1978]. Current designs utilize this analysis in microprocessor-controlled monitors.

True Flux Sensors

As noted previously, the quartz crystal oscillator is often used to determine the deposition rate or, equivalently, the deposition flux. It does so by making film thickness measurements over time, from which a historical rate is calculated. True flux sensors, on the other hand, measure the beam intensity instantaneously by effectively measuring the density of particles in the incident beam (n_i). This quantity is related to the incident flux through

$$j_i = n_i \cdot v_i, \qquad (\text{VIII}.43)$$

where v_i is the average speed of the particles in the beam. For a Knudsen cell at temperature T, it may be shown from kinetic theory that $v_i = (8kT/\pi m)^{1/2}$. For a sputtering source, the average speed might be estimated from the most probable particle energy, as $(U_{sb}/m)^{1/2}$ (see Chapter VII).

In practice, one typically does not need to know the value of v_i. Through a long string of theoretical and empirically determined factors, we obtain

$$j_i \propto MP, \qquad (\text{VIII}.44)$$

where MP is some measured property of the incident flux that is proportional to n_i. The overall proportionality factor between j_i and MP is empirically determined. What measured properties are customarily used?

1. A standard ionization-type vacuum gauge, such as was discussed in Chapter IV, can be employed, as suggested in Figure VIII.22. It must typically have a direct line of sight to the source since evaporant particles are condensible. As long as the ion current due to the evaporant (the MP in this instance) dominates the measured ion current of the gauge, the ion gauge's signal is proportional to n_i. The gauge may be calibrated by using it in tandem with a quartz crystal oscillator or some direct film thickness measurement. An implementation of this technique has been described by Hammond [1975], such a system has been commercially available.
2. Another standard high-vacuum technology that has been adapted to the measurement of deposition flux is the quadrupole residual gas analyzer. Here, the MP is the ion current at the mass of the impinging particles, also represented in Figure VIII.22. A setup optimized for silicon deposition exhibited a rather impressive resolution of 0.05 Å/s

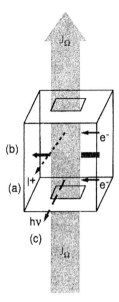

Figure VIII.22 True flux sensors measure some property of the beam of particles coming from the source that is proportional to the density of particles in the beam. This might be (*a*) the ion current I^+ in a vacuum gauge or a residual gas analyzer, (*b*) the attenuation of a transverse light beam, or (*c*) the intensity of light emitted by excited particles of the beam.

[Peter et al., 1991]. Rate control with Si isotopes and Si_2 molecules, as well as ^{28}Si, was found to be possible [van de Leur et al., 1988].

3. Optical absorption measurement is a promising approach to real-time flux monitoring, as represented schematically in Figure VIII.22. The phenomenon is called "atomic beam resonant absorption." By Beer's law, the particle density is related to the optical transmission (T_{opt}) through

$$n_i = -C \cdot \ln T_{opt}, \qquad (VIII.45)$$

where C is a constant. Thus, the MP is the logarithm of the optical transmittance. Since T_{opt} is typically near 100% in practice, there can be a linear relationship between n_i and T_{opt}: $-\ln T_{opt} \approx 1 - T_{opt}$. Chalmers and Killeen [1993] discussed an implementation of this technique, in which the optical transmission signal was utilized to control effusion cells. Atomic beam *fluorescence* can also be induced by the external light beam, and utilized alternatively to absorption.

4. EIES has been developed as a commercially available flux sensor technology [Gogol and Cipro, 1985; Kubiak et al., 1991]. As represented schematically in Figure VIII.22, the principles of operation

are electron impact excitation of neutral flux particles, radiative transition of the excited particles to the ground state, and detection of the emitted radiation. This fluorescence intensity is the *MP* in this instance.

VIII.6 SUMMARY OF PRINCIPAL EQUATIONS

Beam intensity	$J_\Omega = z\,\delta A \cos\theta/\pi$
Differential solid angle	$d\Omega = (\bar{R}\cdot\bar{n})dA_s/R^3$
Deposition flux	$j_c = \alpha_c j_i$
Deposition rate	$v_n = j_c/n_f$
Monolayer formation time	$t_f \approx (n_f)^{2/3}/\alpha_t z_{rg}$
Power flux due to heat of condensation	$j_{thermo} = (\Delta_{con}H°/N_a)\cdot\alpha_c j_i$
Power flux due to kinetic energy of evaporated particles	$j_{evap} = (3kT_{source}/2)\cdot\delta j_i$
Flux of radiant power absorbed by the substrate	$j_{rad} = \varepsilon_{sub} J_{\Omega\,rad} \cos\beta/R^2$
Relationship between incident particle flux and radiant power	$j_{rad} = [(\varepsilon_{sub}\varepsilon_{source}\,\sigma T^4_{source})/z_{source}]\cdot j_i$
Power flux due to kinetic energy of sputtered particles	$j_{sput} = (U_{sb}/2)\cdot\delta j_i$
Power flux delivered to substrate by charged particles	$j_{part} = [3kT^-/2(-q)]\cdot j_e + (V_p - v_{sub})\cdot j_{ion} + (E_i/q)\cdot j_{ion}$
Incident flux	$j_i = J_\Omega \cos\beta/R^2$
Hemisphere	$j_{ih}(\theta) = z\,\delta A \cos\theta/\pi h^2$
Plane	$j_{ip}(\theta) = z\,\delta A (\cos\theta)^4/\pi h^2$
Sphere	$j_{is}(\theta) = z\,\delta A/\pi h^2$
Fraction of particles thermalized	$F = 1 - e^{-y/k\lambda}$
Resonant frequency	$f_r = v_{tr}/2x$
Frequency shift	$\Delta f_r = -S\rho_f \Delta t$
Sensitivity factor	$S \equiv -2f_{ro}^2/v_{tr}\rho_{qz}$
Incident flux and measured property	$j_i \propto MP$

VIII.7 APPENDIX: SOME DEFINITE INTEGRALS

$$\int_0^{2\pi} \frac{d\varphi}{[h^2 + W^2 + r^2 - 2rW\cos\varphi]^2}$$
$$= \frac{2\pi[1 + (r/h)^2 + (W/h)^2]}{h^4[1 + (r/h)^4 + (W/h)^4 + 2(r/h)^2 + 2(W/h)^2 - 2(rW/h^2)^2]^{3/2}} \quad \text{(VIII.46)}$$

$$\int_0^s r\,dr \frac{1 + (r/h)^2 + (X/h)^2}{[1 + (r/h)^4 + (X/h)^4 + 2(r/h)^2 + 2(X/h)^2 - 2(rX/h^2)^2]^{3/2}}$$
$$= \frac{h^2}{4} \cdot \left\{ 1 - \frac{1 - (s/h)^2 + (X/h)^2}{[1 + (s/h)^4 + (X/h)^4 + 2(s/h)^2 + 2(X/h)^2 - 2(sX/h^2)^2]^{1/2}} \right\}$$
(VIII.47)

VIII.8 MATHEMATICAL SYMBOLS, CONSTANTS, AND THEIR UNITS

SI units are given first, followed by other units in widespread use.

$a_{m\ell}$	Monolayer thickness (m)
f	Fraction of atoms resputtered from the substrate
f_r	Resonant frequency (s^{-1})
h	Throw distance (m)
j	Heat flux (W/m^2); current density (A $\cdot$ m^{-2} $\cdot$ s^{-1})
j_c	Condensation flux (m^{-2} $\cdot$ s^{-1})
j_i	Incident flux (m^{-2} $\cdot$ s^{-1})
j_Ω	Incident flux angular distribution (m^{-2} $\cdot$ steradian^{-1} $\cdot$ s^{-1})
k	Ratio of thermalization distance to mean free path (dimensionless); Boltzmann's constant (1.38×10^{-23} J $\cdot$ K^{-1})
m_f	Mass of film (kg)
n	Particle density (m^{-3})
$\vec{n}$	Unit normal vector
q	Magnitude of the electronic charge (1.6×10^{-19} C)
r	Radial position on fixed flat substrate (m)
r'	Radial position on rotating flat substrate (m)
t	Time (s)
v_i	Average speed of particles in source beam (m/s)
v_n	Deposition rate (m/s)
v_{tr}	Acoustic phase velocity of transverse shear wave (m/s)
x	Crystal thickness (m)
y	Distance from sputtering target (m)
z	Impingement rate (m^{-2} $\cdot$ s^{-1})
A	Source area (m^2)
A_{xl}	Area of exposed surface of crystal (m^2)
C	Constant
E_i	Ionization potential (J; eV $= 1.6 \times 10^{-19}$ J)
F	Fraction of sputtered flux that has been thermalized
J_Ω	Beam intensity (steradian^{-1} $\cdot$ s^{-1})
$J_{\Omega rad}$	Radiation beam intensity (W $\cdot$ steradian^{-1} $\cdot$ s^{-1})
MP	Measured property

N_a	Avogadro's number (6.02×10^{23})
P	Point on stationary substrate
P'	Point on rotating substrate
$\overline{R}$	Displacement vector (m)
R	Distance from source to point on substrate (m)
S	Sensitivity factor ($s^{-1} \cdot m^2 \cdot kg^{-1}$; $Hz \cdot cm^2 \cdot g^{-1}$)
T	Period of rotation of substrate (s); temperature (K)
T_{opt}	Optical transmittance
U_{sb}	Surface binding energy (J; $eV = 1.6 \times 10^{-19}$ J)
V_p	Plasma potential (V)
W	Offset distance of rotating substrate (m)
X	Lateral distance on flat substrate (m)
Y	Sputter yield for normal projectile incidence
α_c	Condensation coefficient (dimensionless)
β	Deposition angle (rad)
δ	Trapping probability (dimensionless)
δA	Orifice area (m^2)
δt	Monolayer deposition time (s)
$\Delta_{con} H$	Heat of condensation (J/mol)
ε	Emissivity (dimensionless)
θ	Polar angle (rad)
λ	Mean free path (m)
ρ	Mass density (kg/m^3)
σ	5.67×10^{-8} W $\cdot$ m^{-2} $\cdot$ K^{-4}
φ	Azimuthal angle (rad)
φ'	Azimuthal angle coordinate of rotating substrate (rad)
Ω	Solid angle (steradian)

REFERENCES

Anderson, H. L., 1981, *Physics Vade Mecum*, American Institute of Physics, New York, p. 44.

Behrndt, K. H., 1962, "Films of Uniform Thickness Obtained from a Point Source," *Transact. 9th Natl. Vacuum Symp.*, Macmillan, New York, p. 111.

Chalmers, S. A., and Killeen, K. P., 1993, "Real-time Control of Molecular Beam Epitaxy by Optical-Based Flux Monitoring," *Appl. Phys. Lett.* **63**(23), 3131.

Class, W., and Hieronymi, R., 1982, "The Measurement and Sources of Substrate Heat Flux Encountered with Magnetron Sputtering," *Solid State Technol.* (December), 55.

Dickson, M., Zhong, G. and Hopwood, J. 1998, "Radial Uniformity of an External-Coil Ionized Physical Vapor Deposition Source," *J. Vac. Sci. Technol.* **B16**(2), 523.

Eisgruber, I., 1993, private communication.

Fisher, R. A., and Platt, J. R., 1937, "The Deposit of Films of Uniform Thickness for Interferometer Mirrors," *Rev. Sci. Instr.* **8**, 505.

Glang, R., 1970, "Vacuum Evaporation," in L. I. Maissel and R. Glang, eds., *Handbook of Thin Film Technology*, McGraw-Hill, New York, 1970.

Gogol, C. A., and Cipro, C., 1985, "An Improved Deposition Process Controller for MBE Applications," in J. C. Bean, ed., *Proc. 1st Internatl. Symp. Silicon Molecular Beam Epitaxy*, Proc. Vol. 85-7, The Electrochemical Society, Pennington, NJ, p. 415.

Graper, E. B., 1973, "Distribution and Apparent Source Geometry of Electron-Beam-Heated Evaporation Sources," *J. Vac. Sci. Technol.* **10**(1), 100.

Hammond, R. H., 1975, "Electron Beam Evaporation Synthesis of A15 Superconducting Compounds: Accomplishments and Prospects," *IEEE Transact. Magnetics*, **MAG-11**(2), 201.

Heyman, J. S., and Miller, W. E., 1978, "Effects of Accumulated Film Layers on the Accuracy of Quartz Film Thickness Monitors," *J. Vac. Sci. Technol.* **15**(2), 219.

Hoffman, D. W., 1990, "Intrinsic Resputtering — Theory and Experiment," *J. Vac. Sci. Technol.* **A8**(5), 3707.

Holland, L., and Steckelmacher, W., 1952, "The Distribution of Thin Films Condensed on Surfaces by the Vacuum Evaporation Method," *Vacuum*, **11**(4), 346.

Holland, L., 1963, *Vacuum Deposition of Thin Films*, Chapman & Hall, London.

Knudsen, M., 1915, "Das Cosinusgesetz in den Kinetischen Gastheorie," *Annalen der Physik*, IV, Folge 48, 1113.

Kubiak, R. A., Newstead, S. M., Powell, A. R., Parker, E. H. C., Whall, T. E., Naylor, T. and Bowen, K., 1991, "Improved Flux Control from the Sentinel III Electron Impact Emission Spectroscopy System," *J. Vac. Sci. Technol.* **A9**(4), 2423.

Long, A. R. and MacLeod, A. M., 1979, "A Co-evaporation Control System for Two-Component Alloys," *J. Phys. E: Sci. Instr.*, **12**, 205.

Lu, C. -S., 1974, "Mass Determination with Piezoelectric Quartz Crystal Resonators," *J. Vac. Sci. Technol.* **12**(1), 578.

Macleod, A., 1989, *Thin-Film Optical Filters*, 2nd ed., McGraw-Hill, New York, Chapter 10.

Maier-Komor, P., 1985, "Target Thickness Measurements with Quartz Crystal Sensors of the Third Generation," *Nucl. Instr. Meth. Phys. Res.* **A236**, 641.

Miller, J. G. and Bolef, D. I., 1968, "Acoustic Wave Analysis of the Operation of Quartz-Crystal Film-Thickness Monitors," *J. Appl. Phys.* **39**, 5815.

Pargellis, A. N., 1989, "Evaporating and Sputtering: Substrate Heating Dependence on Deposition Rate," *J. Vac. Sci. Technol.* **A7**(1), 27.

Peter, G., Koller, A. and Vazquez, S., 1991, "A New Electron Beam Evaporation Source for Si Molecular Beam Epitaxy Controlled by a Quadrupole Mass Spectrometer," *J. Vac. Sci. Technol.* **A9**(6), 3061.

Petrov, I., Ivanov, I., Orlinov, V. and Sundgren, J.-E., 1993, "Comparison of Magnetron Sputter Deposition Conditions in Neon, Argon, Krypton, and Xenon Discharges," *J. Vac. Sci. Technol.* **A11**(5), 2733.

Ramsay, J. V., Netterfield, R. P. and Mugridge, E. G. V., 1974, "Large-Area Uniform Evaporated Thin Films," *Vacuum*, **24**(8), 337.

Riegert, R. P., 1968, "Optimum Usage of Quartz-Crystal Monitor-Based Devices," *Proc. 4th Internatl. Vacuum Congress, Part Two*, Institute of Physics and Physical Society, Manchester, April 1968, p. 527.

Rosenblatt, G. M., 1976, "Evaporation from Solids," in N. B. Hannay, ed., *Treatise on Solid State Chemistry*, Vol. 6A, Plenum Press, New York, p. 165.

Rossnagel, S. M., 1989, "Magnetron Plasma Deposition Processes," *Thin Solid Films* **171**, 125.

Rossnagel, S. M., 1998, "Interaction between Gas Rarefaction and Metal Ionization in Ionized Physical Vapor Deposition," *J. Vac. Sci. Technol.* **B16**(6), 3008.

Sauerbrey, G., 1957, *Phys. Verhandl.* **8**, 193.

Somekh, R. E., 1984, "The Thermalization of Energetic Atoms during the Sputtering Process," *J. Vac. Sci. Technol.* **A2**(3), 1285.

Thomas, M. T., 1979, "Vacuum Deposition Techniques," in G. L. Weissler and R. W. Carlson, eds., *Vacuum Physics and Technology*, Vol. 14, Academic Press, New York, Chapter 16.

Turner, G. M., Falconer, I. S., James, B. W. and McKenzie, D. R., 1992, "Monte Carlo Calculations of the Properties of Sputtered Atoms at a Substrate Surface in a Magnetron Discharge," *J. Vac. Sci. Technol.* **A10**(3), 455.

Upadhye, R. S., and Hsieh, E. J., 1990, "A Unified Integrated Model for Sputter Coating Uniformity," *J. Vac. Sci. Technol.* **A8**(3), 1348.

van de Leur, R. H. M., Schellingerhout, A. J. G., Mooij, J. E. and Tuinstra, F., 1988, "Mass-Spectrometer-Controlled Fabrication of Si/Ge Superlattices," *Appl. Phys. Lett.* **52**(12), 1005.

Westwood, W. D., 1978, "Calculation of Deposition Rates in Diode Sputtering Systems," *J. Vac. Sci. Technol.* **15**(1), 1.

INDEX

Abnormal glow regime, *see* Discharge, DC
Activation energy
  calculation, 41
  definition, 20
Activity, thermodynamic, 61, 75–76
Adatom, 46
Adsorbate, 46
Adsorption, 47–53
  ideal model, 46
  isotherm, 46
  mean residence time, 49
  multilayer, 52
  thermal accomodation in, 48, 50
  time for monolayer formation
    expression, 272
    numerical example, 7
Altered layer, 67, 145
Anion, 171
Anode, 171
Arc regime, *see* Discharge, DC
Area-ratio effect, *see* Discharge, RF
Arrhenius plot
  analysis of, 40–41
  definition, 20
Atmosphere, standard, 39
Atomic beam
  fluorescence, 299
  resonant absorption, 299
Atomic layer epitaxy, 53–57
  CVD variant, 56
  growth mechanism, 53
  growth procedure, 54
  process window, 55

Bakeout of vacuum systems, 49, 97
Beam former of E-gun, 130
Beaming effect, *see* Sources
Beam intensity (*see also* various source types)
  differential, 247, 266
  numerical example, 10
  of ideal small sources, 265
Beam steering and focusing, *see* E-gun
Beer can, 8
Billiard(s)
  ball particles, 27
  with atoms, 199
Blind man and elephant, 206
Blocking capacitor, *see* Discharge, RF
Bohm
  criterion, 166
  presheath diffusion, 176
Boltzmann
  distribution, 20
  factor, 20
  probability distribution function, 21
Breakdown, *see* Discharge, DC
Bulk binding energy, 223

Cathode
  definition, 171
  fall, 173

Cation, 171
Chemisorption, 48
Clausing factor, 37, 120
Coevaporation, 67
Cohesive energy, 224
Collision(s)
  binary elastic, 225–227
  cross-section, 27
  rate derivation, 27
Condensation, 62–71
  coefficient
    definition, 63, 269
    in three-temperature method, 69
  definition, 46
  flux, 269
    definition, 62
    expression, 63
  of compounds producing a stoichiometric vapor, 64–65
  steady-state techniques for alloys, 65
  terminology, 50
Conductance
  definition, 37
  for molecular flow through orifice, 38
  for molecular flow through tube, 38
  for viscous flow through orifice, 38
  for viscous flow through tube, 37
  of port, 86
  of series combinations, 93, 95–96
Cosine law
  of emission
    derivation for ideal effusion cell, 118
    derivation for ideal sputtering source, 245
  of impingement
    definition, 23
    derivation, 25
Cryogenic pump, 102

Dalton's law, 26
Dayton theory, 122–124
Debye
  length derivation, 169
  shielding regime, 170
Degassing of ionization gauge tube, 97
Density values for elements, 252–256
Deposition
  flux sensors, 298–300
  modeling, historical development, 267
  monitors, 295–300
  rate derivation, 270–271
Desorption
  process, 96–97
  rate, 50

Diffusion pump, 101–102
Diode sputtering, 153
Diameter(s) of gas and vapor particles
  effective, 28
  tables of values, 28, 42–43
Diffusion pump, 101–102
Diffusivity
  derivation, 31–32
  of dilute impurity, 32
Discharge(s)
  DC
    abnormal glow regime, 188–189
    arc regime, 189
    as fundamental sputtering arrangement, 153
    breakdown
      definition, 187
      explanation, 187–188
    creation and maintenance of, 173
    electrical model, 170–173
    normal glow regime, 188
    ohmic conduction regime, 185
    plasma potential of, 172
    self-sustaining, definition, 185, 188
    sputtering power density, 173
    Townsend discharge regime, 185
    Townsend processes, 190
    voltage-current characteristic, 184–189
  definition, 154
  exceptions for real, 193
  glow
    DC, 190–193
    definition, 154, 190
    distribution of light intensity, 193
    luminous and dark regions, 191–193
    RF, 193
  magnetron, 153
  rarefaction of, 194
  RF (radio frequency)
    area-ratio effect, 181
    as fundamental sputtering arrangement, 153
    blocking capacitor
      purpose, 178
      value, 180–181
    creation and maintenance of, 182
    electrical model, 178–182
    for sputtering insulating materials, 178
    self-bias, 178
    value, 180
    voltage-current characteristic, 189–190
  thermionically assisted, 158
Displacement of target atoms, 224

INDEX   307

Effusion
  cell
    assumptions of ideal, 117–118
    beam intensity, 118
    conical, 124
    definition, 115
    near-ideal, 121–123
    nonequilibrium, 119–121
    open-tube, 123–124
    photograph, 2
  definition, 117
E-gun evaporator
  beam intensity, 131–133
  beam steering and focusing, 130
  deviation from cosine law, 128–129
  Graper's expression for beam intensity
    of, 133, 268
  photograph, 3
  principles of operation, 129–131
  virtual source of, 132–133
Electrical network analogy, see Vacuum systems
Electron
  beam evaporator, see E-gun evaporator
  concentration and temperature
    determination with Langmuir probe,
      167–168
    in plasmas, 154
    in "practical plasma," 160
  current density at probe, 163–164
  energy distribution
    determination with Langmuir probe, 164
    non-Maxwellian, 194
  impact emission spectroscopy (EIES),
    299–300
  impingement rate in "practical plasma," 160
  retardation regime, 163
  saturation regime, 164
  trapping in magnetron, 157
  two populations in discharge, 164
Electronic energy loss, 207
Energy distribution(s), see Electron
Energy transfer
  function, 226
  mass factor, 227
Enhanced ion current density, 165–166
Enthalpies of vaporization for elements,
  table, 78
Enthalpy, 73
Entrainment type pump, 99
Entropies of vaporization for elements,
  table, 78
Entropy, 73
Equilibrium constant, 59, 75–76
Equipartition of energy, 31

Evaporation (see also specific source types)
  congruent, 65
  definition, 46
  of alloys, 65–67, 144–146
  reactive, 70–71
  temperature, 143–144

Film thickness profile(s)
  of ideal small source, 277–278
  of point source, 278
  with large disk source, 291–293
  with large ring source, 293–294
  with off-axis substrate rotation, 290–292
Flash evaporation
  of compounds that dissociate, 65
  pulsed laser deposition as, 65, 133
Flat source, see Sources
Floating potential, see Potential
Flow regime(s)
  choked, 35
  molecular, 34, 36–37
  turbulent, 35
  viscous laminar, 35–36
Flux sensor, 298–300
  atomic beam, 299
  electron impact emission spectroscopy
    (EIES), 299
  ionization gauge, 298
  quadrupole residual gas analyzer, 298
Formation
  quantities, 74
  reaction, 75
Fourth phase of matter, 153
Free evaporation, 127–129

Gas
  expressing amounts of, 39
  versus vapor, 45–46
Gibbs free energy
  criterion for stability or spontaneous
    changes, 73
  definition, 73
  of evaporation, 57
Glow discharge, see Discharge
Graybody radiation, 272–273
Günther technique, 68
Gyro radius, 157

Heat capacity
  derivation, 30
  values, 31
Helmholtz free energy, 73
Hertz–Knudsen equation, 37, 63

Ideal gas law, 26
Ideal solution, 60
Impact ionization, 186
Impingement rate
 definition, 23
 derivation, 24
 expressions, 26
 numerical example, 6–7, 10
Impurity concentration due to residual gas, 71
Incident flux
 definition, 62, 267
 derivation, 267–268
 distribution, 25
 general expression for a small source, 277
 numerical example, 10
Ion
 concentration and temperature determination with Langmuir probe, 167–168
 in "practical plasma," 160
 in plasmas, 154
 current density at probe, 163–164
 enhanced current density, 165–166
 impingement rate in "practical plasma," 160
 pump, *see* Sputter-ion pump
 retardation regime, 165
 saturation regime, 164
Ionization
 cross-section, 186
 fraction in ionized physical vapor deposition, 289
 gauge
  as deposition rate monitor, 298
  principles of operation, 106–107
 impact, 186
Ionized physical vapor deposition
 applications, 286–287
 definition, 287

JANAF tables, 77–78

Kinetic theory of gases (*see also* specific gas properties)
 assumptions, 19
 average relative speed, 29
 importance, 5–6
 speeds, 22–23
Knockons, 232
Knudsen cell, *see* Effusion cell

Lambert's law, 118
Langmuir adsorption isotherm, 49
 for dissociative adsorption, 52
Langmuir-child law, 130
Langmuir probe
 current densities, 163–164
 current-voltage characteristic, 162–165
 definition, 161
 electron retardation regime, 163
 electron saturation regime, 164
 ideal assumptions, 164
 ion retardation regime, 165
 ion saturation regime, 164
 rectifying behavior, 163
 saturation current, 163
Langmuir source
 assumptions, 128
 definition, 115
Large source, *see* Sources
Laws of high-temperature chemistry, 146
Leak in vacuum chamber. *See also* Throughput
 numerical example, 13–14
Linear cascade
 average energy of recoils, 236
 definition, 234
 depth distribution, 237
 directional distribution, 236
 energy distribution, 234
 probability distribution function, 236
 spatial distribution, 236
 termination, 235
 theory, 232
Lorentz force, 129

Magnetron. *See also* Discharge, magnetron; Sputtering, magnetron
 current-voltage relation of DC, 189
 planar, cylindrical, or post, 158
Matsunami formula for sputter yield, 251
 Q values, 252
Mean free path
 derivation, 27–28
 of dilute impurity, 29
Mean residence time, 49
Measured property definition, 298
Mechanical pump, *see* Rotary mechanical pump
Melting points of pure elements, table, 78
Microscopic reversibility, 64
Mole, definition, 39
Molecular beam definition, 117
Molecular solid, 64
Monolayer adsorption time
 expression, 272
 numerical example, 7

INDEX 309

Moving shutter technique, 278–280
Moving target for theoreticians, 205

Nonequilibrium source, 115
Nonideal solution, 61
Normal glow regime, see Discharge, DC
Nozzle-jet source
   description, 125–127
   beam intensity, 125
Nuclear energy loss, 207
   reduced projectile energy, 230
   theory, 229
   Wilson's cross-section, 229
Nuclear stopping power, 229–230

Ohmic conduction regime, see Discharge, DC
Outgassing, 97–98

Permeation, 98
Physical vapor deposition
   definition, 1
   main technologies of, 1
   overview, 1–5
Physisorption, 48
Planar binding model, 224
Planetary
   fixturing, 278
   rotation, 289
Plasma
   definition, 154
   potential, see Potential
Point source, see Source(s)
Poisseuille flow
   conductance, 37
   formula, 36
   numerical example, 14
Poisson's equation, 169
Potential
   floating, 163, 167
   plasma, 162–163, 167
Practical plasma
   definition, 153
   parameters and properties of, 159–161
Presheath, 176
Pressure
   ranges, 1
   SI unit of, 38
   standard, 74
   ultimate, 14, 86, 94–96, 101
   unit conversion, 39
Pulsed laser deposition
   applications, 135
   beam intensity, 140
   definition, 133
   discussion of heating model's assumptions, 139, 141–143
   easiest and best target materials, 135
   heating model, 135–141
   number of vapor particles created, 138
   optical absorption length, 136
   parameters, 133
   photograph, 4
   physical picture, 134–135
   plume, 133
   regimes of laser heating, 137
   thermal diffusion length, 137
   thickness removed per pulse, 138
   threshold, 135, 138
   yield, 135
Pump, see Vacuum pump(s)
Pumping speed, 84–87
   definition, 85
   effective, 94–96
   generic expression for near ultimate pressure, 104
   maximum possible, 85–86

Quadrupole residual gas analyzer as deposition rate monitor, 298–299
Quartz crystal microbalance, 295–298
Quasiequilibrium source, 115

Range of ions in solids, 231
   effective, of recoils in simplified collisional model, 239
   mean projected, 231–232
   of projectiles in simplified collisional model, 239
   radial, 238
Raoult's law, 60
Receiving surface(s), hemisphere, plane, tangential sphere, 277–278
Reduced projectile energy, 230
Reference state, thermodynamic, 76
Reflected neutrals, 282
Refraction of escaping recoils, 226
Residence time, mean, 49
Resputtering, 269
Richardson's equation, 130
Rod-fed sources, 67
Rotary mechanical pump, 99–100

Secondary electron emission
   coefficient, 187
   definition, 173
Self-bias, see Discharge, RF

Sheath
  collisional, 175
  collisionless, 174
  definition, 154
  thickness
    at anode of DC discharge, 176
    at cathode of DC discharge, 174–175
    at a probe, 170
    for RF discharge, 182–183
Simplified collisional model of sputtering, 238
  assumptions, 241
  average energy of recoils at threshold, 239
  combined external flux spectra, 248–250
  combined internal flux spectra, 248
  disturbed volume, 238
  effective number of recoils, 240
  effective range of recoils, 239
  escape depth, 238
  fraction of recoils close enough to surface to escape, 240
  predicted trends, 244
  predictions, 241
  range of projectile, 239
  sample yield calculation, 258
  surface binding energy dependence of yield, 243
  yield expression, 240
Small source, *see* Source(s)
Solid angle definition, 267
Sorption
  definition, 98
  pump(s), 98, 100
Source(s) (*see also* specific types)
  advantages of large, 289
  beaming effect, 122
  dimensionality, 289
  flat, 265
  large versus small, 118, 247, 265
  nonequilibrium, 115
  point
    beam intensity, 129
    ideal model, 129
    symmetry, 118, 129
  quasiequilibrium, 115
  small, 122
  thermal, 1
  virtual, *see* E-gun
Space-charge-limited current flow
  ballistic, 174
  mobility-limited version, 175
Spherical binding model, 224
Spluttering, 205
Sputtering. *See also* Sputter yield
  advantages over evaporation, 201

amorphous target theories, 222
applications, 200
backward and forward, 201
collisional, 201
collision cascade, 201
criteria for nonlinear cascade, 202–203
DC, 155
definition, 200
deposited energy, 237
  density, 201
event, 199
ideal source, 245
  angular yield spectrum, 246
  emission flux angular distribution, 246
ion beam, 159
Kelly's timescale, 201, 203
Kelly's yield expression, 258
magnetron, 157–158
mechanism, 14, 201
nonlinear, 202
Pease's yield expression, 256
photograph, 5–6
projectile energy, 199
reactive, 70–71
reviews, 206
RF, 156
Schwarz and Helms' yield expression, 257
Sigmund's regimes, 201, 202
Sigmund's yield expression, 256
simulations, 204
single crystal targets, 222
source beam intensity, 247–248
theories, 203–205
thermal, 201
Thompson's yield expression, 257
Wehner spots, 222
wind, 194
Sputter-ion pump, 104
Sputter yield. *See also* Sputtering
  angle of incidence dependence, 214–219
  angular distribution, 220–222
  Bohdansky formula, 207
  component, 67
  data sources, 205–206
  definition, 199
  energy distribution, 219
  excess, 202
  Mahan and Vantomme expression, 212
  projectile choice dependence, 214
  projectile energy dependence, 207–212
  Steinbrüchel formula, 212
  surface binding energy dependence, 212–214
  target conditioning, 222–223

INDEX

threshold energy, 207, 227–229
trends, 206
Standard state, thermodynamic, 39, 76
Sticking coefficient, 50
Stoichiometric
  coefficient in chemical equation, 75
  interval in three-temperature method, 68
Sublimation, 46
  pump, see Titanium sublimation pump
Substrate
  bias, 158
  heating mechanisms associated with deposition, 274–278
  rotation, see Planetary; Film thickness profile
Supersaturation
  definition, 64
  numerical example, 12
Surface binding energy, 223–225
  definition, 207
  estimating with heat of sublimation, 224
  planar binding model, 249–250
  spherical binding model, 248
  table of values, 252–256
Symmetric charge exchange, 16, 176

Target parameters, table of, 252–256
Temperature(s). See also Electron; Ion
  practical evaporation, 143
  melting, for elements, 78
Thermal accomodation, 48, 50
Thermal conductivity derivation, 34
Thermalization
  concomitant heating and rarefaction of sputtering gas, 285
  definition, 282
  distance, 283
  fraction, 283
  in pulsed laser deposition, 141
  pressure-distance product, 283
  reduction of incident flux by, 283–284
  summary of practical effects, 282
Thermal sources
  definition, 1
  photograph, 2
Thermal transpiration, 101
Thermochemical data
  discussion of sources for, 77–78
  table for elements, 78
Thermocouple gauge principles of operation, 105–106
Thermodynamic(s)
  criterion for equilibrium, 73
  first law of, 72

formation quantities, 74, 77
potentials, 72–73
reaction quantities, 74–75
second law of, 25, 73
Thickness profile, see Film thickness profile(s)
Thomas–Fermi screening length, 230
Thompson formula, 219
Three-temperature method, 67–70
  condensation coefficient, 69
  procedure, 68
Throughput
  definition, 87, 93
  density, 91
  introduction to, 87–88
  law, 83, 88–93
  numerical example, 14
  of leak, pump, and tube, 88
  of vacuum chamber, 87
  previous definitions, 92, 111
Throughput-type pump, 99
Throw distance, 245, 277
Titanium sublimation pump, 102–104
Tooling factor for flux sensor, 295
Townsend discharge, see Discharge, DC
Trapping
  of electrons in magnetron, 157
  probability, 50
TRIM simulation, 204
Turbomolecular pump, 104

Ultimate pressure, see Pressure

Vacuum gauge
  ionization
    as deposition rate monitor, 298
    principles of operation, 106–107
  thermocouple, 105–106
Vacuum pump(s)
  cryogenic, 102
  diffusion, 101–102
  entrainment type, 99
  rotary mechanical, 99–100
  sorption, 100–101
  sputter-ion, 104
  throughput-type, 99
  titanium sublimation, 102–104
  turbomolecular, 104
  ultimate pressure, 94–95
Vacuum systems
  electrical network analogy, 108–111
  how to analyze, 108
  more realistic model, 94–96
  schematic diagrams, 107–108

Vapor
  definition of, 45
  pressure
    data, 58–60
    derivation, 57–58
    empirical expression, 59
    numerical example, 9
    of alloys and compounds, 60–62
    of gallium arsenide, 61–62
  saturation of, 45
  versus gas, 45–46
Vaporization
  coefficient, 64, 117
  definition, 46
  retarded, 64
Virtual source of E-gun, 129
Viscosity derivation, 32–33